BK.1

D1373912

# PRENTICE-HALL FOUNDATIONS OF MODERN BIOCHEMISTRY SERIES

*Lowell Hager and Finn Wold, editors*

**ORGANIC CHEMISTRY OF BIOLOGICAL COMPOUNDS***
*Robert Barker*

**INTERMEDIARY METABOLISM AND ITS REGULATION**
*Joseph Larner*

**PHYSICAL BIOCHEMISTRY**
*Kensal Edward Van Holde*

**MACROMOLECULES: STRUCTURE AND FUNCTION**
*Finn Wold*

SPECIAL TOPICS

**BIOCHEMICAL ENDOCRINOLOGY OF THE VERTEBRATES**
*Earl Frieden and Harry Lipner*

**THE BIOCHEMISTRY OF GREEN PLANTS**
*David W. Krogmann*

**AN INTRODUCTION TO BIOCHEMICAL REACTION MECHANISMS**
*James N. Lowe and Lloyd L. Ingraham*

**EXPERIMENTAL TECHNIQUES IN BIOCHEMISTRY**
*J. M. Brewer, A. J. Pesce, and R. B. Ashworth*

*Published jointly in Prentice-Hall's *Foundation of Modern Organic Chemistry Series*.

# EXPERIMENTAL TECHNIQUES IN BIOCHEMISTRY

**J. M. BREWER**
*Department of Biochemistry*
*University of Georgia*

**A. J. PESCE**
*Departments of Internal Medicine and Biochemistry*
*University of Cincinatti*

**R. B. ASHWORTH**
*Methods Development Laboratory*
*U. S. Department of Agriculture*
*Beltsville, Maryland*

Prentice-Hall, Inc., Englewood Cliffs, New Jersey

*Library of Congress Cataloging in Publication Data*

Brewer, John Michael (date)
  Experimental techniques in biochemistry.

  (Foundations of modern biochemistry series)
  Includes bibliographical references.
  1. Biological chemistry—Technique. I. Pesce,
Amadeo J., joint author. II. Ashworth, Raymond Bernard
(date), joint author. III. Title. [DNLM: 1. Bio-
chemistry—Laboratory manuals. QU25 B847e]
QP519.7.B73     574.1'92'028     74–8641
ISBN 0–13–295071–5

Dedicated to Professor GREGORIO WEBER

10 9 8 7 6 5 4 3 2 1

Printed in the United States of America

PRENTICE-HALL INTERNATIONAL, INC., *London*
PRENTICE-HALL OF AUSTRALIA, PTY. LTD., *Sydney*
PRENTICE-HALL OF CANADA, LTD., *Toronto*
PRENTICE-HALL OF INDIA PRIVATE LIMITED, *New Delhi*
PRENTICE-HALL OF JAPAN, INC., *Tokyo*

# CONTENTS

v

# PREFACE

This book is intended for graduate student "techniques" laboratory courses, and it attempts to provide experience in most of the more common techniques used in biochemical research. Most students go from these courses to their own research, and, for this reason, the book contains many hints on what is considered "good technique." Because we have tried to describe the techniques in more depth than is found in most laboratory manuals, we feel this book can also be used by people doing postgraduate research.

Biochemistry students enter graduate school with widely varying backgrounds. Consequently, we have tried to provide relatively detailed experimental procedures without providing experiments (since each department makes up its own). However, it is most important that the student understand what he is doing and why, so every chapter contains an extensive theoretical section.

We have also tried to relate the techniques to each other, stressing similarities and differences as much as possible. This has often meant descriptions of processes on a molecular, physical–chemical level. In order to make our descriptions easily understood, we have tried to keep them on a qualitative, verbal level, using equations as concise ways of summary, rather than as the descriptive basis. We feel this will help the student who is not strong in physical chemistry, while providing possible additional insight for those who are.

We thank Dr. R. B. Ballentine of the McCollum-Pratt Institute at Johns Hopkins University, who wrote Chapter 2; Dr. R. Gaver of Bristol-Myers Co., who helped prepare Chapter 3; Dr. Clara Schreiner, who helped prepare Chapter 4; T. E. Spencer, who helped write Chapter 6; and Dr. S. R. Anderson of the Department of Biochemistry and Biophysics of Oregon State University, who helped write Chapter 7. Dr. Charles Todd, of the City of Hope Medical Center, Duarte, California, assisted in the early stages of this project. In addition, several of our colleagues were kind enough to read various chapters: Drs. John Lee, John Wampler and Frank Inman of the University of Georgia, and K. E. Van Holde of Oregon State University were especially helpful. Mrs. Linda Postelnik and Mrs. Virginia Mueller were invaluable in typing and proofreading the manuscript. Finally, the American Heart Association provided financial support to A. J. Pesce (Established Investigator, 1968-1973).

J. M. BREWER
A. J. PESCE
R. B. ASHWORTH

# 1 | GENERAL TECHNIQUES: PREPARATION OF MATERIALS

## 1.1 CLEANING GLASS- AND PLASTICWARE AND CUVETTES

The success of experiments will always depend to some extent on the freedom from contamination of the reagents, and this depends in part on the cleanliness of their containers.

However, we stress that *no* container, be it metal, organic, glass or any plastic, is completely inert. For example, the walls of containers made of plastic such as polyethylene in effect form separate hydrocarbon phases, through which water can slowly diffuse (so that the concentration of aqueous reagents will tend to change with time) or in which nonpolar solutes or solvents can dissolve (1). And anything dissolved in the walls of a plastic container is essentially impossible to remove by washing. Note that some heavy metal cations and inhibitory anions can form nonpolar complexes with some substances, and contaminate a plastic container. Consequently, plastic bottles or other plastic vessels which have come in contact with such compounds should never be used for anything else. Gases such as ammonia (found in increasing concentrations in commercial floor and window cleaners) can also penetrate plastic containers; these should not be used where contamination with ammonia, carbon dioxide, etc., can be troublesome.

### 1.1a  Washing Plasticware

The procedure for cleaning plasticware depends on the plastic—see the Table of Chemical Resistance of Plastics (Appendix 1). For resistant plastics, we soak the object to be cleaned in 8 $M$ urea which was brought to pH 1 with HCl. The acid urea is rinsed off with glass-distilled water, and the object is then washed in 1 $M$ KOH, then water, $10^{-3}$ $M$ EDTA, then several more rinses with water. All suitable (resistant) plasticware should receive such an initial wash. Afterwards, a dilute detergent (usually SDS) wash should be sufficient, though occasional acid urea-base washes should be given the objects.

No plastic should be treated with dichromate or any oxidizing agent, as these will tend to produce free carboxyl groups on the walls of the vessel.

### 1.1b  Washing Glassware

Glass behaves like a mixed-bed ion exchange resin, though one of very low capacity. Glass surfaces will bind a variety of anions and cations, and in working with enzymes, especially metal-sensitive ones, this fact can have unfortunate consequences.

Most treatments of glassware, beyond the usual detergent and distilled water washes (0.5% Alconox is preferred), involve exchanging away unwanted adsorbed ions by rinsing the glass with very strong acid or base, or both. We wash glassware first with ethanolic KOH. This removes any greases or oils, and a mono-layer of glass besides, so a rinse should suffice. The ethanolic KOH is rinsed off with distilled water.

Some substances which should not be exposed to this solution are: glass wool, sintered glass, or glass- or quartzware such as cuvettes which have highly polished optical surfaces. The strong base will etch the polished surfaces or weaken the thin glass fibers or partitions.

A soaking in dichromate cleaning solution will remove any organic material remaining, if sufficiently prolonged. Stubborn material may require that the cleaning solution be heated for greater effect. The dichromate rinse will leave a film of adsorbed dichromate on the glass. After rinsing with distilled water, this film is removed, along with any remaining unwanted cations, by a further rinse with concentrated nitric acid (preferred because it is also an oxidizing agent). Do this in a hood! The glass is then rinsed thoroughly with glass-distilled water. If no organic material is present on the glass, only the ethanolic KOH and nitric acid washes need be used.

According to Westhead (2), the nitric acid wash will remove only the superficial cations. Over a period of days, fresh cations, originally in layers beneath the surface, will migrate to the surface, and the glass will have to be re-rinsed. These cations (such as calcium) are of course present throughout the glass, being in the original material used in the manufacture. Westhead has also found that acetate solutions somehow retard this migration.

The glass may be considered clean when water spreads evenly over the surface and does not pull away from the glass to form patches and drops.

It might be expected that freshly cleaned glass surfaces would interact with proteins, and this has been observed in several instances. Lactic dehydrogenase is inactivated by exposure to glass surfaces at low pH. Small amounts of yeast enolase are adsorbed to glass at neutral pH's. However, these interactions are usually noticeable only when working with very dilute protein solutions. For this reason, it is best to work with concentrated solutions, but, when the use of dilute solutions is unavoidable, two tactics can be employed. An excess of some protein such as bovine serum albumin can be added to the solution, or the glass can be coated with dimethyldichlorosilane. The latter is dissolved in benzene (as a $1\%$ solution), heated to $60°C$, and then the glass is rinsed with it. The glass is dried in an oven, and then one or two additional coatings of the silane are added in the same way.

Sintered glass and glass wool should be cleaned with dichromate cleaning solution and nitric acid.

### 1.1c Washing Cuvettes

Manufacturers recommend washing cuvettes with dilute detergent. Scrub gently using a thin stick with a twist of cotton on it. More stubborn cases may require dichromate cleaning solution and nitric acid—but only if the cuvettes are *fused* at the corners and not cemented.

Your work may require cuvettes with a very low level of fluorescence. For this use fused silica or quartz cuvettes. To clean these you should soak them overnight or heat them to $60°C$ for 30 minutes in $50\%$ nitric acid and $50\%$ sulfuric acid (simply mixing the acids—with care—will give a solution that is warm enough). The cuvettes are then rinsed in doubly glass-distilled water and soaked briefly in ammonium hydroxide. Then they are allowed to dry under an inverted glass beaker on a clean glass petri dish.

## 1.2 PROCEDURE FOR CLEANING DIALYSIS TUBING

The tubing should be soaked about an hour in $1\%$ acetic acid. Then it is allowed to stand, with gentle stirring, for a few minutes in glass-distilled water, which is replaced by $1\%$ sodium carbonate, $10^{-3}$ $M$ EDTA. After a few more minutes of gentle stirring, the alkaline-EDTA solution is replaced by fresh solution and heated, with stirring, to about $75°C$. The hot liquid is poured off, replaced by more alkaline-EDTA, and heated again to $75°C$ or so. The solution is again replaced, this time with glass-distilled water. This is poured off after a few minutes stirring, and more water added. This is heated to about 75°C, poured off, and more water added. After a few minutes, the tubing is placed in glass-distilled water in a plastic bottle and stored in a refrigerator with a few drops of chloroform added to discourage bacteria. Acid-washed glassware should be used for the soaking and heating steps.

**N.B.** Nucleases and other enzymic contamination in dialysis tubing must be removed by *boiling* the tubing once for 30 minutes (no more) in alkaline EDTA. For dialyzing substances with sensitive sulfhydryl groups, the tubing should be heated once in 0.1 $M$ sodium ascorbate $+0.001$ $M$ EDTA.

## 1.3 PREPARATION OF CHROMATOGRAPHIC MATERIALS

### 1.3a Procedure for Washing Substituted Celluloses

The dry cellulose powder is poured into 1 $M$ sodium acetate and thoroughly suspended with a magnetic bar and magnetic stirrer. About 3 liters of solution are used for each 100 g of cellulose.

The suspension is allowed to settle until a distinct zone of settled material has formed (about 10 minutes). A considerable amount of fine cellulose particles may be still in suspension. If so, a pipette attached to an aspirator is used to remove as much of this fine suspension as possible. The remaining sedimented cellulose and solution is poured into a Buchner funnel and filtered nearly dry.

If the cellulose has not been used before, it is then resuspended in the same volume of 0.5 $M$ NaOH, allowed to settle for about 20 minutes, and the "fines" aspirated off, as before. This is filtered, and the cellulose cake is resuspended and washed twice more with fresh 0.5 $M$ NaOH. Do not allow the cellulose to stand for excessively long periods (e.g., overnight) in the strong base or acid. These will hydrolyze the cellulose, producing more "fines," or can actually remove the chemical substituents.[1]

After the third wash with NaOH, the cake is suspended in water (3 liters/100 g) and thoroughly dispersed with stirring. Then HCl is added slowly with stirring to a final concentration of 0.5 $M$. After stirring for five minutes, the suspension is allowed to settle as before, and the aspiration and filtering are repeated. The cellulose is washed again with 0.5 $M$ HCl, once with 0.1 $M$ HCl in 95% ethanol, methanol or acetone, and then washed with water. It is finally resuspended in 0.1 $M$ sodium acetate and $3 \times 10^{-3}$ $M$ EDTA, and the pH of the cellulose suspension adjusted to the pH of use or preference. The EDTA is added because celluloses tend to pick up heavy metal ions. The cellulose is allowed to stand overnight, and the pH is checked again and readjusted if necessary. The final washings should be with 2-3 changes of

---

[1]This is especially true for phosphorylated cellulose.

water. The cellulose is filtered nearly dry and stored with a couple of milliliters of chloroform added.

Glass-distilled water and acid-washed glassware (for vessels in actual contact with the cellulose) should be used throughout.

Once the celluloses have been given this preliminary treatment, further cleanings (for example, after using the cellulose in a protein purification step) should be done in the same way, except 0.1 $M$ acid and base solutions are used. If contaminants are adsorbed which resist this treatment and 0.5 $M$ washes as well, discard the cellulose.

### 1.3b  Washing of Molecular Sizing Materials

Sephadex (plain or substituted) should be cleaned by two successive washes in 0.1 $N$ NaOH, a wash with water, a wash in 0.1 $N$ HCl, then (substituted Sephadexes *only*) with 0.1 $N$ HCl in 95% ethanol. The material is rinsed with more water and $10^{-3}$ $M$ EDTA. The EDTA is washed out with more water before pH adjustment or storage. Chloroform should *not* be added to these substances or to Bio-Gels. Sephadexes may be autoclaved or stored in 0.9% NaCl and 0.2% $NaN_3$. Sephadexes lower than G-50 may be sterilized with 1% NaOH, followed by a sterile water rinse.

Bio-Gels cannot be treated to pH's above 9.5 without some hydrolysis of the amide groups. So Bio-Gels should be washed as with Sephadexes *except* that 1% $NaHCO_3$ should be substituted for the 0.1 $N$ NaOH. Bio-Gels should *not* be sterilized with heat—neurotoxins may be formed.

### 1.3c  Preliminary Preparation of Dowex Resins

The Dowex 1, 2, or 50 should first be suspended in one liter/lb of ethanol (95% or absolute): acetone, 1:1. Then the resin is filtered nearly dry and resuspended in water with a magnetic stirrer and magnetic bar, and solid sodium acetate added to a final concentration of 1 $M$. The Dowex is collected by filtration in a Buchner funnel, resuspended in 0.5 $M$ NaOH, and filtered again. It is washed again by suspension in 0.5 $M$ NaOH and filtration; then a third time with 0.5 $M$ NaOH by pouring one liter/lb through the Dowex cake in the Buchner funnel. The Dowex will have changed color from yellow to a reddish brown because of the methyl orange, an indicator present in all Dowexes. The methyl orange cannot be removed completely. Indeed, Dowex should be rewashed if used after long storage.

After the sodium hydroxide washes, the Dowex is again suspended in water, and HCl added to a concentration of 3 $M$ (about a 1:4 dilution of the concentrated acid) and filtered. An additional 0.5 $M$ HCl wash follows and finally the resin is washed with water until the wash liquid is neutral; that is, until all HCl is removed.

Dowex 1, 2, and 50 are now in the chloride or acid forms, and these can be used immediately if desired. If you want to convert any of the Dowexes to other salts, say Dowex-1 formate, they must undergo further washing.

### 1.3d  Cleaning Other Chromatographic Materials

A wide variety of materials is used for chromatography, and the cleaning procedure depends on the material. DEAE-cellulose for thin layer chromatography, for example, is cleaned like regular DEAE. However, whatever the material used, it should be washed beforehand with the solvent used in the planned chromatography. This is true whether the material is DEAE in a column, paper, or Sephadex. After sufficient solvent has dripped off or percolated through, the material can be washed with water (to remove salts) and dried if the chromatography is done on the dry material.

## 1.4  PREPARATION OF REAGENTS

### 1.4a  Water and Other Solvents

Most reagents you will make up are aqueous ones, and it pays to critically consider the quality of the water you use. While tap water may serve for a few reagents, most will require at least deionized or distilled water. Deionized water has been passed through a mixed-bed ion exchange resin and is consequently very low in concentration of ionic substances. On the other hand, the resins used tend to give off ultraviolet-absorbing substances, and, of course, nonionic substances are not removed. Therefore in work where a very low ultraviolet "background" is required, deionized water should not be used without further treatment. Since some plastic containers also tend to leach ultraviolet-absorbing substances and glass tends to give off (exchange) ions, the storage vessel should be chosen depending on the intended use of the water.

Distilled water tends to be low in ions and other nonvolatile substances. Volatile substances, like carbon dioxide, HCl and acetic acid, will not be easily removed by distillation. Aromatic compounds, which tend to be fluorescent, may also come over, but in general, water distilled in a glass still will be adequate for most experiments.

If really high quality water is desired, it must be distilled from acid permanganate: 1 g of potassium permanganate plus 1 ml of phosphoric acid per liter of water is distilled in a clean glass still. The hot permanganate oxidizes all organic substances, and the resulting distilled water has a very low ultraviolet absorption and fluorescence.

### 1.4b  Chemicals

One must add clean chemicals to clean water for best results. Most inorganic chemicals can be purchased with an analysis that states what major contaminants occur and roughly in what amount. However, many chemicals will not come with any analysis, and in these cases a healthy skepticism as to their purity is the best attitude. Some of these compounds, such as Tris or acrylamide, can and should be recrystallized at least once before using. Others, such as DPNH, can be purified by chromatography (see Chapter 3). Generally, reasonable results can be obtained in most experiments with commercial supplies.

### 1.4c  Accurate Reagent Preparation

In weighing out material for a reagent, it is sometimes convenient to weigh out a relatively large amount of the material and make up a stock solution, from which small amounts may be withdrawn at convenience. Then too, some chemicals are more stable in more concentrated solutions. Stock solutions can also be used to cut down on the amount of pipetting needed and at the same time reduce variability between a number of similar incubation mixtures, assay solutions, etc.

Generally, molar solutions are used—solutions made up to a given volume. Molal solutions are prepared by mixing a certain number of gram formula weights with 1000 grams of water or other solvent. Molal solutions are seldom used in biochemistry. "Normality" refers to the molar concentration in equivalents, especially of acid or base. A frequently used term of concentration is the "percent of" some compound. This is either: weight percent— grams of compound in 100 grams of solution (not solvent); or volume percent —milliliters per 100 milliliters of solution.

For best accuracy, the student should learn to differentiate between TC ("to contain") and TD ("to deliver") containers. The former container will *contain* say 100 ml if the liquid is at the 100 ml mark. Such vessels must be rinsed out with the solvent to get all the reagent. TD containers will *deliver* 100 ml of liquid into another container when tipped, but contain slightly more than 100 ml.

Temperature effects on the volume of solutions are not large in the case of water, but can be significant with other solvents. Effects of temperature on the pH of buffers vary with the buffer. The pH of Tris buffers (and probably all primary amine buffers as well) is markedly affected by temperature, decreasing almost one pH unit with an increase of 30°C in temperature. The pH of phosphate buffers shows less temperature dependence.

The pH of a buffer can change with dilution as well. Generally, large dilutions of buffers should not be made without checking the final pH. Excessively dilute (say $10^{-4}$ $M$) buffers will probably be ineffective, due to residual ions in the water, especially ammonium and bicarbonate ions.

Various methods are used to avoid contamination of reagents. The most obvious has been mentioned: storing them in properly cleaned containers. Keeping fingers away from contact with reagents and surfaces in contact with reagents is absolutely necessary, since fingerprints contain not only fats and oils, but metal ions, phosphate, amino acids and at least one enzyme, a nuclease. Sealing the container against vapors such as carbon dioxide, ammonia or hydrochloric acid can be done temporarily with Parafilm. Parafilm itself, however, is a wax and is susceptible to attack by toluene or benzene vapors.

Bacterial contamination is not confined to reagents containing organic chemicals. We have observed bacteria growing on the inside of polyethylene glass-distilled water containers and in glass jars of magnesium chloride solutions. Presumably, they live off ammonia and carbon dioxide in the water. This may be retarded by storage in a freezer or refrigerator, by adding a drop of chloroform to the reagent before sealing, or by filtration through a Millipore filter (the 0.45 micron pore size is sufficient). Growth also seems inhibited in more concentrated solutions.

One minor precaution against contamination is not to pipette excess reagent back into the bottle. Of course, this is only if the reagent is not expensive or hard to replace.

## 1.5  PIPETTES AND PIPETTING

Graduate students in biochemistry should have learned how to pipette reagents before entering graduate school, but in fact some have not. Several types of pipette are commonly in use: volumetric (not graduated), serological (graduated to the tip), measuring (not graduated to the tip), long tip (these may be serological or measuring), and microliter ("lambda").

For adding large volumes of solution with the greatest accuracy, volumetric pipettes are used. These are filled to above the mark ("fill line") while held vertically, the outside is carefully wiped off, and the excess solution released to the fill line. While still held vertically, the specified volume of solution is released into the desired receptacle until the flow stops. Wait about ten seconds for solution on the walls of the pipette to collect in the tip and withdraw the pipette with a twisting motion, holding the tip against the inside of the receiving vessel. Do not blow out the remaining liquid.

For adding intermediate volumes, use long tip pipettes (or measuring pipettes, if long tip pipettes are not available). Use the appropriate size of pipette; do not try to measure out say 0.2 ml samples with a 5 ml pipette. Draw the reagent up past the "0" mark, wipe off the outside of the pipette, release the reagent to the "0" mark while holding the pipette vertically, and touch off the tip end. Then move the pipette to the receiving vessel and allow the liquid in the pipette to drop to the desired level. Keep the pipette vertical and the flow slow—take twenty seconds or so.[2] Then touch the tip again to the inside of the receiving vessel and remove the pipette.

---

[2]Longer for viscous solutions.

For adding small volumes, the tip of the pipette should be wiped carefully and thoroughly after loading the pipette. Then the cleaned tip of the pipette is inserted *into* the receiving solution, the desired volume released, and the pipette withdrawn. Be sure to clean off the tip of the pipette again before reinserting into the reagent.

Any pipettes used should drain cleanly; if they do not they are dirty and should not be used. Some pipettes are calibrated "to contain"; these must be washed out into the receiving vessel to get all the contents transferred. Serological and Mohr pipettes are generally considered to be calibrated less accurately than measuring pipettes. However, after emptying them, they must be held vertical about twenty seconds,[2] and then the last drop of solution collected in the tip is blown out into the receiving vessel.

For some experiments, you will want to transfer solutions with the greatest accuracy—the least real deviation from the "correct" value. For others, precision (reproducibility) rather than accuracy may be of prime importance.

## REFERENCES

1. Hamner, C. E., *University of Virginia Medical School Biochemistry Manual* (unpublished).
2. Westhead, E. W. and G. McLain, *J. Biol. Chem.*, **239**, 2464 (1964).

# 2 | TREATMENT OF GAUSSIAN MEASUREMENT DATA

## 2.1 INTRODUCTION

The economist, agriculturist, and geneticist frequently employ experimental designs in which a sample consisting of a large number of data (greater than thirty) is collected for the calculation of the estimates of the *population* parameters.[1] The biochemist is usually not so fortunate, since the cost and scarcity of his experimental materials and the arduousness of his analyses limit him to a small-number sample for statistical analysis. Therefore, he is forced to

---

*Copyright 1969 by R. B. Ballentine. Reprinted by permission of the author.

[1]It should be emphasized that the term "population" has a special meaning to the statistician which is different from the casual meaning given it by the experimental scientist. As used statistically, a "population" is an infinite set of numbers; for example, a set of numbers constituting a *specific population* having the mathematical properties of a *normal* Gaussian distribution. Such a *normal population* is completely defined by the statistical parameters, the *mean* and the *standard deviation*. Thus, an infinite number of *normal populations* is possible, each differing from every other normal population in that one or both of its parameters have different values from those of any other population.

utilize the small-sample statistics which were developed largely by R. A. Fisher (1). As a result of the research of this great statistician and mathematician, the small-sample statistics are simpler, if anything, to use than the more conventional large-sample statistics, and they are of no less rigor in their derivation. In this section we shall limit ourselves to the utilization of small-sample statistics in the treatment of data.

A collection or *sample* of data from a population can be of two general kinds. Typical of these kinds is the counting data which are represented by a table of frequencies such as would be obtained in counting a cell suspension with a cytometer, or in determining the number of radioactive atoms decomposing in a unit of time. This type of data is *discontinuous*, since the frequencies assume discreté integral values; i.e., we do not have fractional cells in a hemocytometer field count, nor do we have a fractional part of an atom undergoing radioactive decay. The other kind of data sample, the one most often obtained when we perform an analysis, is a collection of measurements whose values ideally form a *continuous* distribution; i.e., we can ideally subdivide the volume delivered by a burette into an infinite number of smaller volumes and are only practically limited to the smallest division on the burette which can be read. This practical limitation on the measurements does not affect the ideal concept of a potentially infinite and continuous distribution of measurements and hence does not alter the mathematics. Gauss showed in his classical contribution to the theory of observations that the individual values of measurements drawn from an infinite population of such measurements are normally distributed about the mean (2). Consequently, the Gaussian *normal error curve* forms the mathematical basis for the treatment of measurement data. It is this type of continuously variable measurement-data samples which we shall be concerned with here.

## 2.2  SOME DEFINITIONS

### 2.2a  Samples and Populations

When we perform a series of replicate analyses upon a material what we are actually doing is drawing a *sample* of data from an ideal, infinite *population* of such analytical values pertaining to the material being studied. Obviously, if it were possible to collect all the data in this infinite population, we would be able to calculate the "true" values for the population parameters representing the quantitative characteristics of this population. We shall represent such population parameters by lower case Greek letters; for example, we shall let $\mu$ stand for the weighted average or "true" *mean* of the *population*. Since it is obviously impossible to collect all the data in an infinite population, we actually draw a sample from it in performing our measurements and then inquire, statistically speaking, how well our sample represents the "true" population values. We shall represent these statistics or estimates derived from the sample of the population parameters by lower case English letters, such as $m$ for the mean of the sample. (For reasons which will develop shortly, it is more usual to represent the *sample mean* by $\bar{x}$).

A statistical concept which we will encounter frequently is that of *degrees of freedom*, d.f., this being the number of variants contributing to a statistic which may be independently varied. To demonstrate, consider the equation

$$Y = X_1 + X_2 + \ldots + X_n$$

$Y$ has associated with it $n - 1$ degrees of freedom. If the value of $Y$ is fixed, we can assign any of an infinite variety of values to $X_1$ thru $X_{n-1}$, but this fixes $X_n$ to a value given by $X_n = Y - \sum_1^{n-1} X_i$ if the equation is to balance. In general, every time we calculate a statistic—for example, the mean of the sample—we reduce the degrees of freedom by one. The calculation and apportionment of the degrees of freedom in statistical calculations is one of fundamental importance, especially in the analysis of variance and experimental design. With this in mind let us now proceed to the definition and mode of calculation of the various useful statistics for a sample of measurement data.

### 2.2b  The Mean, $\bar{x}$

In general, the purpose of most experiments with their attendant analyses is to determine the arithmetical average value, the *mean*, of the variable being measured; for example, the hemoglobin content of the blood of a group of control animals. If we have $n$ determinations of $X$ in our sample of data (where $X$ is the value of each individual determination), the *mean*, $\bar{x}$, is given by

$$\bar{x} = \frac{\sum X}{n} \tag{2-1}$$

Since $\bar{x}$ is determined by $n$ independent values, it has associated with it $n$ degrees of freedom.

### 2.2c  The Median

A moment of consideration will reveal that the mean is very subject to bias by extreme values in the sample. Figure 2.1 illustrates this. To avoid this bias, some of the older statistical treatments used the median as a measure of the central moment of the sample. The *median* is the value of the middle datum in an array of the data arranged according to magnitude. However, while the median does reduce the biasing effect of extreme values, it is not a particularly useful statistic, mathematically speaking. If the data are numerous, determination of the median is an arduous task. Therefore, it is little used today.

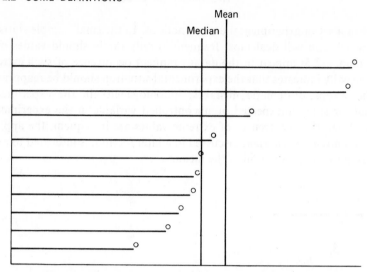

**Figure 2.1**   Showing how extreme values bias the mean more than the median.

### 2.2d   Standard Deviation, $\pm\sigma$ and $\pm s$

The standard deviation of the observations (represented by $\pm\sigma$, with respect to the *population*, and by $\pm s$ with respect to the *sample*) is one of the most useful of statistics. It is especially valuable for the control and the analysis of errors encountered in the development of an analytical procedure. It may be calculated by the equation:

$$s = \pm\sqrt{\frac{\sum (x')^2}{n-1}} \qquad (2\text{-}2)$$

$$\text{where}\quad x' = X - \bar{x} \qquad (2\text{-}3)$$

Or, expressed in words, it is the square root of the sum of the squares of the deviations of the observed values from their mean, divided by the degrees of freedom. Actually, for calculating machines, a more convenient formula is

$$s = \pm\sqrt{\frac{\sum X^2 - n(\bar{x})^2}{n-1}} \qquad (2\text{-}4)$$

If the sample of data being analyzed is drawn from a Gaussian normal population of errors, then the following percentage of individual values $X$ will be expected to lie, on the average, within the intervals of $\pm s$ shown in Table 2.1. Examination of the *deviations* of your experimental values from their *mean*, $(x')$, in the light of this table, will frequently reveal when an experimental process has gone out of control, or when there is an unexpected inhomogeneity in a set of experimental determinations. In the small-sample statistics (with which you will deal most frequently) only rarely should values of $x'$ larger than $\pm 2.5s$ appear in the data. Frequent occurrence of such extreme values usually indicates that the experimental situation should be reappraised for aberrations, systematic mistakes, inhomogeneity in the experimental material, or some unexpected or uncontrolled variable in the experimental design. When the occurrences of extreme values are infrequent, the application of *Chauvenet's criterion*, discussed in a later section, is indicated to obviate bias in the determination of the mean.

**TABLE 2.1**

| | |
|---|---|
| 0.5$s$ | 38.3% |
| 1.0 | 63.2 |
| 1.5 | 86.6 |
| 2.0 | 95.4 |
| 2.5 | 98.8 |
| 3.0 | 99.7 |

### 2.2e    Relative Standard Deviation or Coefficient of Variation, C

The fact that the standard deviation is a dimensioned number reduces its utility, since its magnitude depends upon the values of the sampled data. For example $\pm s$ computed for a sample of weights of infant animals will be of different magnitude from that for a similar sample of weights of adults. Similarly, in comparisons between the enzyme content of the livers of a control and an experimental set of animals, the values of both the mean and the standard deviation may be of different orders of magnitude if there is a definite effect of the treatment under consideration. Therefore, it is convenient to reduce the standard deviation to a relative or dimensionless number, i.e., a percentage. When so reduced by the following equation, it is called the *coefficient of variation, C.*

$$C = \pm \frac{s}{\bar{x}} \cdot 100\% \qquad (2\text{-}5)$$

Since this now expresses the variation of the observations in terms of a percentage of the mean, it is now possible to compare the variation between two sets of data whose means may be widely different. Indeed, we may compare the variation in widely different types of data; for example, the relative error in the determination of an enzyme by a spectrophotometric method as opposed to the error inherent in a radioactive tracer assay. This is of value in designing experimental procedures, since one may now discover which specific step in a sequential series of experimental operations constitutes the major source of error.

### 2.2f  Standard Deviation of the Mean, $s_{\bar{x}}$

After an experimental procedure is established and we are conducting the experiment proper, the statistic of greatest interest is the *mean*. Since $\bar{x}$ is determined from a sample of the populations, we are concerned with how accurately it represents the "true" value, $\mu$ (the mean of the infinite population). If we were to draw repeated samples of data from the population and determine their means, we would find that the means were also distributed according to a Gaussian error function, provided the original population was normal. A corollary of this is that no matter how one subdivides a normal population, the subdivisions are also normal provided that $n$ is reasonably large.

Assuming that the means are distributed normally, we may validly calculate a *standard deviation of the mean*

$$s_{\bar{x}} = \pm \frac{s}{\sqrt{n}} \tag{2-6}$$

With this value we may roughly estimate two means are significantly different by noting whether the difference between them is greater than twice the sum of their standard deviations. This is only a rough indication; the *t*-test, to be

**TABLE 2.2**

| $n$ | $s_{\bar{x}}$ |
|-----|------|
| 1   | 1.00 |
| 2   | .70  |
| 3   | .58  |
| 4   | .50  |
| 6   | .41  |
| 8   | .35  |
| 10  | .32  |
| 15  | .26  |
| 20  | .22  |
| 25  | .20  |
| 30  | .18  |
| 50  | .14  |

discussed subsequently, should be applied if the differences are considered *significant*. Nevertheless, this rough estimate is very useful for controlling overenthusiasm of the experimenter who might otherwise be tempted to regard the differences between his experimental and control groups as significant.

It is this statistic, $s_{\bar{x}}$, which is disheartening to the analyst, since the precision of measurement of the mean does not increase linearly with the increase in the number of replications but rather as the square root of the number of replications. Table 2.2 reveals some very useful information in designing an analytical experiment with reference to the relationship between precision, sample size and economy of effort. Thus, going from a single determination to 3 or 4 replicates in determining the *mean* increases our precision by 42% and 50%, respectively. On the other hand, increasing the number of replicates from 10 to 20 only increases the precision of its estimate 10%, while a further increase from 20 to 50 determinations leads to additional precision of only 8%. Thus it becomes statistically obvious why most analysts choose 3 or 4 replicate determinations for the analysis of an unknown, and even indicates why it is possible to do most analytical experiments at all.

## 2.3  CONFIDENCE OR FIDUCIAL LIMITS

In our discussion of the standard deviation of the mean in the previous section the suggestion was made that this statistic might be utilized to judge the reliability or the precision of estimate of a *sample mean*. Table 2.2 provided a semiquantitative mechanism for judging the significance of the observed differences between the means of two samples of data—or, to be more specific, of forming some opinion as to whether the difference arose simply from the errors of sampling a normally distributed population and therefore should not be considered significant or whether we should judge them to be significant of a response to experimental treatment. Since such decisions with respect to significance form the crux of experimental investigation, we wish to examine at this point in greater detail the development by "Student" and Fisher of a test to place estimates of significance on a firmer theoretical and quantitative basis, namely the *t-test*. We shall limit ourselves first to the use of the *t*-test in assessing the reliability of estimate of the mean of a *single* sample and shall defer to a subsequent section its use in drawing conclusions about the more important, but also more complex question of the significance of *difference* between the means of two samples.

Before entering into a discussion of the *t*-test, it may be well to discuss some aspects of populations and distributions. When a statistician speaks of a population, he is speaking of an infinite set of numbers which may have certain definable parameters. In the case of measurement data, as was shown by Gauss, an *infinite normal population* has a distribution given by the Gaussian error function, and as a result of the mathematics, a normal population is completely defined when the two parameters, the population *mean*, $\mu$, and the population *standard deviation*, $\sigma$, are given. When we draw a sample of observations from such a normal population, we can make estimates of these population parameters from the sample statistics, and then as experimentalists we should inquire how well our statistics evaluate the parameters. There is a basic but rather elusive difference in concept between a statistic and a parameter which must be kept in mind in this discussion. By virtue of the fact that the population is infinite and obeys definable mathematical laws, each parameter for any given population has a single value. For example, there is only one value of the mean and one value of the standard deviation associated with any specific normal population. On the contrary, when samples are drawn from a specific normal population, due to the random nature of the sampling, the statistics estimating the parameters will have a range of values distributed about the unitary value of the population parameter. Thus the sample statistics will also obey a distribution law, and, as Gauss showed (2), as the number of samples drawn from the population is indefinitely increased their distribution approaches that of the ideal normal distribution. This approach to the normalcy in the distribution of statistics does not occur until the number of samples becomes quite large, and as a consequence we must seek a more fitting distribution than the Gaussian one to apply in judging the reliability of these estimates.

Such a distribution is that of $F(t)$, which was derived on a mathematical basis first by "Student" (3) and subsequently by Fisher (1). This "t" distribution, which can in a qualitative sense be considered an adaptation of an infinite normal population distribution to small finite populations, forms the basis for establishing the *confidence* or *fiducial limits* of the sample mean, and, as we shall see in a later section, it will permit us to make probability statements about the difference between the means of two separate samples.

You will note that by the use of the $t$ distribution we have limited ourselves to statements about the mean. This limitation is imposed by the fact that the sample *standard deviation* becomes an integral part of the $t$ distribution in its operational form, the $t$-test. A different frequency distribution, Pearson's chi-square, is used to judge how correctly the standard deviation of the sample represents the standard deviation of the population.

The great strength of the $t$ distribution with its accompanying $t$-test is that it allows us to inquire in a quantitative manner how accurately our sample mean represents the *true mean* of the population. We must recognize that these cannot be absolute statements, but will always have associated with them a qualifying probability. For example, we can make the statement that we can be 95% confident that the true mean lies within certain fiducial limits, but we cannot say with absolute certainty that this is true.

Let us now return to the actual utilization of the $t$ distribution with its attendant $t$-test for calculating the fiducial limits of the sample mean. Fisher (1) and "Student" (3) have provided us with tables of "$t$" which they calculated by integrating within definite percentage confidence limits the distribution function, $F(t)$. Also entering into this calculation are the number of degrees of freedom associated with "$t$" and derived from the sample size, that is $n - 1$. The complementary relationship for $t$, derivable from our sample statistics, is given in the equation

$$\pm t = \frac{\bar{x} - \mu}{s_{\bar{x}}}, \text{d.f.} = n - 1 \tag{2-7}$$

Since $\bar{x}$ and $s_{\bar{x}}$ can be directly calculated from the $n$ observations constituting our sample, we have only to choose our percentage confidence limits to read a value of $t$ from the table and calculate the limits for the value for $\mu$. These limits are called the *fiducial limits*, $l_1$ and $l_2$, and may be expressed by the pair of equations:

$$l_1 = \bar{x} + (t_{p,n-1})s_{\bar{x}}$$
$$l_2 = \bar{x} - (t_{p,n-1})s_{\bar{x}} \tag{2-8}$$

Turning to a numerical example (below), we choose a value of $t$ from Fisher's tables (Appendix 8b) corresponding to nine degrees of freedom and a percentage confidence limit at 0.05. Calculation of the fiducial limits allows us to say that we are 95% confident that the true population mean falls between 126.7 and 120.5, while our observed sample mean is $123.6 \pm 1.38$. (Note we use the value of $s_{\bar{x}}$.)

If we wish even greater reliability, we calculate $l_2$ and $l_2$ for $t_{0.01}$, and thus are able to state that we are 99% confident that the true mean lies in the interval 128.1 to 119.1. If we wish to be 99.9% confident, then the limits become 130.2 to 117.0. These quantitative statements about the precision of our estimates are based upon the assumption of a normal distribution in the population. However, it has been shown that the population may deviate considerably from the normal and still permit analysis by the $t$-distribution without introducing a major unreliability in the conclusions. In plotting data, such as amount of growth obtained at various growth factor concentrations, it is good practice to represent the fiducial limits of each point by vertical bars. This procedure presumes that each plotted point is the mean of a number of observations.

We are given 10 values of $x$: 125.6, 117.9, 130.1, 117.0, 127.1, 127.5, 122.1, 120.0, 122.9, and 124.0. The calculations are

$$n = 10$$
$$\bar{x} = 123.62$$
$$\sum X^2 = 152{,}990.26$$
$$n\bar{x}^2 = 152{,}819.04$$
$$s^2 = \frac{X^2 - n\bar{x}^2}{n - 1} = 171.22/9 = 19.024$$

$$s = \pm 4.362$$
$$C = \pm 3.53\%$$
$$s_{\bar{x}} = \frac{\pm 4.362}{\sqrt{10}} = \pm \frac{4.362}{3.162} = \pm 1.38 \ (\pm 1.12\%)$$
$$n - 1 = 9; \ t_{0.05} = 2.262$$
$$l_1 = \bar{x} + t_{0.05}s_{\bar{x}}$$
$$= 123.62 + 2.262 \times 1.38$$
$$= 126.74$$
$$l_2 = 120.5$$

## 2.4 ERRORS AND THE REJECTION OF DATA

One of the most difficult problems facing the experimentalist is the general question of mistakes in a collection of data and the rejection or elimination of the defective datum. It is axiomatic that we do not exclude a value from the report of our experimental results simply because it does not agree with our preconceived ideas or does not fit in the table of values. However, there are occasions when a value's inclusion in the mean, for example, will give a *biased* result. Under these conditions the datum should be excluded from the calculation of the mean (but not from the table of data), and the conditions and criteria for this exclusion will be considered shortly.

First a word about definite mistakes in an analysis or experimental determination. A safe rule to follow is to reject such a datum as soon as it is known to be in error, up to but *not* including the attainment of the final numerical result. For instance if one suspects that there has been a mismeasurement or that a mistake in weighing or pipetting has occurred, the analysis should be terminated *at that point*. One should not wait to see if the value for the analysis is in agreement with the expected results, for then the value *must* be retained, and if a true mistake in operations has occurred a bias will be entered into all the subsequent calculations. *Data from experimental mistakes should be avoided by terminating the specific determination at the moment of recognition of the mistake, provided this is prior to reading the final value of the determination.*

There is another kind of "error," however, which can introduce bias in the final results. As we have discussed previously, Gauss showed that, in a set of data from a large or infinitely continuous population of values, the error is distributed normally. Consequently, if we draw a small sample from this population—say we perform ten analyses—there is a small but finite and calculable probability of including in the set a value lying in the tail of the normal distribution. As you remember, a characteristic of the mean is that it is very sensitive to extreme values, and consequently if an extreme value is included in the calculation the value of the *sample mean* will be biased and will show a greater divergence from the "true" or "*population mean*" than if the single extreme datum were excluded.

Various techniques are available for avoiding or adjusting this bias. The one most commonly used is *Chauvenet's criterion;* an excellent discussion of this, along with other adjustment techniques, can be found in Chapter 8 of Smart's *Combination of Observations* (4). Chauvenet's criterion cannot be applied under all circumstances. Let us therefore consider some advantages and limitations of this method of adjustment and also when one is justified in applying it.

From a strict mathematical point of view, this method is not on the firmest foundation, and there are many purist statisticians who reject its application under all circumstances. Others feel that when it is wisely and judiciously used it leads to a value more closely approaching the "true" value than if no method of adjustment were used. (Incidentally, the use of this criterion frequently results in the rejection of aberrant values introduced by an analytical mistake, although this is not its purpose.) Further, one of the main values of the technique is that it is objective. There is one stipulation, however, which must always be observed in applying this technique: the method to which it is being applied must be a statistically tested one, and one of recognized reliability. Use of the Chauvenet criterion in the development of a new method would quite likely obscure and delay the recognition of a lack of control in the method.

Let us now examine the statistical justification of the method. Chauvenet's criterion defines mathematically the conditions which a value must satisfy to be considered sufficiently extreme to be eliminated from the calculation of the mean. Obviously if the number of data constituting the sample is large (several hundred), extreme values would have little effect on the mean, and there would be neither justification nor reason for excluding them. Therefore the criterion is applied only to small samples. The mathematical weakness of Chauvenet's criterion lies in its derivation from the theory for infinite populations. If you are interested in this point, reference 4 will supply you with the mathematical arguments. In my own opinion, while I am rather bothered by the lack of mathematical rigor in Chauvenet's criterion, it does bridge the gap between the unjustifiable intuitive discarding of data because they appear to be wrong and the arbitrary inclusion of data in the final results even though it is apparent they will lead to a more incorrect value than if they were rejected. Therefore the method will be presented here.

First calculate the mean $(\bar{x})$, the standard deviation $(s)$ and the standard deviation of the mean $(s_{\bar{x}})$. Regardless of whether the criterion results in a rejection of one or more values, the standard deviations remain unaltered (i.e., they are calculated from the full set of values in the data sample), and only the mean is adjusted. The rejected values are indicated by an asterisk (*) in the table of data.

Next calculate the deviations from the *mean*

$$x' = X - \bar{x} \qquad\qquad (2\text{-}3)$$

and tabulate them. Select the single value which has the greatest deviation from the mean and test it for rejection. The steps are:

1. Calculate

$$\text{erf } hx' = \frac{2n - 1}{2n} \qquad (2\text{-}9)$$

where $n$ = number of observations.

2. Obtain the value of $hx'$ from tables of erf$(hx')$ (Appendix 8c). erf$(hx')$ is the error function of $hx'$ given by

$$\frac{2}{\sqrt{\pi}} \int_0^{hx'} e^{-(hx')^2} \, d(hx')$$

and is simply the probability that an error will exceed the value, $x'$. Appendix 8c gives erf$(t)$ for arguments of $t$ (in our case $t = hx'$), and, since erf$(hx') = (2n - 1)/2n$, one simply reads off the value of $hx'$ $(t)$.

3. $h$ is the *modulus of precision* and is also given by

$$h = \pm \frac{1}{s\sqrt{2}} \qquad (2\text{-}10)$$

4. Since we know $x'$ and we have found $hx'$ in step 2, we can readily calculate the *observed value* of $h$. If $h$, calculated in step 3, exceeds this, the observation is to be excluded from the calculation of the mean.

Actually the calculations are very simple and can be performed rapidly. If the value is rejected, then one proceeds to examine the next largest $x'$ after recalculating the mean and the deviations from the mean. Remember that the new $n$ is now one smaller and that the *standard deviations* are the same as calculated with the complete set of data. It is infrequent that a second value is found to merit rejection. When second and third values are frequently rejected, it should be suspected that the method is out of control, and the experimental techniques should be carefully examined for operational error or mistakes.

This objective test is very useful to remove the bias from one's results without recourse to intuition. An excellent example of the use of Chauvenet's criteria was that of Ryan and Brand (5) in their *Neurospora* bioassay for leucine. It eventually turned out that their aberrant values were actually drawn from a different population of samples since they were caused by back mutation of the leucineless mutant in the presence of a limiting concentration of its required growth factor. In this case the use of Chauvenet's criteria resulted not only in a more accurate assay of proteins for the amino acid, but initiated a whole study of back-mutation by the Ryan school.

Let us examine another set of data to see how Chauvenet's criterion is applied. Given the data:

| $x$ | $x'$ |
|-------|-------|
| 172.4 | 13.8 |
| 154.9 | −3.7 |
| 162.6 | 4.0 |
| 150.6 | −8.0 |
| 152.7 | −5.9 |
| 163.3 | 4.7 |
| 160.8 | 2.2 |
| 156.0 | −2.6 |
| 164.0 | 5.4 |
| 149.0 | −9.6 |

$$n = 10$$
$$\bar{x} = 158.63$$
$$\sum X^2 = 252108.71$$
$$n\bar{x}^2 = 251634.77$$
$$s^2 = 473.94/(10 - 1) = 52.66$$
$$s = \pm 7.257$$
$$s_{\bar{x}} = 2.295$$

*Chauvenet's Criterion Test*

$$x' = 13.8$$

$$\text{erf}\,(hx') = \frac{2n - 1}{2n} = 0.9500 \qquad h = \pm \frac{1}{|s|\sqrt{2}} = \pm \frac{1}{(7.257)(1.414)}$$

$$hx' = 1.39$$
$$h = 0.1007 \qquad\qquad\qquad h = \pm 0.0975$$

Therefore, since $0.1007 > 0.0975$, the observation is retained.

## 2.5  COMPOUNDING STANDARD DEVIATIONS

Frequently in handling experimental data we wish to obtain the standard deviation either for a particular step in a process or for a sequence of steps in an analysis. Sometimes these values can be determined by direct measurement, but we may wish to conserve the labor which would be involved. Again we may wish to compare the measured error (*coefficient of variation*) in a complete analysis with that calculated from the individual steps. Finally it is often impossible to measure directly the error of a particular operation in an analytical scheme, although both the total error of the process and the errors of each of the other contributing steps are known. Any of these may be calculated from the following relationship. For a process:

$$A = B + C + \ldots + N$$

the *standard deviation* of $A$ is given by

$$s_A = \pm\sqrt{s_B^2 + s_C^2 + \ldots + s_N^2} \qquad (2\text{-}11)$$

and

$$s_{\bar{x}_A} = \frac{\pm s_A}{\sqrt{n}} \qquad (2\text{-}12)$$

By suitable algebraic rearrangements of these equations any of the standard deviations, together with fiducial limits, may be calculated. This can be illustrated using another numerical example.

In determination of total nitrogen by the Kjeldahl procedure, the sample is first oxidized; then the ammonia which is formed during the ashing process is steam distilled into a standard acid solution and titrated. We wish to inquire about the apportionment of the variation to the individual steps in this analytical sequence. However, it is impossible to measure directly the error associated uniquely with either the ashing or the distillation. Since we are interested in the relative errors, but the $s$'s have been determined on samples with different *means*, we shall use the *coefficient of variation* for our calculations. We have the experimental values:

$$C_{\text{titration only}} = \pm 0.056\%$$

$$C_{\text{determination of ammonia}} = \pm 0.132\%$$

$$C_{\text{determination of total nitrogen}} = \pm 0.785\%$$

We may now set up the following equations:

$$(C_{\text{ammonia}})^2 = (C_{\text{distillation}})^2 + 2(C_{\text{titration}})^2$$

The factor 2 enters because the analysis is based on the difference between a standardizing and a sample titration

$$(C_{\text{total}})^2 = (C_{\text{ammonia}})^2 + (C_{\text{ashing}})^2$$

Substituting, rearranging, and extracting the square root, we obtain the following relative variations:

titration = $\pm 0.079\%$
distillation = $\pm 0.105$
ashing = $\pm 0.774$
total procedure = $\pm 0.785$

It is thus obvious that the ashing process contributes virtually all the error to the procedure, and if we wish to improve the precision of the analysis it is this step upon which we must concentrate our attention, at least until its error is of the same order of magnitude as that of the distillation step.

## 2.6  COMPARISON BETWEEN GROUPS; TESTS OF SIGNIFICANCE

Up to this point we have been primarily concerned with the analysis of a single sample of data drawn from a given population. For example, we have seen how to calculate the various statistics for estimating the parameters of the population, the mean and the standard deviation, how to use the standard deviation of the mean with the $t$-test, and have set up fiducial limits to enable us to judge how well our estimates represent the parameters. These are important statistical operations which are used in the estimation of the precision and accuracy of various biochemical analyses and techniques.

However, as experimentalists we are frequently more interested in the comparison between two samples of data, one of which we derive from measurements on a control group and the other from a group which has been subjected to certain experimental manipulation. Each of these samples will be characterized by a set of statistics calculated as described in the preceding sections. These are the numerical data which we may use to make a statistical inference as to which of the following two hypotheses to accept: (1) that both samples, control and experimental alike, were drawn from the same statistical population and therefore the difference between their means is simply a reflection of the normal *random variation* of sampling, or (2) each sample has been drawn from a separate statistical population having uniquely different parameters, and consequently the difference between the means reflects a *difference in the response* of the two groups to their respective treatments in addition to the normal variation of random sampling.

Let us again emphasize that in making statistical inferences about the sample populations, we can say that we are $x\%$ confident that this is the situation, but we can never make absolute inferences concerning randomly drawn samples. Nevertheless, if we can be $99\%$ or $99.9\%$ confident that the *mean* of a control and the mean of an experimental sample represent separate statistical populations and if we have used wisdom and judgement in the design and the execution of our experiment, we have a better basis than that of intuition alone on which to accept or reject the hypothesis.

A moment's consideration concerning statistical inference based on tests of significance will reveal that we can always be subject to one of two types of mistakes. Type I error is that in which we reject a hypothesis when it is true. Type II error arises when we accept a hypothesis which is actually false. For any given degree of freedom in our samples, it is possible to minimize the chance of making either a Type I or Type II error—but not both at the same time. Thus in experimental design it is important to consider which of these two types of error will have the most serious consequences; then we must design the experiment and select the percentage confidence and test of significance which will achieve the preferred reliability.

Having learned the necessary statistical calculations, the experimenter must still be cautioned that statistical tests of significance cannot replace good experimental design, accurate and meticulous execution of the laboratory operations, nor the use of common sense in the analysis of the experimental complex. For example, the statistician frequently speaks of *confounding*: the distortion of the tests of significance by uncontrolled variables in the experimental situation. To cite a trivial example, suppose the liver lactic dehydrogenase was carefully determined in a group of "control" rats. At some subsequent time a second group of rats maintained under different conditions, such as a different basal diet, is fed a toxic level of zinc and the liver dehydrogenase level is determined. In all likelihood the difference between the means of these two samples will be found to be highly statistically significant. However, the effect of the zinc toxicity has now been *confounded* with the effects of many other experimental differences, and the experiment, in spite of its tests of significance, is essentially meaningless. It would be a rare experimentalist who would be guilty of such poor experimental design, but subtle confounding is one of the pitfalls most difficult to avoid in experimental work, and statistics are no substitute for wisdom in the planning and interpretation of an experiment.

However, in a well-designed experiment these tests of significance provide an objective bulwark against reaching the wrong conclusion. The following quotation from Snedecor (6) sums up the relationship between statistics and experimental intuition.

It is sometimes not understood that statistical methods can bring out only that information which has been incorporated into the data by careful design and execution of sampling. Elaborate statistics are no substitute for meticulous experimentation. Population inferences are futile if dependent on carelessly collected data. It is equally often overlooked that extensive and conscientiously done measurements may contain little worthwhile information if the experimental design is faulty. It is only by a combination of appropriate design, skilful conduct of the experiment and suitable statistical methods that the investigator is assured of reliable evidence upon which to base his decisions.[2]

With these considerations in mind let us now turn to the actual calculations themselves. To conduct these tests we use the *t*-distribution with its attendant

[2]From Snedecor, G. W., *Statistical Methods Applied to Experiments in Agriculture and Biology*, 5th ed., Iowa State College Press, Ames, Iowa (1956), p. 100. Reprinted by permission of the publisher.

$t$-test which was discussed in connection with fiducial limits. The question we wish to answer with an $x\%$ confidence is whether the two samples come from the same or different populations. To simplify the mathematics, we set up the hypothesis (a *null hypothesis*) that both samples were randomly drawn from the same population, and hence the population mean and the standard deviation are the same for both samples. Under these conditions we may write

$$t = \pm \frac{\bar{x}'}{s_{\bar{x}'}} \qquad (2\text{-}13)$$

where $\bar{x}' = \bar{x}_1 - \bar{x}_2$ and $(s_{\bar{x}'})^2$ is the variance of the mean estimated from the *pooled variances*, $s^2$, of both samples. The quantity $s^2$ may be calculated by any of the following

$$s^2 = \frac{\sum X_1^2 - n_1(\bar{x}_1)^2 + \sum X_2^2 - n_2(\bar{x}_2)^2}{n_1 + n_2 - 2} \qquad (2\text{-}14a)$$

$$s^2 = \frac{\sum (x_1')^2 + \sum (x_2')^2}{n_1 + n_2 - 2} \qquad (2\text{-}14b)$$

$$= \frac{s_1^2(n_1 - 1) + s_2^2(n_2 - 1)}{n_1 + n_2 - 2} \qquad (2\text{-}14c)$$

and $(s_{\bar{x}'})^2$ is then given by

$$(s_{\bar{x}'})^2 = s^2 \frac{(n_1 + n_2)}{n_1 n_2} \qquad (2\text{-}15a)$$

which simplifies to

$$(s_{\bar{x}'})^2 = \frac{2s^2}{n}, \quad \text{when } n_1 = n_2 \qquad (2\text{-}15b)$$

As in the calculation of the fiducial limits, we establish the degree of confidence with which we wish to accept or reject our null hypothesis and obtain from the tables of $t$ the value corresponding to this percentage and to the number of *degrees of freedom* which is given by

$$\text{d.f.} = n_1 + n_2 - 2 \qquad (2\text{-}16)$$

If the value for $t_{p,\text{d.f.}}$ calculated from Equation 2-13 is greater than the tabular value, the hypothesis is rejected and we say with a *given $\%$ of confidence* that the samples were drawn from different populations, that the difference between their means is significant, and that the experimental variable under test has affected the response of the experimental group. Contrariwise, we accept the null hypothesis and reject the hypothesis that the experimental treatment was effective.

In this formulation of the *t*-test, it was assumed that the *variance* of the two samples was the same, as it would be if the samples were from the same population. However, this assumption biases our judgement in favor of making a Type I error, and in the case of the proposed null hypothesis such a bias is undesirable; i.e., we are apt to set up a program to investigate the effect of this experimental variable when actually it is without significant effect upon the test system. This bias can be avoided by employing a slightly more involved version of the *t*-test. Instead of using the pooled variances calculated by the Equations 2-14 above, we calculate

$$(s_{\bar{x}'})^2 = (s_{\bar{x}_1})^2 + (s_{\bar{x}_2})^2 \tag{2-17}$$

and use this in Equation 2.13 for calculating the value of *t*. The value of *t* derived from the table must be adjusted for the *two variances* used in the above calculation. First look up the value of *t* for d.f. $= n_1 - 1$, calling it $(t)_1$, and similarly $(t)_2$ with d.f. $= n_2 - 1$ (Appendix 8b). The adjusted value of *t* for comparison with that derived directly from the samples is given by

$$t = \frac{(t)_1(s_{\bar{x}_1}) + (t)_2(s_{\bar{x}_2})}{(s_{\bar{x}_1}) + (s_{\bar{x}_2})} \tag{2-18}$$

Although this procedure for the *t*-test involves slightly more calculation, it is the one preferred.

The above methods for calculating the tests of significance are those adopted when there is no *a priori* reason for pairing the experimental observations. However, there are circumstances in which we can greatly decrease our experimental variability. For example, we would anticipate that litter mates would respond more similarly than would animals drawn at random from a colony. Consequently, in setting up an experiment we would be wise to divide the litter mates equally between the control and the experimental groups, and then to test for significant differences between the data pertaining to litter mates. To conduct this test of significance, tabulate the difference between the pairs and treat this as a new sample, calculating the mean and standard deviation of the mean. The degree of freedom is $n - 1$, where *n* is the number of pairs. These values are substituted directly into Equation 2-13 and the table of *t* is entered for the chosen *p* and d.f. The conclusions follow the same pattern as for unpaired data.

This is best illustrated with another numerical example.

|        | Sample 1          |         | Sample 2          |
|--------|-------------------|---------|-------------------|
|        | 10                | $n$     | 10                |
|        | 158.63            | $\bar{x}$ | 123.62          |
|        | $473.94/9 = 52.66$ | $s^2$  | $171.22/9 = 19.024$ |
|        | $\pm 7.257$       | $s$     | $\pm 4.362$       |
|        | $\pm 2.295$       | $s_{\bar{x}}$ | $\pm 1.38$    |
|        | $\pm 4.57$        | $C$     | $\pm 3.53$        |

1. Simple $t$-test, $\sigma_1^2 = \sigma_2^2$ assumed.

(2-16) $$\text{d.f.} = n_1 + n_2 - 2 = 18$$

(2-14) $$\text{Pooled variance } s^2 = \frac{473.94 + 171.22}{18} = 16.818$$

(2-15) $$(s_{\bar{x}'})^2 = \frac{2(16.818)}{20} = 1.68$$

$$(s_{\bar{x}'}) = \pm 1.30$$

$$\bar{x}' = 158.63 - 123.62 = 35.01$$

(2-13) $$t = \pm \frac{35.01}{1.30} = 26.93$$

$$t_{0.01, \text{d.f.} 18} = 2.878$$

Therefore the means are very significantly different:

$$p > 0.01$$

2. *t-test without assumption of equal variances.*

(2-17) $$s_{x'}^2 = (2.295)^2 + (1.38)^2 = 7.171$$

$$s_{x'} = \pm 2.68$$

(2-13) $$t = \frac{35.01}{2.68} = 13.06$$

(2-18) $$t = \frac{(t_{0.01,\text{d.f.}9})(s_{\bar{x}_1}) + (t_{0.01,\text{d.f.}9})(s_{\bar{x}_2})}{s_{\bar{x}_1} + s_{\bar{x}_2}}$$

$$= \frac{(3.250)(2.29) + (3.250)(1.38)}{(2.29) + (1.38)} = 3.25$$

The differences are still very significant, but note the relative decrease in $t$.

## 2.7 SUMMARY

### 2.7a Summary of Formulas

*Mean*

$$(2\text{-}1) \qquad \bar{x} = \frac{\sum X}{n}$$

*Standard deviation of observation*

$$(2\text{-}2) \qquad s = \pm\sqrt{\frac{\sum (x')^2}{n-1}}$$

$$(2\text{-}3) \qquad \text{where } x' = X - \bar{x}$$

and

$$(2\text{-}4) \qquad s = \pm\sqrt{\frac{\sum X^2 - n(\bar{x})^2}{n-1}}$$

*Coefficient of variation (relative standard deviation)*

$$(2\text{-}5) \qquad C = \pm\frac{s}{\bar{x}} \cdot 100\%$$

*Standard deviation of the mean*

$$(2\text{-}6) \qquad s_{\bar{x}} = \pm\frac{s}{\sqrt{n}}$$

*t-test of "Student"*

$$(2\text{-}7) \qquad t = \pm\frac{\bar{x} - \mu}{s_{\bar{x}}}, \quad \text{d.f.} = n - 1$$

$$(2\text{-}8) \qquad l_1 = \bar{x} + (t_{p,n-1})(s_{\bar{x}}) \qquad l_2 = \bar{x} - (t_{p,n-1})(s_{\bar{x}})$$

*Chauvenet's Criterion*

$$(2\text{-}9) \qquad \text{erf } hx' = \frac{2n-1}{2n} = \frac{2}{\sqrt{\pi}}\int_0^{hx'} e^{-(hx')^2} \, d(hx')$$

$$(2\text{-}10) \qquad h = \pm\frac{1}{s\sqrt{2}}$$

*Compounding standard deviations*

$$(2\text{-}11) \qquad s_A = \pm\sqrt{s_B^2 + s_C^2 + \ldots + s_N^2}$$

$$(2\text{-}12) \qquad s_{\bar{x}_A} = \pm\frac{s_A}{\sqrt{n}}$$

*Test of significance*

(2-13)
$$t = \pm \frac{\bar{x}'}{s_{\bar{x}'}}, \quad \text{where } \bar{x}' = \bar{x}_1 - \bar{x}_2$$

(2-14a)
$$s^2 = \frac{\sum X_1^2 - n_1(\bar{x}_1)^2 + \sum X_2^2 - n_2(\bar{x}_2)^2}{n_1 + n_2 - 2}$$

(2-14b)
$$s^2 = \frac{\sum (x_1')^2 + \sum (x_2')^2}{n_1 + n_2 - 2}$$

(2-14c)
$$s^2 = \frac{s_1^2(n_1 - 1) + s_2^2(n_2 - 1)}{n_1 + n_2 - 2}$$

(2-15a)
$$(s_{\bar{x}'})^2 = s^2 \frac{(n_1 + n_2)}{n_1 n_2} \quad \text{when } n_1 \neq n_2$$

(2-15b)
$$(s_{\bar{x}'})^2 = \frac{2s^2}{n}, \quad \text{when } n_1 = n_2 = n$$

(2-16)
$$\text{d.f.} = (n_1 - 1) + (n_2 - 1)$$

(2-17)
$$(s_{\bar{x}'})^2 = (s_{\bar{x}_1})^2 + (s_{\bar{x}_2})^2$$

(2-18)
$$t = \frac{(t_{p,n_1-1})(s_{\bar{x}_1}) + (t_{p,n_2-1})(s_{\bar{x}_2})}{(s_{\bar{x}_1}) + (s_{\bar{x}_2})}$$

### 2.7b  Symbols Used

| | |
|---|---|
| $C$ | coefficient of variation |
| d.f. | degrees of freedom |
| erf $(t)$ | error function of $(t)$. (Note: this is not "Student's".) |
| $h$ | modulus of precision |
| $l_1, l_2$ | fiducial limits |
| $n$ | number of observations |
| $\pm s$ | standard deviation of the observations with respect to the sample |
| $s_{A,B,\text{etc}}$ | standard deviation of observations with respect to the sample A, B, etc |
| $s_{\bar{x}}$ | standard deviation of mean with respect to sample |
| $s_{\bar{x}_A}$ | standard deviation of the mean of $s$ with respect to the sample A |
| $(s_{\bar{x}'})^2$ | variance of the mean estimated from the pooled variances |
| $t$ | "Student's" test |
| $X$ | observation or observed value |
| $\bar{x}$ | the mean of the sample |
| $x'$ | deviation of the observed value, $X$, from the mean |
| $\bar{x}'$ | the difference between sample means |

$\mu$       the mean of the population or true mean

$\sigma$       standard deviation of the observation of the population

## REFERENCES

1. Fisher, R. A., *Statistical Methods and Scientific Inference*, Hafner, New York (1956).
2. Gauss, C. F., "Theoria Combinationis Observatorium, Supplement," *Werke* (1826).
3. "Student," *Biometrika*, **6**, 1 (1908).
4. Smart, W. M., *Combination of Observations*, Cambridge University Press, New York (1958), Ch. 8.
5. Ryan, F. J. and E. Brand, *J. Biol. Chem.*, **154**, 161 (1944).
6. Snedecor, G. W., *Statistical Methods Applied to Experiments in Agriculture and Biology*, 5th ed., Iowa State College Press, Ames, Iowa (1956).

# 3 ‖ CHROMATOGRAPHY

## 3.1 INTRODUCTION

Biochemistry has been extensively concerned with isolation and analysis of substances, and isolation has almost always involved some form of chromatography.

Chromatography involves separation of the components in a mixture, using a medium through which a flow of gas or liquid is passed which causes a differential migration of the components (1). The flow is generally driven by pressure or gravity or some other mechanical action.

Any procedure that effects separations of substances can be carried out as a boundary method or a zone method. A *boundary method* involves analysis of the distribution of a moving material at its boundary with pure solvent. This is strictly an analytical method to determine the degree of heterogeneity of the material. A *zone method* starts with the material concentrated in a thin band or zone—the thinner the better—to produce complete separation of the components, if there are more than one, into subzones. Zone methods can consequently be used either for analytical or for preparative purposes. Chromatographic procedures are normally carried out as zone methods.

Chromatography can also be divided into partition and adsorption chromatography depending on the interaction between the medium and the components. In *partition chromatography* the medium can be another solvent or liquid immobilized on a solid such as paper or Sephadex. It acts essentially by providing another phase; that is, separation is achieved by means of solubility differences. Specific interactions between the components to be separated and the medium or solvent are minimal. In *adsorption chromatography* the medium is designed or selected to interact more or less specifically with some or all of the components, and the solvent is selected to increase or decrease these interactions.

Essentially, partition chromatography is based on ideal thermodynamic behavior of all the substances involved, while the opposite is true of adsorption chromatography. In practice of course, no such sharp division can exist in nature since all substances interact with each other to some extent. Adsorption effects can assist or interfere with separation by partition methods and vice versa. Consequently, the selection of chromatographic materials for a particular separation is of great importance. Another consequence is that irrespective of the particular chromatographic technique used, both partition and adsorption processes can be of importance in effecting a particular separation.

In this chapter we discuss the more common chromatographic methods, with particular emphasis on column chromatography. Because chromatographic techniques vary tremendously in the principles involved in separation, the instrumentation, etc., each major technique is dealt with separately, and arranged in order of increasing importance of adsorption effects in the technique.

## 3.2 CHROMATOGRAPHIC METHODS BASED PRIMARILY ON PARTITION

### 3.2a The Partition Coefficient

The partition coefficient is the ratio at equilibrium of the amounts of a substance dissolved in two immiscible solvents which are in contact. That is, it is the ratio of the solubilities of the substance in the solvents.

For the partition coefficient to be constant over a range of solute concentrations, Raoult's law and the gas law must hold; that is, the activity coefficient of the solute must be fairly constant over the concentration range of interest. Extensive adsorption effects can lead to widely varying partition coefficients so these effects must be minimal.

Raoult's law states that the partial pressure of a substance in a mixture is proportional only to the concentration of that substance (in mole fractions, or fractional purity) and the vapor pressure or volatility of the pure substance, and is otherwise independent of any other materials present (2).

If an ideal substance $A$ is dissolved in two ideal immiscible solvents, 1 and 2, and the solvents are left in contact until equilibrium is reached, the partition coefficient can be shown to be a constant (1):

$$\text{partition coefficient} = k = \frac{(A)_1}{(A)_2} = \frac{(C_A)_1}{(C_A)_2} \tag{3-1}$$

According to the gas laws

$$PV = nRT \tag{3-2}$$

$$\frac{n}{V} = C = \frac{P}{RT} \tag{3-3}$$

so

$$k = \frac{P_1}{(A)_2 RT} \tag{3-4}$$

According to Raoult's law

$$P_1 = XP_A^\circ \tag{3-5}$$

where $X$ is the mole fraction of substance $A$ in solvent 1 and $P_A^\circ$ is the vapor pressure of $A$. $X$ is proportional to $(C_A)_2$ or $b(A)_2$ ($b$ is a constant),

$$k = \frac{b(A)_2 P_A^\circ}{(A)_2 RT} = \frac{bP_A^\circ}{RT} \tag{3-6}$$

This is of course constant at a given temperature. Non-ideality can be accommodated by inclusion of the activity coefficient; if this is constant over the concentration range covered—no adsorption effects—the partition coefficient remains constant.

### 3.2b   Countercurrent Distribution

This involves the direct partitioning of a solute between two solvents (3). We assume here and in the following sections that the solute distribution between the two phases is at equilibrium.

### (i) Separation of solutes by partitioning between two solvents

This might be done with separatory funnels. For example, consider the successive extraction of a solvent, $U$, containing two substances $A$ and $B$, with a second solvent, $h$, and suppose that the respective solubilities of $A$ and $B$ in $U$ and $h$ are 1/4 and 4/1, and that the partition coefficients of $A$ and $B$ are constant, whatever the volumes of $U$ and $h$ [Figure 3.1(a)]. Note that we have effectively a binomial distribution of the two solutes [Figure 3.1(b)].

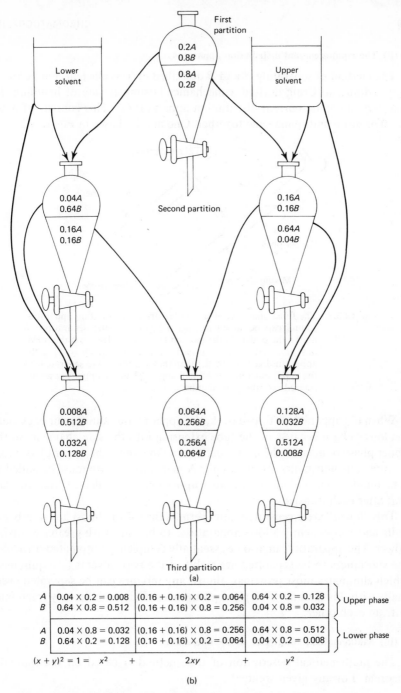

**Figure 3.1** (a) Representation of separation of two substances, $A$ and $B$, by successive extraction of a solvent ($U$) with equal volumes of a second solvent $h$, using separatory funnels. The partition coefficients of $A$ and $B$ in $U$ and $h$ are 0.25 and 4.00. (b) Distribution of $A$ and $B$ in upper and lower phases after three extractions. Note that a binomial distribution of each is obtained.

### (ii) The countercurrent distribution apparatus

The method of separating $A$ and $B$ described in (i) is efficient, but becomes very tedious, so Craig devised a mechanically simpler way of handling the transfers: a countercurrent distribution apparatus (3). This consists of a set of 30 or more units connected together. One unit is shown in Figure 3.2.

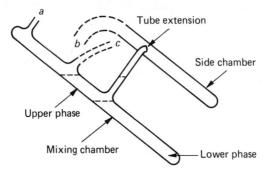

**Figure 3.2**   The functional unit of a countercurrent distribution apparatus. Liquid can be added to inlet $a$ before or after separation; $b$ marks the outlet, to the next unit; $c$ is the inlet, from the previous unit. Two phases in each unit are mixed by rocking the apparatus back and forth; then they are allowed to separate with the apparatus tilted at a small angle. Then the upper phases are poured into the side chambers.

When the apparatus is tipped back, the upper phase cannot pour back onto the lower phase because of the tube extension into the side chamber, so the upper phase pours into the next unit. At the same time, the upper phase from the previous unit pours into this unit. A new upper phase must be added to tube number one and an equilibrated upper phase is collected from the last unit after each transfer.

This "linear" arrangement of units permits the use of almost any number of units and consequently allows apparatuses to be built with greater resolving power. The apparatus can also be essentially completely automated. Provided the substances to be separated are stable in the two solvents (a requirement which eliminates most proteins), almost any mixture can be separated eventually with this method, even if the partition coefficients differ only fractionally from each other.

### (iii) Mathematical description of the operation

The mathematical description of this method is based on the binomial theorem. For any given solute

$$(x + y)^n = 1 \qquad (3\text{--}7)$$

$$\text{partition coefficient} = k = \frac{x}{y} = \frac{\text{fraction in upper phase}}{\text{fraction in lower phase}} \qquad (3\text{-}1a)$$

If we expand $(x + y)^n$, the terms of the resulting equation will each represent the amount in one tube (both phases) where the first term corresponds to tube 0, and $n$ is the number of successive extractions ("transfers").

$$(x + y)^n = x^n + nx^{n-1}y + \frac{n(n-1)}{2!}x^{n-2}y^2$$
$$+ \frac{n(n-1)(n-2)}{3!}x^{n-3}y^3 + \ldots + y^n \qquad (3-8)$$

As $n$ becomes large this equation begins to describe a Gaussian distribution, where the fraction $f_r$ in the $r$th tube is

$$f_r = \frac{1}{\sqrt{2\pi nxy}} \exp \frac{(r - nx)^2}{2nxy} \qquad (3\text{-}9)$$

This calculation can be further simplified if one regards $x$—the fraction of solute in the upper phase—as the probability that any one molecule will move forward at a transfer. It can be shown that

$$M = xn \qquad (3\text{-}10)$$

where $M$ is the tube of maximum material after $n$ transfers. From the position of the maximum, the total amount of material may be determined if it were not previously known. The fraction of material in the maximum tube ($f_M$) will be

$$f_M = \frac{1}{\sqrt{2\pi nxy}} \exp \left( \frac{-(M - nx)^2}{2nxy} \right) \qquad (3\text{-}9a)$$

$$= \frac{1}{\sqrt{2\pi nxy}} \qquad (3\text{-}11)$$

For a tube displaced $i$ tubes from the maximum tube, the fraction of material present can also be calculated.

$$f_{M+i} = \frac{1}{\sqrt{2\pi nxy}} \exp \left( \frac{-(M + i - nx)^2}{2nxy} \right) \qquad (3\text{-}9b)$$

$$= \frac{1}{\sqrt{2\pi nxy}} \exp \left( \frac{-i^2}{2nxy} \right) \qquad (3\text{-}12)$$

The area under the peak—proportional to the amount of solute—is easily calculated if the tube (transfer) numbers of the maximum and half maximum concentrations of solute are known (see Figure 3.3):

$$A = \text{area} = f_M \cdot \frac{2a}{0.94} \qquad (3\text{-}13)$$

where the maximum concentration of solute is $f_M$, and $a$ is the number of tubes between tube $f_M$ and the tube at which the concentration has dropped 50%.

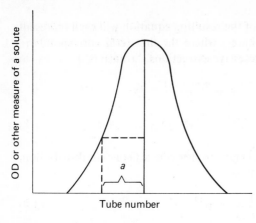

**Figure 3.3**
Measurement of the area under the peak of a solute eluted from a countercurrent distribution apparatus. *a* is the distance in fractions between the tube of highest solute concentration and one containing half that concentration.

### (iv) Sample calculations

Representative data from a countercurrent distribution experiment after 24 transfers are given in Table 3.1.

**TABLE 3.1**

| Tube No. | 0 | 1 | 2 | 3 | 4 | 5 | 6 | 7 | 8 |
|----------|------|------|------|------|------|------|------|------|------|
| OD | 0.00 | 0.07 | 0.26 | 0.68 | 1.27 | 1.92 | 2.16 | 2.06 | 1.52 |

| Tube No. | 9 | 10 | 11 | 12 | 13 |
|----------|------|------|------|------|------|
| OD | 1.03 | 0.55 | 0.25 | 0.09 | 0.05 |

If we take the tube of maximum concentration to be 6 (it is actually 6.35), we find

$$M = nx \qquad (3\text{--}10)$$

So

$$6 = 24(x) \qquad x = \frac{M}{n} = \frac{6}{24} = 0.25$$

Since $x + y = 1$, $y = 0.75$

So

$$\frac{x}{y} = k = 0.33 \qquad (3\text{-}7a)$$

As a test of homogeneity, the theoretical amount of material two tubes from the peak is

$$f \text{ at} (M + 2) = 2.16 \exp\left[\frac{-(2)^2}{2(24)(0.25)(0.75)}\right] = 2.16 \cdot \exp(-4/9)$$

$$= 2.16 \cdot 0.641 = 1.38$$

SEC. 3.2 CHROMATOGRAPHIC METHODS BASED PRIMARILY ON PARTITION 39

The actual values are 1.27 and 1.52. Depending on the accuracy of these measurements and the effect of taking tube 6.0 instead of 6.35 as the maximum fraction, these values may or may not be close enough to indicate a true Gaussian distribution. This would be the case if the material was homogeneous. A comparison of each measured value with the corresponding theoretical value is often performed to see if any discrepancies occurring are random or if they suggest an impurity at the leading or trailing edge of the peak.

From the data in Table 3.1, $a = 2.5$, so the area under the peak is

$$A = f_m \cdot \frac{2a}{0.94} = \frac{2.16(2)(2.5)}{0.94} = 11.5 \text{ OD units} \tag{3-13}$$

### (v) Significance of the number of transfer units ("plates")

Let us now assume we have a countercurrent distribution apparatus of $p$ units or "plates," where $p$ is a large number. We will put a marker in the upper phase of the first tube with $x = 1$, $k = 1.00$, and we will pour the effluent continuously through an optical density recorder. Note that the upper phase is transferred successively from one unit to the next, so it is referred to as the "mobile" phase, while the lower phase is called "stationary." We assume equal volumes of upper and lower phases (3) for simplicity. See Figure 3.4.

The place where the maximum emerges is proportional to the distances (or time, or volume) $(b + c)$. Therefore the number of transfers $n$ is proportional to $(b + c)$. We can define, in fact,

$$n = b + c \tag{3-14}$$

Now $x$, which we have redefined as the probability that a molecule of solute will move forward at a transfer, can be thought of as being analogous to the chromatographic $R_f$ of the solute; as the probability of moving forward increases, so does the $R_f$. In fact, we can define

$$x = R_f = \frac{b}{b + c} \tag{3-15}$$

Note that as the probability of transfer becomes one, the solute emerges with tube one after passing through the instrument, $c$ is zero and the $R_f$ of the compound is one. From Equations 3-10 and 3-7a,

$$M = nx = (b + c) \cdot \frac{b}{b + c} = b \tag{3-16}$$

and

$$M = n(1 - y) = n - ny \qquad y = \frac{n - M}{n} \tag{3-17}$$

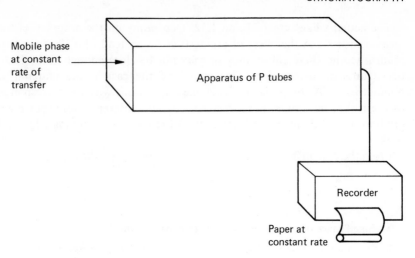

Mobile phase at constant rate of transfer

Apparatus of P tubes

Recorder

Paper at constant rate

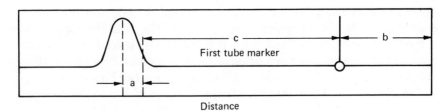

c

First tube marker

b

a

Distance

**Figure 3.4**  Elution of a solute from a countercurrent distribution apparatus containing $p$ "plates." The lower part of the figure is the recording of the elution; increasing distance (time or volume) is toward the left.

Also

$$n - M = c \qquad (3\text{–}18)$$

Now:  $i = \Delta n \cdot x = \Delta n \cdot M/n =$ number of tubes between the    (3–19)
maximum tube $M$ and the tube of half-maximal solute concentration and, where $\Delta n$ is proportional to $a$, the number of transfers between $f_M/2$ and $f_M$.

From Equation 3-12, at the half-maximum of the peak where

$$\frac{f_m + i}{f_m} = \frac{1}{2} = \exp\left(\frac{-i^2}{2nxy}\right)$$

we can substitute the above relationships for $i, n, x, y$:

$$\ln 2 = \frac{i^2}{2nxy} = \frac{\left(\Delta n \dfrac{M}{n}\right)^2}{2n\dfrac{M}{n}\dfrac{n - M}{n}} = \frac{(\Delta n)^2 M}{2n(n - M)}$$

Upon rearranging and solving for $M$, we obtain

$$M = \frac{2n(n - M)\ln 2}{(\Delta n)^2} \tag{3-20}$$

Since $\Delta n$ is proportional to $a$, we obtain an expression relating the half-width of the solute peak $a$ and the elution parameters $b$ and $c$ to the number of units or "plates" in the countercurrent distribution apparatus

$$p = \frac{5.54(b + c)c}{(2a)^2} \tag{3-21}$$

In other words, the greater the number of plates, the better the resolution in the sense that the peaks are sharper ($a$ is smaller) for a given elution volume ($b + c$) and different substances are better separated (if $a$ is constant, $b + c$ is larger).

#### (vi) Applications to other systems

The partition process is involved in gas and liquid partition chromatography, and similar formulas are found to apply to other systems. In liquid chromatography, the relation is

$$p = \frac{5.54(Vm)(Vm - V)}{2\left(Vm - \dfrac{Vm}{2}\right)} \tag{3-22}$$

$Vm$ is the effluent volume, and $V$ is the volume of the mobile phase in the column. In gas chromatography, $b + c \cong c$, so we have

$$p = \frac{5.54(c)^2}{(2a)^2} \tag{3-23}$$

#### 3.2c Liquid-Liquid Partition Chromatography (4)

#### (i) Theoretical considerations

Liquid-liquid partition chromatography is similar to countercurrent distribution except that the "plates" are not distinct. This type of chromatography includes paper and thin-layer chromatography; as a first approximation, there is a mobile phase and a liquid phase adsorbed to the paper or other carrier material. Separation depends on differences in solubility; the stationary phase in paper chromatography behaves like a concentrated carbohydrate solution (5). The mobile phase can be an aqueous or nonaqueous solvent. If the solute (s) is charged, the pH of the solvent will be an important consideration (see Appendices 9 and 10 for examples).

The $R_f$ is the counterpart of $x$; that is

$$M = nx \qquad\qquad (3\text{--}10)$$

(distance traveled by solute) = (distance traveled by solvent)$\cdot R_f$   (3–24)

The $R_f$ of a particular substance in a particular system will be character-istic of that substance, so $R_f$'s, or sometimes $R_r$'s—the migration of a sub-stance relative to some standard substance rather than the solvent—are customarily given.

However, the movement of a solute will be influenced by any adsorption to the solid phase. If adsorption is not involved and the amount of solute is not sufficient to change the distribution between the two phases (i.e., $x$ is constant), symmetrical peaks will be obtained (see Sec. 3.3). Since equilibrium must be relatively rapidly attained between the two phases and this involves diffusion in liquids, the flow rate is important. It must be kept low but not too low (cf. Sec. 3.2e).

### (ii) Practical applications

Partition chromatography can be carried out with columns as well as sheets. The carrier material should be inert, i.e., nonadsorbing, and of the proper mesh size for good balance between flow rates and separation. Some popular materials are Kieselguhr ("Celite" or diatomaceous earth), silica gel or cellulose powder. The support can be coated with a hydrophobic liquid such as benzene for chromatography of nonpolar materials (the mobile phase is a hydrophilic solvent such as methanol or formamide), or it can be coated with a hydrophilic liquid such as butanol for separations of polar substances (with water as the mobile phase). See Figure 3.5. Another possi-bility is to coat the support with silicone or ester groups and then coat this with a hydrophobic solvent. Using another hydrophobic solvent as the mobile phase gives a generally effective system for chromatography of less polar substances, since more polar solutes will not move as fast.

In general, partition chromatography is not used today for proteins. How-ever, for examples of its use for bovine $\gamma$ globulin and rabbit antibody separa-tion, see references 6 and 7. The major employment of this technique is in separation of small molecular weight compounds. Since these diffuse more rapidly than large molecules such as proteins, resolution is even more of a problem. A great deal of effort is required to keep or make the initial zone of sample as narrow as possible. A common technique used in paper chro-matography and especially in paper electrophoresis of peptides, is to apply the sample to the paper and then wet the paper with a sufficient amount of solvent around the sample zone so that the solvent advances from all sides to the center of the sample zone. This concentrates the sample by capillary

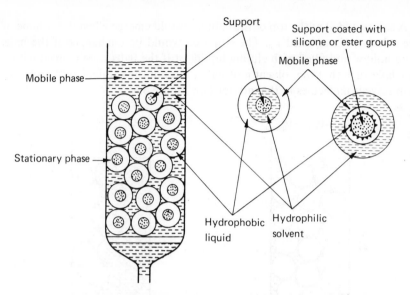

**Figure 3.5** Types of support used in liquid–liquid partition chromatography. The support, which may be Celite, is coated with the stationary phase, which may be hydrophilic or hydrophobic, packed into a column, and the solute eluted with an immiscible solvent.

action. Then the paper can be dried (if necessary) and chromatography begun. Obviously it is best if the solutes have a reasonably high solubility in the solvent used. If the paper is to be dried, the solvent should be volatile, and it needn't be the same as the solvent used in the chromatography.

Another point which is applicable to all chromatographic systems is that, if possible, the chromatographic material—paper, coated Celite or whatever —should be washed with the solvents used in chromatography before sample application to remove possible interfering substances.

### 3.2d   Gel Filtration (8)

This is considered a partition method but partitioning is on the basis of the size of the molecules which are to be separated, rather than their solubility or volatility. Again, the method employs a mobile liquid phase, with a stationary phase which consists of the same liquid trapped in a lattice.

#### (i) Mechanism of "molecular sieving"

Consider a column of total volume $V_t$ packed with spheres occupying a volume $V_p$. The interstitial volume $V_0$ would be

$$V_0 = V_t - V_p \tag{3–25}$$

A solute added at the top of the column would emerge when the volume of the effluent, $V_e$, equaled $V_o$. The situation would be unchanged if the balls were hollow and filled with eluting liquid, or if the balls were pierced with a tiny hole such that the solvent molecules, but not the solute, could exchange with solvent molecules in the interior (see Figure 3.6).

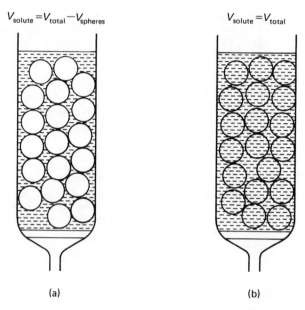

$$V_{solute} = V_{total} - V_{spheres} \qquad\qquad V_{solute} = V_{total}$$

(a)                                                     (b)

**Figure 3.6** A column packed with spheres, showing the effective column volume available to a solute if the spheres are (a) impenetrable or (b) porous to the solute.

Now assume that the holes in a fraction $z$ of the balls are sufficiently large to permit solute molecules to move back and forth. The fraction $x$ of solute molecules remaining in the mobile volume $V_o$ is reduced by permitting solute to move into an additional volume of $zV_p$:

$$x = \frac{V_0}{V_0 + zV_p} \tag{3-26}$$

neglecting the volume occupied by the walls of the spheres. The $x$ used here can be considered analogous to the $x$ (the fraction in the mobile phase or the probability of migration) in countercurrent distribution. Thus, the solute molecules will emerge from the column (ignoring zone spreading caused by diffusion) when

$$V_e = \frac{V_0}{x} = V_0 + zV_p \tag{3-27}$$

Now if the molecules were small enough to enter all the balls (i.e., $z = 1.0$),

$$V_e = \frac{V_0}{x} = V_0 + V_p = V_t \qquad (3\text{–}27a)$$

In this case all the molecules (ignoring diffusion) should emerge when a volume of eluant equal to the column volume has passed.

Note that in such a column, the larger the molecule the sooner it will emerge, up to the point where the molecules are sufficiently large that all are excluded from the nonmobile phase (the spheres). All molecules of this size and larger will emerge at $V_0$.

While this picture has been developed using as a physical model a trap or cage which the solute molecules cannot penetrate, we need not restrict our thinking to such a model. Indeed some separation can be obtained using a column packing of very small glass beads (9). In this system there is a mobile phase which is available to all the molecules and an immobile phase consisting of solvent adsorbed onto the beads. Only fractions of the adsorbed volume are available to the solute, according to the size of the molecules (10).

A variety of chromatographic material with "molecular sieving" properties is now available (see section iii).

### (ii) Quantitative aspects

It was discovered that a plot of elution volume from a column of "molecular sieving" material was a roughly linear function of the log of the molecular weight of a number of globular proteins (10,11). This relationship is shown in Figure 3.7.

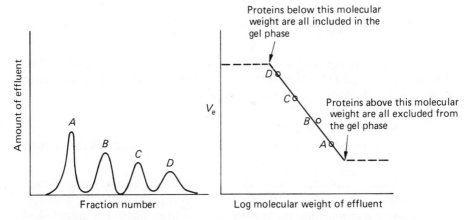

**Figure 3.7**   Representation of a plot of the elution volume of several proteins versus the logarithm of their molecular weight. The column packing material was Sephadex G-200.

To explain this, Laurent and Killander (12) considered the matrix as a series of rods and then considered what volume is available to a sphere of radius $r_s$, using the relation developed by Ogsten (13) for a series of infinitely long straight rods which are distributed randomly throughout the matrix. They found that

$$K_{av} = \exp[-L(r_s + r_r)^2] \tag{3-28}$$

where $K_{av}$ is that fraction of the matrix which can be occupied by a sphere of radius $r_s$. $L$ is the concentration of rods in solution as centimeters of rod per cubic centimeter, and $r_r$ is the radius of these rods.

There are two parts to the matrix, one interstitial and penetrable by any size particle and the other consisting of rods whose penetrability is characterized by Equation 3-28.

The elution volume, $V_e$, of a particle of radius $r_s$ will then be

$$V_e = V_0 + K_{av} \cdot V_x; \qquad K_{av} = \frac{V_e - V_0}{V_t - V_0} \tag{3-29; 3-29a}$$

where $V_x$ is the total volume of the column chromatographic material or "bed", $V_0$ is now called the "void volume" of the column and $V_t$ is the volume available to a small (unexcluded) molecule. A plot of $(-\ln K_{av})$ versus $r_s$, therefore, should be linear for spherical molecules.

An elipsoidal particle will tumble in solution because of thermal agitation (see Chapter 6). The average swept volume of any nonspherical particle is related to the cube of its Stokes radius. The model proposed by Ackers (11) provides a relationship of the form:

$$K_{av} = \left[1 - \left(\frac{a}{r}\right)^2\right]\left[1 - 2.104\left(\frac{a}{r}\right) + 2.09\left(\frac{a}{r}\right)^3 - 0.95\left(\frac{a}{r}\right)^5\right] \tag{3-30}$$

where $r$ is the average pore size of the bed material and $a$ is the Stokes radius. Monty and Siegel (14) showed that the Stokes radii of proteins are better correlated with elution volumes than are their molecular weights.

### (iii) Materials used for gel filtration

There are several commercially produced "sieving" materials in common use today.

*1. Sephadex.* Sephadex is made of dextran. This is a polysaccharide of high molecular weight (originally $10\text{-}300 \cdot 10^6$) which is prepared by the action of *Leuconostoc mesenteroides* on sucrose. The dextran fibers are cross-linked with epichlorohydrin (see Figure 3.8). Originally the cross-linking was done in bulk, but it is now done in emulsions to get beads which give better flow

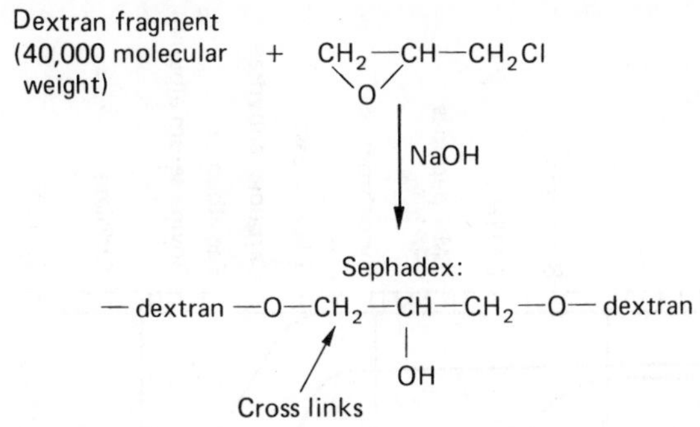

Dextran fragment
(40,000 molecular
weight)
$+$  $CH_2$—$CH$—$CH_2Cl$
\ O /

NaOH

Sephadex:
— dextran —O—$CH_2$—$CH$—$CH_2$—O— dextran
|
OH

Cross links

**Figure 3.8**  Production of Sephadex by reaction of dextran with epichloro-hydrin.

rates. By varying the ratio of dextran to epichlorohydrin, varying amounts of cross-links are introduced, which leads to differences in the size of molecules which will be totally excluded (Appendix 11). Several mesh sizes of each grade are also sold, e.g., coarse beads, medium, fine, etc.

2. *Bio-Gels.* Acrylamide gels, sold as "Bio-Gel," are another "molecular sieve" chromatographic material. The gels result from the polymerization of acrylamide with varying percentages of methylene bisacryamide (see Chapter 5), which provides cross-links to the polyacrylamide strands. Bio-Gels are insoluble in water, salt solutions, and common organic solvents, and can be used within the pH range 2 to 10. Lower pH's can be tolerated for short periods, but very strong bases should be avoided due to the possibility of hydrolysis of the amide groups.

Bio-Gels are supplied as dry beads which swell readily in water, salt solutions, and some polar organic solvents like ethylene glycol. Appendix 11 gives the hydrated bed volume per dry gram and the water regain per dry gram for the different Bio-Gels in water. The bed volume of polyacrylamide gels is negligibly affected by changes in ionic strength, the volume of Bio-Gel P-100 decreasing only 2% in going from zero to 0.4 $M$ phosphate buffer.

As with Sephadex, the degree of crosslinking determines the size of the molecules excluded. The higher the degree of cross-linking, the smaller the average pore size, and the lower the exclusion limit (Figure 3.9). The Bio-Gel beads also become harder, giving good flow rates.

Essentially one can do anything with these that one can do with Sephadex. They appear to have some advantages over Sephadex: greater range and subdivisions, lower price, better flow rates, no mold growth, and less change in volume with changes in salt concentration. Also, Sephadexes leach glucose into the eluent; this can be troublesome if you are measuring carbohydrate content in a sample chromatographed on Sephadex.

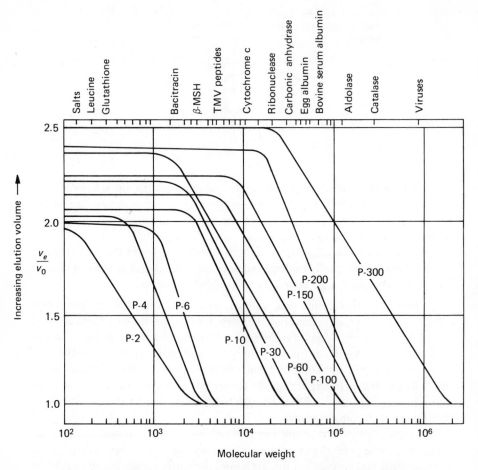

**Figure 3.9**  $R_f$'s of proteins versus their molecular weights on various Bio-Gels. [Redrawn from catalogue W of the Bio-Rad Corporation (Richmond, Calif.) with their permission.]

On the other hand, a disadvantage of Bio-Gels appears to be a somewhat lower resolving power. Curiously, the void or exclusion volumes, $V_0$, of both Bio-Gels and Sephadexes are about 35% of the column volume.

*3. Ago-Gel.* The third variety of "molecular sieve" material is a gel made from agar. Agar is a mixture of agarose (the major component) and agaropectin. Most of the sulfate and carboxyl groups in agar are in agaropectin. Preparation of agarose is described in reference 15. It can now be purchased from Bio-rad, General Bio-Chemicals and Mann Research Laboratories. Ago-Gel (agarose) granules are sold as suspensions in $10^{-3}$ $M$ EDTA and 0.02% sodium azide. They usually have very low flow rates, e.g., 5 ml/hour, and are relatively expensive. Their major feature is much higher exclusion limits (Appendix 11), and they are used for separation of molecules of 400,000 or more in molecular weight. The same relations between elution volumes and Stokes radii appear to hold for agarose gels as for Sephadexes or Bio-Gels.

*4. Other materials.* Two other types of molecular sieve materials have been introduced: glass beads with a closely controlled pore size for separations of proteins and even viruses (16), and polystyrene beads for separation of compounds of molecular weights below 3000 in nonaqueous media. Both have as yet found only limited application.

### (iv) Experimental applications: (7)

*1. Concentration.* Solute may be concentrated with dry Sephadex. Sephadex G-25 (coarse beads) is recommended for this since the beads settle rapidly and do not contain fine particles. Sephadex G-25 excludes the large solute molecules, taking up only the solvent, resulting in a concentration of solute.

For example, the sample is in a volume of 100 ml and concentration to 50 ml is desired. The water regain for G-25 is 2.5 ml of water per gram of dry gel, so it will take 20 grams of G-25 to take up 50 ml of water. Add 20 grams of dry Sephadex to the solution and allow it to swell for 20-30 minutes.

The system now consists of four parts: the supernatant, 16 ml, contains solute. In addition, as in a column bed, the swollen Sephadex consists of three parts: the volume of dry dextran gel; the solvent which has been taken up by the dry gel—the 50 ml which were taken up by the 20 grams of gel; and water and solute which is in $V_0$, the space between the Sephadex beads (24 ml). The material in $V_0$ must be recovered, since two-thirds of the solute is there. This can be recovered by centrifugation, or filtration. Thus, a macromolecule solution can be concentrated without changing the pH or ionic strength. This procedure can be repeated to concentrate the solution further.

*2. Desalting.* Molecular sieves are also used for changing buffers, desalting, removal of small molecules, etc. This is done simply by passing the protein solution through a column, usually of Sephadex G-25 or Bio-Gel P-2, prepared in the replacement buffer. The protein will emerge in this buffer. Unfortunately, the sample is always diluted somewhat by this procedure.

*3. Analysis of ligand binding.* Gel filtration can be used to study binding of ligands to proteins (17, 18). A column of some material such as Sephadex G-25 is equilibrated with a solvent containing the ligand. Then the protein is added and chromatographed, eluting with the same solvent. The protein will take up a certain amount of ligand, and comes off in the void volume of the column. The total concentration of ligand is greater in the protein fractions, but drops to much less in a volume equal to $V_t$, the total (available) column volume (see Figure 3.10). From the changes in ligand concentration

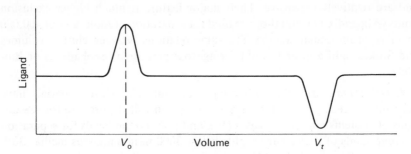

**Figure 3.10**   Measurement of ligand binding by a protein using chromatography of the protein on a Sephadex G-25 column equilibrated with the ligand.

and knowing the concentrations of protein and ligand, the stoichiometry of binding and the binding constant(s) can be determined (see Chapter 7).

*4. Purification of materials.* The widest use of these chromatographic materials is for separations of substances on the basis of their size. Sephadex, Bio-Gel and Agarose allow this to be done for molecules of molecular weight up into the millions. Andrews has presented data on several proteins, chromatographed on various concentrations of agar gel. At an agar gel concentration of less than 5%, thyroglobin (MW 670,000) was separated from hog gastric mucoid whose molecular weight is very high (19). Separation of cellular particles and even viruses has been obtained with agar gels. Hjerten (20) separated $T_2$ phage (particle weight $200 \cdot 10^6$) from $R$ phycoerythin (MW 290,000) and 50S ribosomes from protein contaminants. Steere and Ackers (21) separated tobacco mosaic virus from southern bean mosaic virus (MW $6.6 \times 10^6$) on a 4% agar column.

A technique called "zone precipitation" has been developed. This involves a Sephadex column made up in an ammonium sulfate concentration gradient (22), as shown in Figure 3.11. Proteins tend to travel faster than the salt, so they are migrating into a gradient of increasing ammonium sulfate concentration. Consequently, they tend to precipitate out of solution. If the column is not plugged by this, the precipitated protein travels faster yet. This column consequently separates proteins not only by their size, but concentration and solubility in ammonium sulfate as well.

*5. Analysis of molecular size.* Sephadex chromatography can be used to determine molecular weights of enzymes even in crude extracts (13, 23). Since this method actually gives relative Stokes radii rather than direct molecular weights, the method relies on the assumption that the asymmetry and extent of hydration of the proteins being analyzed and used as markers are the same. This assumption appears to hold reasonably well for globular

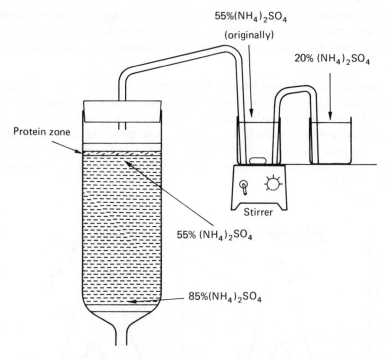

**Figure 3.11**  "Zone precipitation" using a Sephadex G-100 column (22). The column is poured; then a gradient of decreasing ammonium sulfate concentration is applied. The gradient is stopped after the salt concentration applied to the column drops to 55%. A protein solution is applied, and this is eluted with the remainder of the salt gradient (55% to 20%).

(i.e., most of the soluble) proteins, but not very well for proteins containing large amounts of carbohydrate.

Measurements of the molecular weight of polynucleic acids are currently done by comparing their mobilities during electrophoresis (Chapter 5) in a column of polyacrylamide gel. The gel sorts the molecules according to size by offering greater fractional resistance to movement of larger molecules than smaller (24).

Polson (25) stated that hemoglobin (MW 68,000) ($D = 5.6 \cdot 10^{-7}$ cm²/sec) can be separated from bovine serum albumin (MW 68,000, $D = 6.0 \cdot 10^{-7}$) and attributed this to a difference in diffusion constant. However, it was later determined that hemoglobin dissociates at low concentrations, so he was measuring the average Stokes radius of a partially dissociated system. Attempts have been made to use gel filtration measurements to give information on association-dissociation phenomena in proteins (26).

Cytochrome c (MW 13,000) movement on agar columns has been found to be a function of salt concentration (19), so adsorption (ion exchange) phenomena may be important in this case. This should be minimal on agarose columns, since the sulfate- and carboxyl-containing agaropectin has been removed.

6. *Analysis of molecular purity.* Measurement of some specific property of an eluate—its activity if it is an enzyme—across its elution peak is often done. The ratio of the property to the amount of elute, such as the specific activity, is determined to see if it is constant. Systematic variation in the ratio across the peak suggests contamination with molecules of a different size. Constancy of the ratio is often used as a criterion of purity of a substance. See Figure 3.12.

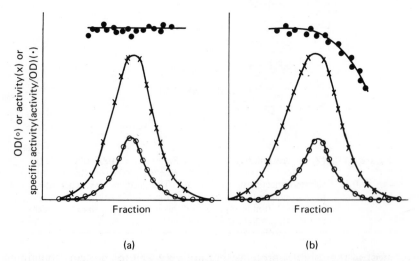

Figure 3.12    Analysis of purity of size of a protein preparation using constancy of specific activity across the peak, after elution from a
molecular sieving column: (a) constant specific activity indicates
purity with respect to molecular size; (b) drop at the tailing
edge of the peak suggests a contaminant of smaller size is
present.

### 3.2e  Gas-Liquid Chromatography (GLC) (27-30)

This can be considered as analogous to countercurrent distribution. There are, however, some differences. As in countercurrent distribution, the mixtures are partitioned between a mobile and a stationary phase. However, the mobile phase ("carrier gas") is gaseous, and the stationary phase is a liquid adsorbed on a solid. The solid is immobilized on the walls of a tube—the column. The "plates," though real, are not visibly distinct; and the number of "plates" is much higher, e.g., as high as $10^6$. The partition coefficient always favors the gas phase, approaching 1.0 as a limit. Also, the sample size is usually smaller.

### (i) Theory

*1. "Theoretical Plate" theory.* A GLC column is considered analogous to a distillation column, but it has a number of "theoretical plates" instead of actual ones.

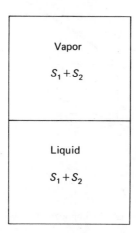

**Figure 3.13**
A closed vessel containing two volatile components, $S_1$ and $S_2$.

Consider a solution of two volatile components, 1 and 2, which partly fill a closed vessel, as shown in Figure 3.13. Raoult's law states that the vapor pressure of a substance $P_n$ is

$$P_n = x_n P_n^\circ \qquad (3\text{-}5a)$$

where $P_n^\circ$ is the vapor pressure of pure $s_n$ at a temperature $T$; and

$$x_n = \frac{W_n}{M_n} \bigg/ \sum_n \frac{W_n}{M_n} \qquad (3\text{-}31)$$

which is the mole fraction of $s_n$: $W_n$ is the weight of $s_n$ in the liquid phase, and $M_n$ is the molecular weight of $s_n$. If $S_1$ is more volatile than $S_2$ at $T$, then $P_1^\circ$ is greater than $P_2^\circ$, or the vapor phase is richer in $S_1$ (see Figure 3.14). If we now condense the vapor phase at $T_1$ by lowering the temperature to $T_2$, we have completed one distillation or one "theoretical plate." This new liquid $l_1$, composed of $S_1$ and $S_2$, will now be enriched in $S_1$ because $P_1^\circ$ is greater than $P_2^\circ$. This $l_1$ may now be distilled at a lower temperature $T_2$, as indicated in the diagram. From Figure 3.14, it would require three "theoretical plates" to go from a liquid of composition $a_1$ to a distillate of composition $a_3$, where $a_3$ was greatly enriched in $S_1$ (2).

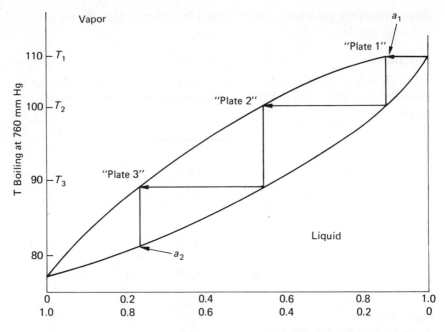

**Figure 3.14** Partial purification of one volatile substance (component 1) from nearly pure component 2 by distillation in three stages or "plates" (2).

The theoretical plate theory was originally developed for liquid-liquid chromatography (27, 30), and like the "rate theory" is valid for conventional (as opposed to capillary) columns.

The theory gives good agreement with experimental results. It also enables one to calculate the column efficiency from distribution theory, and to evaluate and compare a number of experimental column parameters. To evaluate column efficiency, several parameters must be defined.

$$\text{Partition coefficient} = k = \frac{p}{q} = \frac{\text{fraction of total solute in liquid phase}}{\text{fraction of total solute in gas phase}}$$

(3-1b)

$$p + q = 1$$

Writing $p$ and $q$ in terms of $k$ gives

$$\frac{k}{(k+1)} + \frac{1}{(k+1)} = 1$$

(3-32)

This relation is also used in countercurrent distribution.

Partition ratio $= K = \dfrac{C_L}{C_M}$

$$= \dfrac{\text{concentration of solute in the liquid phase}}{\text{concentration of solute in the gas (mobile) phase}} \qquad (3\text{--}33)$$

$K$ is independent of gas and liquid volumes, but is temperature dependent. It is related to $k$ by

$$K = k\dfrac{V_M}{V_L} \qquad (3\text{--}34)$$

where $V_L$ and $V_M$ are the volumes of liquid and gas, respectively, in the column.

$$\text{Retardation factor} = R = \dfrac{\dfrac{L}{t_R}}{\dfrac{L}{t_M}} = \dfrac{t_M}{t_R} = \dfrac{V_M}{V_R^{\circ}}$$

$$= \text{the ratio of the solute band velocity to}$$
$$\text{the carrier gas velocity} \qquad (3\text{--}35)$$

$L$ is the length of the column; $t_M$ is the time required for the carrier gas to pass through the column; $t_R$ is the time required for the maximum of the solute band to pass through the column; and $V_R^{\circ}$ is the "corrected retention volume." Note that $R$ is temperature dependent and varies with the ratio of gas to liquid phase.

Let $C_L$ and $C_M$ be the concentrations of the solute in the liquid and gas phases at the point of appearance of the peak maximum. Half of the solute has been eluted in a retention volume $V_R^{\circ}$, and half remains behind in the column gas (volume $V_M$) and liquid (volume $V_L$).

Let $n$ be some multiple of $V_M$ which equals $V_R^{\circ}$:

$$nV_M C_M = V_R^{\circ} C_M = V_M C_M + V_L C_L \qquad (3\text{--}36)$$

Divide both sides by $C_M$:

$$V_R^{\circ} = \dfrac{V_M}{R} = V_M + \dfrac{V_L C_L}{C_M} = V_M + KV_L \qquad (3\text{--}37)$$

These are the fundamental relations employed in deriving the distribution of a solute from the theoretical plate concept. To simplify the theory, it is assumed that the equilibrium distribution of solute between the two phases is obtained instantaneously, that all of the solute is initially present on the first "plate" of the column, and that the distribution coefficient is constant and independent of the amount of solute in a "plate."

The result, if plotted, gives a curve which may be expressed as a binomial expansion of $(p + q)^n$. This is a Gaussian curve, subject to treatment by probability theory. From probability theory we can derive

$$C_{\max} = \left(\frac{N^{1/2}}{V_R^\circ}\right)\left(\frac{w}{(2\pi)^{1/2}}\right) \tag{3-38}$$

where $N$ is the number of "theoretical plates," and $w$ is the initial weight of solute. Compare Equation 3-38 with Equations 3-21 to 3-23.

Thus, early peaks are sharp, while late peaks are low and broad.

For a constant $V_R^\circ$, a more efficient column will yield higher peaks, the height being proportional to $N^{1/2}$. The width of the peaks increases directly with $V_R^\circ$ and inversely with $N^{1/2}$.

The number of theoretical plates ($N$) can be calculated from the actual sharpness of a peak produced by a substance and the distance (in time or volume) between the appearance of the solvent and the emergence of the substance.

In Figure 3.15, $x$ is the distance from the solvent peak, and $y$ is the distance at the baseline between the tangents to the curve. $N$ is calculated from

$$N = 16\left(\frac{x}{y}\right)^2 \tag{3-39}$$

The "height equivalent to theoretical plate" (HETP) is inversely pro-

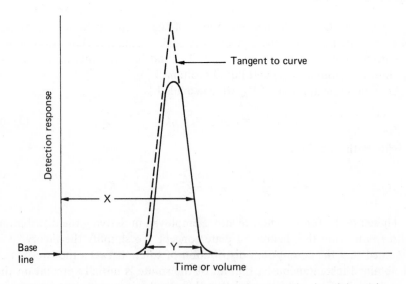

**Figure 3.15**  Calculation of the amount of a solute from the size of the peak from gas–liquid chromatography.

portional to the column efficiency. It is defined as

$$\text{HETP} = \frac{L}{N} \tag{3-40}$$

The smaller HETP is, the more efficient is the column. This is the definition of theoretical efficiency in gas chromatography.

*2. Rate theory.* This considers the detailed physical phenomena governing the exchange between the gas and liquid phases (31). Consider the passage of a solute through a column. There are two basic processes occurring: zone spreading by diffusion, and exchange. The problem is to obtain complete exchange with minimum zone spreading.

Exchange is provided by transport of the solute by the carrier gas to a point in the column where the solute dissolves in the stationary or liquid phase. The solute then evaporates and is again transported by the carrier gas to another point in the column where the process is repeated.

However, the stationary phase is not a homogeneous film spread over smooth supporting particles; there are pores and channels with varying lengths within and between the supporting particles. These simultaneously greatly increase the surface area available for exchange and greatly increase the frictional resistance (particularly in the smaller channels[1]) to carrier gas flow. The extent of increase in these quantities is related to a number of factors: the irregularity in packing, denoted by a function $\lambda$ (Equation 3-41, below), and the sinuosity of the gas path, denoted by $\gamma$. $\lambda$ and $\gamma$ together constitute a measure of the effective porosity of the packing particles. Related to these quantities is the diameter of the support particles $d_p$.

Diffusion processes are of two types: molecular and turbulent. The former is a function of the absolute temperature, the viscosity of the medium, and the molecular radius of the solute, and occurs irrespective of whether there is any gas flow at all. In the stationary phase, this is the only type of diffusion occurring. Turbulent diffusion occurs because of turbulence of the gas flow and naturally is increased by a more irregular path and higher carrier gas velocities.

The extent to which exchange actually occurs will be a function of several factors. The gas flow rate together with the diffusion constant of the solute in the mobile and stationary phases are the most important parameters. Too rapid a flow rate relative to the diffusion constant of the solute will not allow complete exchange, and too slow a flow rate will reduce the resolution by allowing too much diffusion to occur[2].

---

[1] Poiseulle's law states that the flow through a tube decreases as the *fourth* power of the radius of the tube (Chapter 5). ·
[2] The same can of course be said for molecular sieve or any other type of partition chromatography.

The diffusion constant of a given solute is different in the mobile (gas) and stationary phases—denoted by $D_M$ and $D_f$, respectively. The gas velocity at a particular point in the column is called $u_M$.

Other factors important in regulating exchange are the effective thickness, $e_f$, of the liquid around the supporting particles and the effective partition coefficient, $k'$. The latter is equal to $KV_L/V_M$.

The combination of these factors is expressed by the Van Deemter equation (31), which like plate theory gives a Gaussian distribution of solute concentration as a function of time or volume. The column efficiency parameter, HETP, is related to the variables described above in Equation 3-41:

$$\text{HETP} = 2\lambda d_p + \frac{2\gamma D_M}{u_M} + \frac{8}{\pi^2} \cdot \frac{k'}{(1+k')^2} \cdot \frac{e_f^2}{D_f} \cdot u_M \qquad (3\text{–}41)$$

Much work has been done to verify the equation. The object is to find conditions for the highest efficiency; that is, when HETP is a minimum. The particle size of support $(d_p)$ is related to $\lambda$ in such a way that as $d_p$ is reduced, $\lambda$ increases; the product is a constant. Since a fine grained support would resist the passage of the carrier gas, somewhat larger meshed particles are better—110-120 mesh or 120-160 mesh material is generally used. Although this is not indicated by the equation, the particles should all be of the same size to reduce "channeling."

It is best that the diffusion of the solute in the gas $(D_M)$ be kept as low as possible. Carrier gases of higher density are also of higher viscosity and hence retard solute diffusion.[3] Since the density of gas increases with its molecular weight, $CO_2$ or argon is preferred to helium or hydrogen. Higher gas pressures in the column also improve separations partly by reducing solute diffusion, because of higher viscosity.

The amount of the stationary (liquid) phase $V_L$ determines both the film thickness $(e_f)$ and the effective partition coefficient $k'$. It is found experimentally that the column efficiency is greatest at some percent liquid phase. Unfortunately there is no consensus as to what "percent load" is best; at present, 5-10% loads are preferred for analytical columns, and somewhat higher percentage loadings for preparative work.

The gas velocity $(u_M)$ is the most important factor. If all the other factors are constant, the Van Deemter equation assumes the form of a hyperbola:

$$\text{HETP} = a + \frac{B}{u_M} + Cu_M \qquad (3\text{-}43)$$

A similar hyperbola is obtained as a function of the temperature. One can calculate the column efficiency from $16(x/y)^2$ and plot it against the gas flow

[3] $$D = RT/Nf = RT/6N\pi\eta r_0 = (R/6N\pi r_0)(T/\eta) \qquad (3\text{-}42)$$

$D$ is the diffusion constant of the solute, $T$ is its absolute temperature, $r_0$ its molecular radius, $\eta$ is the viscosity of the solvent (Chapter 6), and $f$ is the frictional coefficient of the solute.

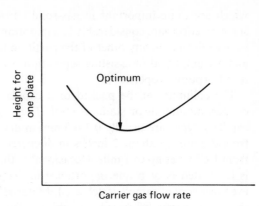

**Figure 3.16**
Dependence of column efficiency on
gas flow rate.

rate or the temperature. This curve should have a maximum. However, conditions for the best column efficiency are not necessarily the conditions for the best separation of one solute from another. That must be determined empirically.

For capillary (as opposed to conventional) columns, a separate theory of operation has been developed. This will not be covered here.

#### (ii) Instrumentation

The basic components of a gas chromatograph are: a carrier gas supply and flow control, a sample port and means of introducing the sample, a column with a packing coated with the liquid phase, a detector, a recorder, and temperature controls.

*1. Gas.* The gas can be helium, hydrogen, argon or $CO_2$, delivered at a rate of 10-400 ml/min.

*2. Column and packing.* The stationary phase depends on the separation desired. Our treatment of this subject deliberately ignores adsorption effects,

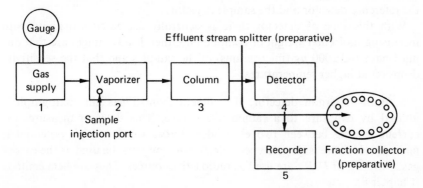

**Figure 3.17** Schematic representation of a GLC apparatus.

which are quite important in gas-solid chromatography (GSC). While adsorption effects are considerably less important in GLC (though perhaps more important than in any other of the partition techniques covered), they can be and are employed in assisting separation, by choosing the solid support for its adsorption properties (1).

The columns can be packed or a capillary type, made of glass, plastic, copper, aluminum or stainless steel. They can be U-shaped or coiled. The capillary types are usually 0.2-0.3 mm in diameter. The packed columns are from 2-6 mm to about 2 inches in diameter. Either type can vary in length from 1-6 inches up to 1 mile. Occasionally, the material from which a column is fabricated is of prime importance in determining whether separation or even recovery of sample is achieved. Materials such as phenols are so reactive that no recovery of sample is achieved from any column fabricated from aluminum or copper, no matter what packing or stationary phase is used. In such cases, stainless steel, Teflon, or glass columns are used.

*3. Detector.* A detector should respond to all compounds and have a high sensitivity, intrinsic stability and a low dead volume. It should also provide a rapid response, which ideally should be independent of the pressure and flow rate of the gas and the chemical nature of the solute and have a response which is linearly proportional to the concentration of solute. There are several types of detector in common use (32).

a. A thermal conductivity detector operates on the principle that the electrical resistance of a heated wire varies with its temperature. With a gas of constant composition flowing over the wire, it gets constant cooling, and thus has a constant resistance. If the gas composition changes, a change in cooling results, causing a change in resistance. For this detector, a gas of high thermal conductivity relative to that of most organic vapors is needed. Helium is commonly used. A heated metal filament or thermistor is used as the sensing device. A reference detector is also needed. The instrument measures an unbalanced bridge voltage, the result of the difference in resistance between the reference detector and the sample detector.

With this type of detector there is generally decreased sensitivity with increasing molecular weight of compound. Other disadvantages are that one must have 100-2000 $\mu$g (micrograms) of the samples and that the sensitivity drops off at higher temperatures.

b. With an argon detector, the effluent from a gas chromatograph is ionized by exposure to a radioactive source. This detector measures the current passing between two electrodes, across which a high potential is maintained. There is no reference detector. Argon must be used as the carrier gas. $^{90}$Sr, $^3$H, or $^{226}$Ra are used as radioactive sources. They are beta emitters (Chapter 8).

Most organic compounds have ionization potentials of 9-11 ev. However, some compounds having higher ionization potentials (i.e., noble gases, inorganic gases, and fluorocarbons) are not detected. The response of this detector consequently varies with the structure of the compound. The operating range is 0.1 to 100 $\mu$g. However, this detector is linear only at certain voltages.

c. The schematic of a "flame ionization" detector is shown in Figure 3.19. The principle of the hydrogen flame or "flame ionization" detector is not

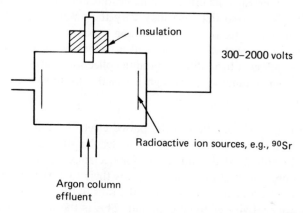

Figure 3.18   Schematic representation of an argon detector.

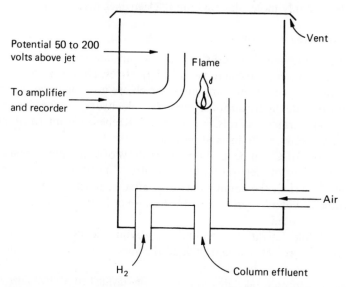

**Figure 3.19**   Schematic representation of a flame ionization detector.

thoroughly understood. Organic material is burned in a hydrogen flame, and electrons are produced, allowing a current to be measured. The sensitivity reaches $10^{-10}$ gm. The operation range is 0.01 $\mu$g to 5 mg, which are lower and higher limits than the thermal conductivity or argon detectors. The linear range is almost infinite. Any gas can be used, but helium or nitrogen are generally favored. However, the sample is totally destroyed in analytical instruments, and the detector is insensitive to air, $CO_2$, the noble gases, and relatively insensitive to water. The sensitivity of the detector depends on the number of carbon-carbon bonds.

In preparative instruments fitted with flame ionization detectors, an effluent stream splitter is used so that one may vary the amount of sample which passes through the detector. The remaining amount of the sample peak passes out through a heated collector head, to which is affixed a sample collecting device such as a turntable containing collection bottles. The collector is operated either manually or through a signal sensing switch attached to the recorder.

d. The principle of operation of the pulsed electron capture detector (33) is based on the affinity of certain types of molecular structures, such as polyhalogenated or polynitrated compounds, for electrons. The detector provides a constant source of electrons and measures their loss as the sample flows through the electron field. Consequently the sensitivity is determined by the relative electronegativity of the compound. This is the most sensitive of all detectors, up to 1000 times more sensitive than the flame ionization detector, but it detects only molecules containing electronegative groups.

### (iii) Experimental procedure

A major disadvantage of GLC is that the substances to be separated must be volatile, or be convertible to volatile derivatives. However, in biochemical applications, the classes of compounds which are now analyzed or separated by GLC include hydrocarbons, alcohols, aldehydes, amines, gases, fatty acids, carbohydrates,[4] sterols, amino acids, Krebs cycle intermediates, bile acids, barbiturates, vitamins and pesticides.

To see how GLC can be used for the analysis of a mixture of related compounds, consider the separation of a mixture of fatty acids. In order to be separated by GLC, the components of a mixture must be either volatile or convertible to volatile derivatives. In the case of fatty acids, the methyl ester derivatives are used.

Once suitable derivatives are obtained, we must next know what operating conditions to use. This would include the specific instrument; column size

---

[4]Trioses to tetrasaccharides are converted to trimethylsilyl ether (TMS) derivatives.

(I.D., length) and composition (glass, copper); the type and mesh size of the column packing; the type and percent liquid phase; the column and flash heater temperature, the type of detector and its temperature; the appropriate range and attenuation settings of the instrument; the chart speed; the carrier gas and its flow rate; the flow rate of hydrogen and air (if a hydrogen flame detector is used) and the amount of sample injected. The optimal operating conditions are determined by using a mixture of known compounds and adjusting the various parameters until a suitable separation is achieved.

Often a column must be prepared for the separation desired. We include specific directions for preparing a column for fatty acid methyl ester separation, since preparation of most columns will follow similar lines.

A suitable column is a 10 feet long, $\frac{3}{8}$ inch diameter aluminum coil packed with 30% (w/w) cyanosilicone coated on 60-70 mesh Anakrom SD (a silanized diatomaceous earth support). DEGS or EGS or similar polyester columns can also be used (see iv (2), below). They give better separation of fatty acid esters of the same length but different degrees of saturation.

Approximately 30 gm of prepared support is needed to fill the column. Since the percentage coating used here is 30% (w/w), 21 gm of the Anakrom SD is weighed out along with 9 gm of cyanosilicone. The Anakrom SD is placed in a round bottom flask and the cyanosilicone (XE-60) is dissolved in 100 ml of chloroform, the coating material is poured in, mixed well with the solid support and allowed to stand 15 minutes. The flask is then placed on a rotary evaporator and the solvent evaporated off at 40°C. When all visible solvent has evaporated, increase the bath temperature to 60°C and continue drying until all of the powder falls freely away from the walls of the flask. When all lumps have disappeared and the powder is dry to the touch, it is ready to pack into the column. Any excess should be stored in a tightly stoppered jar to avoid moisture uptake.

Cut a 10 ft length of aluminum tubing with a tubing cutter and straighten it. The bottom end of the column is stopped with a 1 cm plug of glass wool. The column is supported on as nearly vertical a position as convenient, and a powder funnel is attached to the top with a short length of plastic tubing. The packing is poured in the funnel and sifted into place by tapping the column gently on the floor or by using a vibrator.

The column is filled to within 1 cm of the top and sealed with a plug of glass wool. It is then bent to the required shape or spiraled around a mandril. For the former, a tube bender is very convenient. Connection to the chromatograph is usually made with Swagelok fittings.

With either glass columns or those with low percentage loadings, another approach is to precoil the column and then pack, using vacuum and pressure.

Any newly packed or long-stored column must be conditioned in an oven overnight at a temperature 50° higher than the highest anticipated operating temperature. If no oven is available the column can be conditioned in the column oven of the gas chromatograph. The effluent end of the column is left disconnected during the conditioning period.

Inject into the sample port the unknown fatty acid esters or the model mixture of standard fatty acid esters in chloroform. Use a 1-10$\lambda$ (microliter) sample, depending on the concentration of the standard. A concentration range of 0.05 to 0.1 mg/component is convenient. Note and record the time required for each ester to come off the column. This will help identify the fatty acids in the biological samples. Change attenuation settings if necessary to keep the peaks on scale. Record any attenuation changes alongside the individual peaks.

### (iv) Treatment of data

The recording obtained may show a series of peaks. These are normally Gaussian curves if the substances are pure, and the areas are usually proportional to the amounts of material present. We will discuss primarily methods for identification of unknown substances.

1. *Identification of components of a single homologous series.* A homologous series of fatty acid methyl esters (lauric, myristic, palmitic, stearic) come off the column in order of increasing chain length. A graph of the logs of the absolute retention times (in cm or mm) versus the number of carbon atoms in the fatty acids ($C_{12}$, $C_{14}$, $C_{16}$, $C_{18}$) gives a straight line. Likewise a graph of a homologous series of mono-unsaturated fatty acids would also give a straight line. However, $C_{16}$ unsaturated esters would come off the column either before or after the saturated esters, depending on the specific liquid phase used.

Absolute retention times will vary if there is any change in the operating conditions, such as a slight change in column temperature or gas flow. This effect of column temperature on retention time can be used to advantage. If a peak comes off very quickly, the temperature can be lowered to give a longer retention time. If a peak takes a long period of time to be eluted, raising the column temperature will decrease the retention time (the log of the retention time is proportional to $1/T$). To avoid the variability of absolute retention times an internal standard is used. For example, a 17-carbon fatty acid methyl ester could be added to a mixture to be analyzed and the retention times of all the other components expressed relative to the retention time of the $C_{17}$ ester (Table 3.2).

Once the appropriate conditions have been found for the complete separation of a known mixture of standards, mixtures of unknown composition can be examined. After the unknowns have been run alone, a suitable internal standard is added to the unknown mixture and the retention times are calculated relative to the extra peak occurring on the chromatogram. These

**TABLE 3.2**

| | Retention Times of Known Fatty Acid Methyl Esters Relative to $C_{17}$ | Relative Retention Times of Unknowns |
|---|---|---|
| $C_{14}$ | 0.35 | — |
| $C_{15}$ | 0.50 | 0.50 |
| $C_{16}$ | 0.71 | 0.71 |
| $C_{17}$ | 1.00 | — |
| $C_{18}$ | 1.41 | 1.41 |

relative retention times are then compared to those of known standards. If an extra peak does not occur one of the unknowns should have increased in size. In any analysis of a homologous series one must be careful that short-chain members of the series are not missed by using too high a column temperature. Likewise, small quantities of the long-chain members of a series may be missed if too low a temperature is used.

An approximation of the weight percentage composition is obtained by determining the area of each peak (by multiplying peak height times peak width at half height) and calculating what percentage of the total peak areas is represented by this specific peak. To make an accurate quantitative determination of the amount of a particular substance, an internal standard of known concentration must be added to the unknown and the two chromatographed together. In addition, a correction factor must be determined to correct for differences in the response of the detector to the two components, and the load limits of the column and detector must be known (34).

*2. Identification of components of several homologous series.* Although relative retention times can be used to identify a member of a single homologous series the analysis of fatty acids from biological tissues is more complex, since several homologous series are represented. For example, consider a mixture of saturated, mono-, di-, and tri-unsaturated fatty acids. The analysis of such a mixture is carried out by chromatography on two separate columns, one coated with a polar liquid phase such as diethylene glycol succinate (DEGS), the other coated with a relatively nonpolar phase such as methyl silicone (SE-30). As shown in Figure 3.20 each column gives a different elution pattern for each homologous series. On the nonpolar column (SE-30) the saturated fatty acids appear after the unsaturated fatty acids of the same chain length. For example, methyl linoleate ($C_{18}^{\Delta 9, 12}$) comes off before methyl oleate ($C_{18}^{\Delta 9}$) which precedes methyl stearate ($C_{18}$). On the polar column (DEGS) the pattern is reversed, with the saturated fatty acids coming off before the unsaturated ones. In addition, di- and tri-unsaturated can be separated on the polar column, but they are eluted together on the nonpolar column.

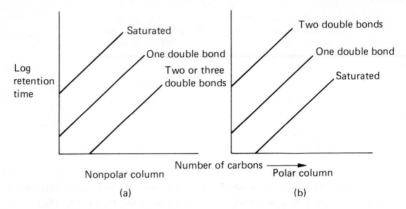

Figure 3.20   Analysis of mixture of several homologous series on two GLC
columns: (a) with a nonpolar packing material (SE-30); (b) with
a polar packing material (EGS or DEGS).

A tentative identification of each component in the complex mixture can
be arrived at by assigning each peak an "equivalent chain length" (ECL)
number (35). This number is obtained by drawing a plot of log retention time
versus carbon number for the saturated standards for each column, and
using it to convert the retention times of the unknowns to a corresponding
carbon number. As shown in Figure 3.21 the unknown has an ECL of 17.2.
In other words, all the fatty acid methyl esters are converted to hypothetical

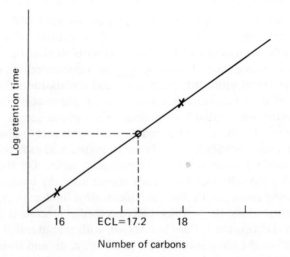

Figure 3.21   Calculation of the "equivalent chain length" (ECL) of an
unknown fatty acid methyl ester.

saturated fatty acids. The ECLs of several known fatty acids, on both polar and nonpolar columns, are given in Table 3.3. Note that saturated and branched chain fatty acids have the same ECL on both polar and nonpolar columns, while the unsaturated acids have different ECL'S.

**TABLE 3.3**

USE OF ECL TO IDENTIFY UNKNOWN FATTY ACIDS

|  | Column A | Column B |
|---|---|---|
| $C_{18}$ | 18.00 | 18.00 |
| $C_{19}\Delta 9$ | 17.71 | 18.51 |
| $C_{18}\Delta 9, 12$ | 17.53 | 19.30 |
| $C_{18}\Delta 9, 12, 15$ | 17.51 | 20.40 |
| $C_{18}Br$ | 17.50 | 17.60 |

Although this sort of GLC analysis can indicate tentatively what structures might be represented by the peaks, and certainly what the peaks cannot be, retention times or ECL alone do not prove a structure. However, one can hydrogenate the fatty acids and rechromatograph them to determine how many carbon atoms are present, and one can oxidize the unsaturated fatty acids and analyze the fragments by GLC to determine the positions of the double bonds. Finally GLC can be used to obtain highly purified samples which can then be analyzed by mass spectometry, all on a microgram scale.

### (v) Precision and advantages of GLC

Quantitative accuracy by GLC depends on the proper combination of columns for achieving separation, the detectors used for measuring eluates, and conditions under which neither is overloaded. Therefore, the load limits of column and detector, the linear range of the instrument, the response of the instrument to homologs of widely differing molecular weights and double bond character, the separating efficiency of the column, and the effect of changes in operating conditions (gas flow, voltage in argon ionization detector) must all be determined. If this is done, high reproducibility is generally achieved.

The advantages of GLC are many. The separation achieved is excellent—one can separate mixtures of closely related compounds such as fatty acids. The technique is versatile: one can use analytical instruments for small samples or preparative instruments for large samples; GLC is used to prepare microgram to gram quantities of ultra pure compounds as well as obtain information on the structure of unknown molecules. The instruments have high sensitivity—to $10^{-12}$ gm. The time required for analysis ranges from seconds to hours, with resolution of six components per second possible. The automatic character of the instrument eliminates errors and allows ease of operation. The simplicity of the instrument makes it easy to change columns and to repair or test various components. The sample can also be collected and analyzed by other methods: infrared or mass spectra or radioactive counting.

## 3.3 ADSORPTION CHROMATOGRAPHIC METHODS

In practice, the interactions producing adsorption effects can include anything from van der Waal's interactions to electrostatic effects to the complex set of hydrophobic, steric, and electrostatic forces involved in substrate binding by enzymes. The various types of adsorption chromatography are classified on the basis of these interactions. Initially, we will consider only the most nonspecific ones.

### 3.3a  Theory of Adsorption Chromatography

#### (i) Adsorption isotherms

Consider the detail of a solid surface. Bonds have been broken, and there are electron rich (basic) and electron poor (acidic) regions. The nature and number of these regions capable of adsorbing solute will be a function of the particular solid, its previous history, and the nature of the material to be adsorbed. The point is, there will be many types of adsorption sites available for interaction with a given substance. As we increase the concentration of solute in the liquid phase ($c$) and measure the amount adsorbed, $m$, we can have at least three possible adsorption isotherms, as shown in Figure 3.22. By far the most common is the convex, and this is the one we will consider.

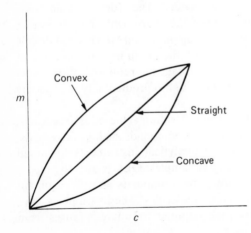

Figure 3.22
Possible types of adsorption isotherms.

It arises because the binding centers with greater affinities tend to be populated first, so additional increments of solute are less firmly bound.

### (ii) Consequences of heterogeneity of adsorption sites

Let us assume we wish to separate on a column filled with adsorbent, a mixture of two substances, $A$ and $B$, with concentrations $C_A$ and $C_B$, and with the adsorption isotherms shown in Figure 3.23. The quantity of importance is the ratio of the concentration of substance to the amount adsorbed ($c/m$).

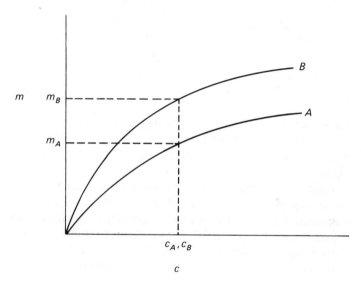

**Figure 3.23**   Adsorption isotherms of two substances with different affinities for the adsorbent.

This is a measure of the affinity of the substance for the adsorbent. Note that $B$ has a greater affinity for the adsorbent than $A$, at equal concentrations of $A$ and $B$.

At $c_A = c_B$,

$$\frac{c_A}{m_A} > \frac{c_B}{m_B} \qquad (3\text{--}44)$$

since the greater the affinity, the smaller is ($c/m$). In this case, $A$ will emerge first, since it is bound less firmly on average than $B$, and is retarded less. The effluent from this column will give the pattern in Figure 3.24.

$B$ will not be contaminated as much by $A$ as might be thought, since the tail of the $A$ peak will be displaced by the front of the $B$ zone because $B$ has a greater affinity for the adsorbent than $A$. The long "tail" on $B$ may be said to

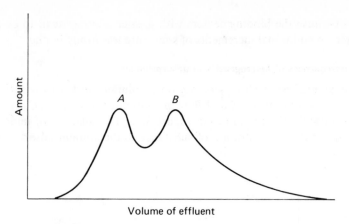

**Figure 3.24**   Elution of two substances with different affinities from a column of adsorbent.

arise because the "$R_f$" of $B$ is not constant. As the concentration of $B$ decreases, so does "$R_f$," since $B$ is left bound to the sites of greatest affinity as its concentration goes down. This "tailing" is a major problem of adsorption chromatography.

Now assume that the material is distributed on a column; the zone of $B$ will lie atop the zone of $A$, since $B$ will have displaced $A$ to sites further down the column. A new eluant with a stronger affinity for the binding sites of both $A$ and $B$ is introduced. It will "chase the tail" of $B$ (as $B$ does for $A$) removing the most firmly adsorbed molecules, and the effluent pattern will be partly artifactual (see Figure 3.25). $B'$ is what would otherwise be the tail

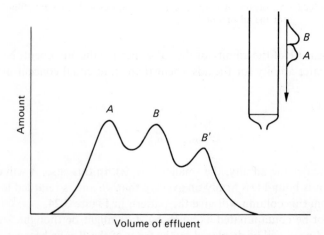

**Figure 3.25**   Artifactual peak produced by stepwise elution of $A$ and $B$ from an adsorption column with an eluant of greater affinity for the adsorption sites.

of $B$, only concentrated into another peak, so there can be three peaks for two substances. To avoid this, it is better to use a continuous elution *gradient*. This reduces tailing, since the most strongly adsorbed molecules encounter an eluant of steadily increasing effectiveness. A proper choice might give two well separated, nearly symmetrical peaks.

### 3.3b Adsorption Chromatography (37, 38)

#### (i) General description

This technique employs a mobile liquid phase and a stationary solid phase. One solid phase commonly used is alumina. Others used are silica, florisil or carbon. This procedure is not much used for protein separation, but some of its applications are instructive.

#### (ii) Applications of adsorption chromatography

*1. Frontal analysis.* Frontal analysis for purity of a material can be done by the continuous application of sample and solvent, and measuring the total concentration of material leaving the column. This is a boundary method, in other words. Each succeeding front serves to compete for binding sites with the preceding material, thus speeding the movement of any part of the preceding material which is not as strongly adsorbed. For a mixture of substances, a series of "steps" (boundaries) is obtained. Note that the height of the "step" will not be proportional to the amount of substance represented by front. Thus, frontal analysis only counts components. See Figure 3.26. If the material were homogeneous, only a single "step" would be obtained, since the material displaced would be identical in amount and nature to the material adsorbed.

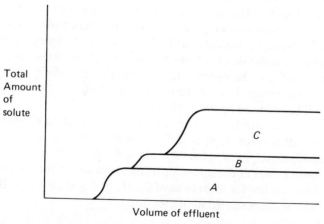

Figure 3.26   Analysis of purity of a substance by "frontal analysis" chromatography.

 2. *Displacement analysis* (39, 40). A limited amount of material containing
*A* and *B* is placed on a column, so that both are adsorbed. Then a displacer,
*C*, which is very strongly adsorbed to the immobile phase, is added. Essentially all the material will then move down the column at about the same
speed or "$R_f$." See Figure 3.27.

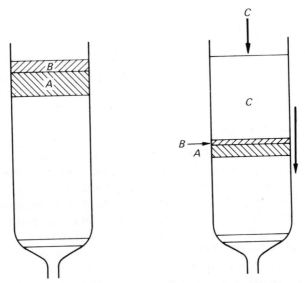

**Figure 3.27**  Representation of "displacement analysis" of a mixture. Modified from S. Dal Nogare and R. S. Juvet, Jr., *Gas–Liquid Chromatography*, Wiley, New York (1962).

If *B* is so low in concentration that only the sites for which it has the highest
affinity are filled, *C* will displace some of *B*, increasing its concentration so
that it adsorbs to sites of lower affinity in the same section of column. That is,
the zone of column containing *B* will become narrower. Since a molecule of
*B* can displace a molecule of *A* from any site, *A* will be eluted first in a band
whose width will also be decreased, depending on its affinity ($c/m$) for the
column packing material. *A* will be immediately followed by the band of *B*,
and that will be followed by the displacer.

 This process can also be described in terms of adsorption diagrams. If the
adsorption isotherms of *A*, *B*, and *C* are given in Figure 3.28, and we call
$C_c/M_c$ [actually $C_c/(C_c + M_c)$] the "$R_f$" of the displacer, then all the material
on the column moves at the same rate as the displacer. If the concentrations
of *A* and *B* are too low for them to have $C_A/M_A$ or $C_B/M_B$ equal to this "$R_f$",
then they will be concentrated until they do.

$$\frac{C_c}{C_c + M_c} = \frac{C_B}{C_B + M_B} = \frac{C_A}{C_A + M_A} \qquad (3\text{--}45)$$

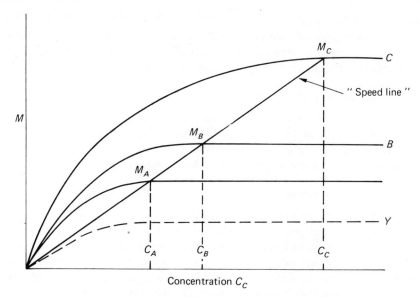

**Figure 3.28** Adsorption isotherms of *A*, *B*, and *C*. The "speed line" gives the final concentrations of *A* and *B* obtained on elution with a certain concentration of *C*. Modified from S. Dal Nogare and R. S. Juvet, Jr., *Gas–Liquid Chromatography*, Wiley, New York (1962).

The component *A* cannot have a lower concentration, corresponding to point $C'_A$, because then it would move down the column more slowly than the displacer.

The effluent pattern from such a column is shown in Figure 3.29. The concentrations of *A* and *B* which are eluted from the column can be determined from the "speed line" (Figure 3.28). The volume in which the sample is eluted is determined by $C_A$ and $C_B$ and the amount of sample originally adsorbed on the column. If a component *Y* is part of the sample and its

**Figure 3.29**
Elution pattern from the situation in Figures 3.27 and 3.28. Modified from S. Dal Nogare and R. S. Juvet, Jr., *Gas–Liquid Chromatography*, Wiley, New York (1962).

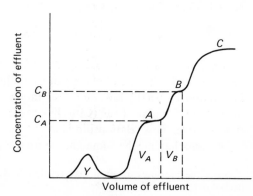

adsorption isotherm does not cross the "speed line," it will be eluted in front of $A$ as a discrete component. If we increased the concentration of the displacer so that the speed line crossed the $Y$ adsorption isotherm, it again would be concentrated and move at the same speed and precede $A$ off the column.

### 3.3c  Ion-exchange Chromatography

Adsorption with these materials is based primarily on ionic charge, though hydrogen bonding and hydrophobic interactions can also affect separation. The resins are partly or entirely synthetic, which provides more uniform adsorption sites. Consequently, there is less tailing.

#### (i) Theory of ion exchange

Ion exchangers are solids that can exchange ions with aqueous solutions. They consist of charged groups attached to a resin network. These charges have to be balanced by ions of opposite charge that are relatively small and mobile. The resins are submicroscopic networks like three dimensional fishing nets. When placed in water, they swell in order to maintain osmotic equilibrium with internal and external ions and take water molecules into their interior, hydrating the ionic groups and making the inside of the resin beads into something like a drop of concentrated salt solution.[5] The exact concentration depends on the extent of chemical cross-linking between resin chains, but it is high, about 5 molar. If the resin beads are placed in solutions containing other ions, some of the ions inside and outside the beads can exchange:

$$RSO_3^-Na^+ + K^+ \xrightarrow{(+Cl^-)} RSO_3^-K^+ + Na^+(+Cl^-)$$

This is ion exchange.

Of course the inside of the bead must stay electrically neutral. For every $Na^+$ lost by the bead in this example, one $K^+$ is gained. However, $Cl^-$ can penetrate the bead to some extent and every $Cl^-$ that enters must bring a positive ion with it. The concentration of $Cl^-$ in a bead of cation exchanger stays extremely small, however. Consider $RSO_3^-Na^+$ placed in a solution of $Na^+Cl^-$. Thermodynamics requires that

$$(Na^+)_r(Cl^-)_r(\text{in resin}) = (Na^+)_s(Cl^-)_s(\text{in solution}) \qquad (3\text{-}46)$$

Using concentrations instead of activities in this equation presumes ideal behavior, but the general effect can easily be seen. The NaCl concentration in solution is usually low, about 0.1 molar. In the resin, however, $(Na^+)_r$ is high about 5 molar. Thus $(Cl^-)_r$ has to be very small, about 0.002 molar in this example. This relationship is called the Donnan equilibrium.

The ions having the same charge as the resin framework (here $Cl^-$) are

---

[5]Since the pressures involved may reach 160 atmospheres, columns should *never* be packed dry.

called *co-ions*; those having opposite charge to the framework (here Na⁺) are called *counterions*. The Donnan equilibrium causes a partial exclusion of co-ions from the resin beads.

### (ii) Materials

Resin ion exchangers are also used primarily for substances other than proteins (41). We will use "Dowex" resins as an example. Other trade names for closely similar products are "Amberlite," "Duolite," "Permutit," etc. (Appendix 11). Most important are the ion exchange resins based upon cross-linked polystyrene (see Figure 3.30). The group $X$ represents a pair of ions: if this is $-SO_3^-H^+$, we have a cation exchanger, for the H⁺ can now be

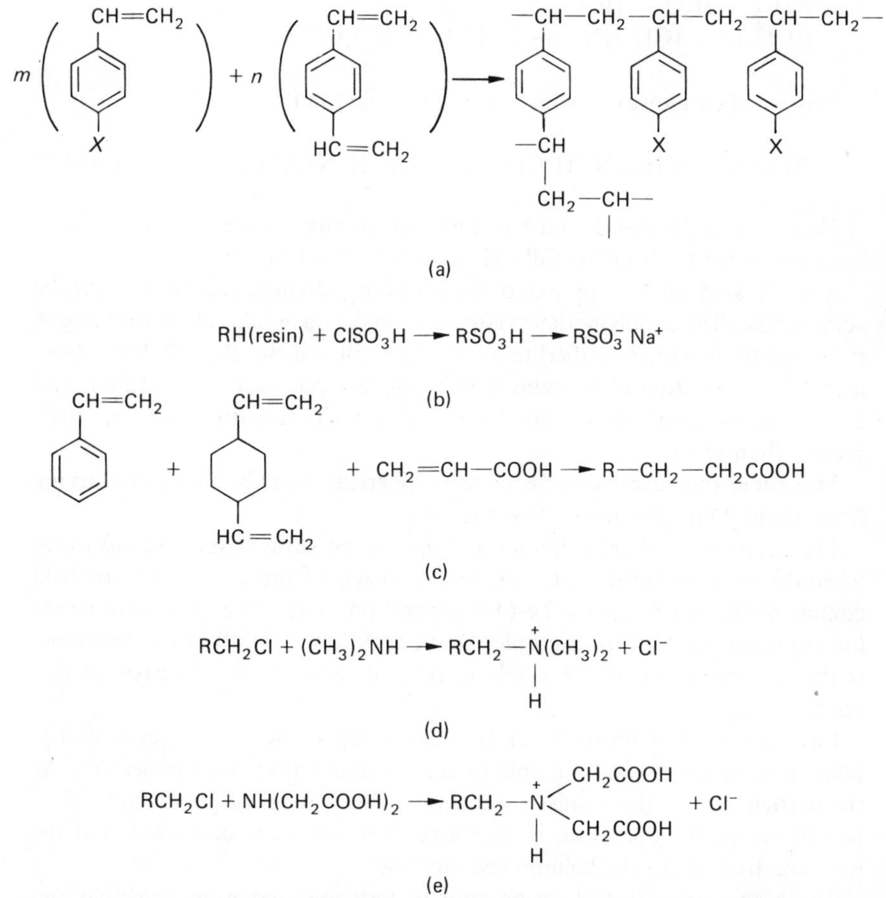

(a)

$$RH(resin) + ClSO_3H \longrightarrow RSO_3H \longrightarrow RSO_3^- Na^+$$

(b)

$$+ \ CH_2{=}CH{-}COOH \longrightarrow R{-}CH_2{-}CH_2COOH$$

(c)

$$RCH_2Cl + (CH_3)_2NH \longrightarrow RCH_2{-}\overset{+}{N}(CH_3)_2 + Cl^-$$
$$\underset{H}{|}$$

(d)

$$RCH_2Cl + NH(CH_2COOH)_2 \longrightarrow RCH_2{-}\overset{+}{\underset{\underset{H}{|}}{N}}{\overset{CH_2COOH}{\underset{CH_2COOH}{\diagup\diagdown}}} + Cl^-$$

(e)

**Figure 3.30** Chemical reactions for production of: (a) cross-linked substituted polystyrene; (b) sulfonic acid resins; (c) polymethacrylate resins (weak acid type); (d) weakly basic resin ("Amberlite IR 45"); and (e) a chelating resin ("Dowex-A-1", "Chelex 100").

replaced by any other positive ion such as $Na^+$. Two $H^+$ can be replaced by $Ca^{2+}$ and so on. If $X$ is $-CH_2-N(CH_3)_3{}^+Cl^-$, we have an anion exchanger, in which $Cl^-$ can be replaced by other negative ions, such as $OH^-$, or two $Cl^-$ by $SO_4^{2-}$. These two kinds of exchanger suffice for at least 90% of the laboratory and commercial applications of ion exchangers.

Many types of ion-exchange resins are available (Appendix 11). The major ones are described in Figure 3.30.

Strong acid resin is produced by treating the resin with chlorosulfonic acid (Figure 3.30). This is called Dowex-50 when produced by the Dow Company, and functions over the full pH range.

Two main types of strongly basic exchanger are produced.

a.  $RH + CH_2O + HCl \longrightarrow RCH_2Cl$
    $RCH_2Cl + (CH_3)_3N \longrightarrow RCH_2N^+(CH_3)_3Cl^-$

This is called Dowex-1, which is more stable than Dowex-2.

b.  $RCH_2Cl + (CH_3)_2NCH_2CH_2OH \longrightarrow RCH_2N^+(CH_3)_2(CH_2CH_2OH)Cl^-$

This is called Dowex-2, and has a greater affinity for negative ions. These both also function over the full pH range for hydroxyl ion.

A weak acid resin is prepared from styrene, divinylbenzene and acrylic acid. It is called a polymethacrylate resin and is a weak cation exchanger. It has the trade names Amberlite XE-64, IRC-50 and Bio-Rex 70. It has been useful for separation of proteins with isoelectric points near neutrality, and for adsorbing antibiotics from broth. This resin functions only at pH's greater than seven.

These acid and base resins are made as spherical beads of diameters varying from about 2 mm down to a few microns.

The proportion of cross-linking groups can be varied over a broad range when the resin is synthesized. Proportions varying from 2% to 16% are sold commercially, but 8% is the best for general purposes. The amount of cross-linking is specified as, for example, Dowex 50X8; 8% DVB (divinylbenzene) is the concentration of cross-linking reagent used in the synthesis of this resin.

Low cross-linked resins generally have a higher degree of permeability, take up more water, and are able to accomodate larger sized molecules. As the particle size of the beads is decreased, the volume of resin required to perform a specific operation is decreased, flow rates are decreased, and the pressure drop across the column bed increases.

The higher cross-linked resins tend to be somewhat more selective and exhibit molecular sieving effects. They also have greater adsorption capacity per unit weight or bed volume. High cross-linking also produces less swelling when placed in water, greater physical stability, more difficulty in introducing the ionic groups during the synthesis of the resin, and increased time to attain equilibrium.

In amino acid analyzers, the trend is to higher cross-linking (e.g., 12%) with a smaller particle size to compensate for increased equilibrium time and pressure elution to compensate for increased flow resistance. For peptide chromatography, a low percentage (e.g., 2%) cross-linked resin is used to allow increased diffusion to the resin interior.

### (iii) Applications

In general any soluble ionic compound will participate stoichiometrically and reversibly in ion exchange if it is small enough to penetrate the resin. Since the reaction is stoichiometric, the amount of resin needed can be calculated from its capacity. Generally speaking, the greater the valency the more strongly the material is bound, and at equal valency, the binding affinity increases with atomic number.

*1. Exchanging of counterions.* Ion exchangers are almost always used in columns. The solution traveling down the column continually meets fresh resin as it proceeds, and in this way reactions that are reversible can be carried to completion in any direction desired. A sodium salt, for example, passed through a column of cation exchange resin carrying replaceable hydrogen ions is converted completely into the corresponding acid, provided the column is long enough and the flow of the solution is slow enough to give time for reaction.

Similarly, conversion of a resin from one counterion to another is effected by treatment with an excess of the ion wanted as counterion, preferably also in a column.

*2. Concentration.* Concentration of a substance can be done by adsorbing it on a resin and eluting with a substance with a higher affinity, either on a column or batchwise (see Sec. 3.3b).

*3. Separation of similar substances.* If we have two ions, $A^-$ and $B^-$, and $B^-$ is more strongly adsorbed to a cation exchanger than $A^-$ at the same pH and ionic strength, it is clear that if ion $A^-$ is adsorbed on a column it would form a layer right at the top of the bed. If a solution containing $B^-$ is now applied, $B^-$ acts like an eluting ion, that is, ion $A^-$ is displaced downward on the column until all of ion $B^-$ is adsorbed. The same process occurs even when the ions are applied in mixed solution.

Chromatography with a resin such as Dowex 50 can separate closely related ions such as calcium and magnesium (see Figure 3.31). Elution is with sodium ion, applied stepwise or in a gradually increasing concentration gradient. Frontal analysis or displacement analysis may be used here (see Sec. 3.3b).

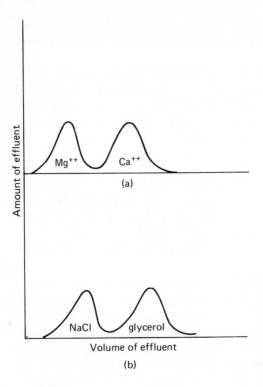

**Figure 3.31**
(a) Separation of calcium (atomic weight 20) and magnesium (atomic weight 12) by elution from an ion-exchange resin. (b) Separation of salt (NaCl) and glycerol using an ion-exchange resin.

Purification of proteins may be attempted by adsorbing the protein on the resin as the counterion. It can then be eluted with some neutral salt, such as sodium chloride. However, proteins are easily denatured by these types of ion-exchange resins, though not by ion-exchange celluloses (see Sec. 3.3d). *Never* commit an enzyme to treatment with a resin without first checking the effect on its activity.

*4. Ion exclusion.* Ion exclusion is a property of these resins that is sometimes useful. Consider the chromatography of a mixture of sodium chloride and some nonionic substance such as glycerol on a column of Dowex-50 ($Na^+$).

If a chloride ion moves in to a bead of resin a sodium ion must follow. If sodium ion follows, the sodium ion activity in the interior increases and the activity in the exterior drops. Thus the movement of sodium chloride into the resin is repressed. The interior volume available to NaCl is thus less than that available to glycerol; i.e., the ions are excluded. Consequently, they will emerge first from the column. See Figure 3.31b.

*5. Partition chromatography on ion-exchange resins.* Nonionic separations can also be performed. In the case of a mixture of ethylene glycol and phenol, the ethylene glycol is equally partitioned between the resin phase and the interstitial phase. The phenol prefers the resin phase, so the ethylene glycol comes out first. In the case of methanol and glucose, the glucose tends to be excluded from the interior of the beads because of its greater size and emerges first. Thus, both "molecular sieving" and solubility partition effects can occur and be useful with resin ion exchangers.

### 3.3d  Cellulose Ion Exchangers

As indicated above, large molecules such as proteins are excluded from the interior of resin ion exchangers limiting their effective exchange capacity. In addition, some proteins seem to react with the resins or be denatured by them. In 1956 a new approach was opened by the development of ion exchangers made from cellulose (42). The disadvantages of the resin exchangers as far as protein separations are concerned are not nearly so marked with the cellulose exchangers. Currently the latter are used largely for protein purification, though they can also be used for separations of small molecular weight compounds (43).

**(i) Theory and applications**

Since proteins are complex amphoteric materials, and moreover may have regions which are predominantly acidic while other regions in the same protein are predominantly basic, it is difficult to give a precise picture of their interactions with cellulose ion exchangers. In general, however, the major attractive force between proteins and these celluloses is thought to be electrostatic (44). The attractive forces may consist of many such interactions, but the net charge on the protein in the case of cation exchangers should be positive; that is, the pH should be below the isoelectric point of the protein. The protein can in theory be eluted by changing the pH so that either the protein loses its net charge or the resin loses its charge (see Figure 3.32).

Elution by decreasing the pH should be avoided; with anionic celluloses, (cation exchangers) decreasing the pH to near or below the pK of the exchange group would produce poor selectivity since all the adsorbed proteins would tend to be released at once. A similar situation would occur with cationic celluloses (anion exchangers); in this case, the best procedure would be to elute adsorbed proteins by decreasing the pH. Increasing the pH enough would neutralize the charge on the cellulose, releasing all adsorbed proteins.

With these precautions, proteins may be progressively eluted by a gradually increasing or decreasing pH gradient. The major limitation remaining is that you must stay in a pH range where the protein is stable.

There is also a "damping" effect of ionic strength on the interactions between ion-exchange celluloses and proteins. Thus one can elute proteins at a constant pH by increasing the ionic strength. This seems to produce sharper elution patterns than changing the pH, so it is most favored (see 3-4c below).

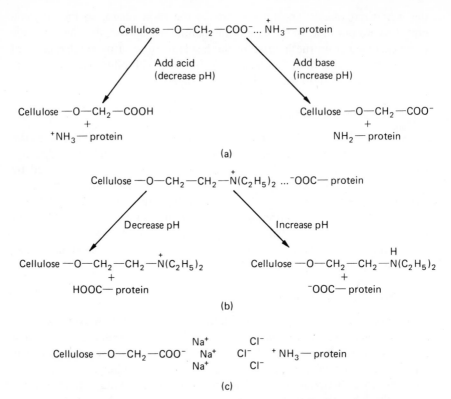

Figure 3.32  Elution of a protein from: (a) cellulose cation exchanger by changing the pH; (b) anion exchange cellulose by changing the pH; (c) cation exchange cellulose by increasing the ionic strength.

**(ii) Materials.**

A variety of types are available. The most common are:

    1. DEAE (diethylaminoethyl) cellulose is an anion exchange resin used primarily with neutral and acidic proteins.

    2. CM (carboxymethyl) cellulose is used primarily for separation of neutral and basic proteins; it is a cation exchanger.

    3. ECTEOLA cellulose is used for separation of nucleotides, nucleic acids and viruses.

Other types which may find special uses are shown in Figure 3.33.

The final concentration of chemical groups is usually 0.2-0.9 meq/gm. Above 1.0 meq/gm the cellulose becomes gelatinous, giving poor solvent flow rates.

Another chromatographic material which has found wide application is the resin DEAE-Sephadex; i.e., Sephadex which has been substituted with

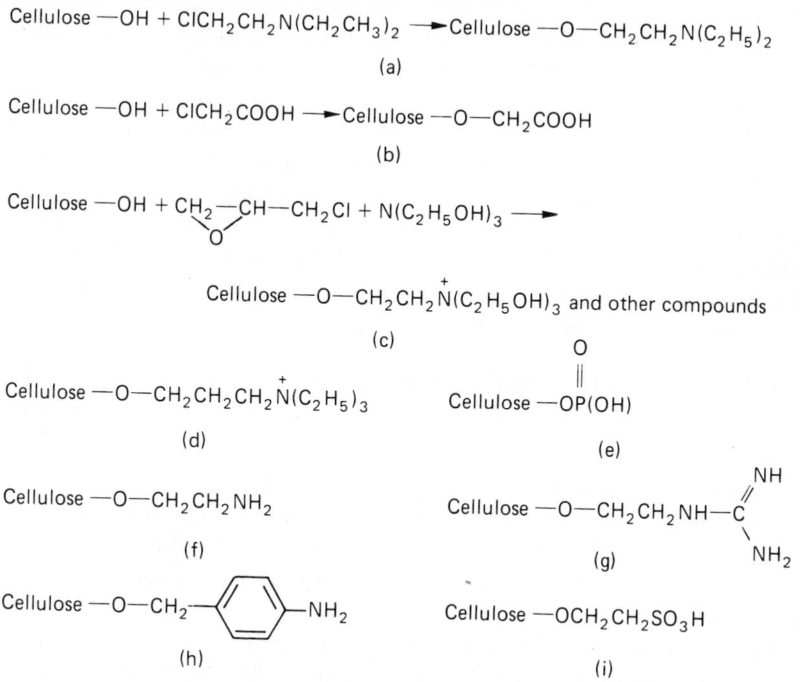

Cellulose $-OH + ClCH_2CH_2N(CH_2CH_3)_2 \longrightarrow$ Cellulose $-O-CH_2CH_2N(C_2H_5)_2$

(a)

Cellulose $-OH + ClCH_2COOH \longrightarrow$ Cellulose $-O-CH_2COOH$

(b)

Cellulose $-OH + CH_2-CH-CH_2Cl + N(C_2H_5OH)_3 \longrightarrow$

Cellulose $-O-CH_2CH_2\overset{+}{N}(C_2H_5OH)_3$ and other compounds

(c)

Cellulose $-O-CH_2CH_2CH_2\overset{+}{N}(C_2H_5)_3$

(d)

Cellulose $-OP(OH)$
$$\overset{O}{\underset{\|}{}}$$

(e)

Cellulose $-O-CH_2CH_2NH_2$

(f)

Cellulose $-O-CH_2CH_2NH-C\overset{NH}{\underset{NH_2}{\diagdown}}$

(g)

Cellulose $-O-CH_2-$〈benzene ring〉$-NH_2$

(h)

Cellulose $-OCH_2CH_2SO_3H$

(i)

Figure 3.33 Chemical formulas and synthetic reactions for: (a) DEAE-cellulose; (b) CM-cellulose; (c) ECTEOLA-cellulose; (d) TEAE-cellulose; (e) Phospho-cellulose; (f) aminoethyl-cellulose; (g) guanidinylethyl-cellulose; (h) PAB-cellulose; (i) sulfoethyl-cellulose.

DEAE groups. Because it has a cross-linked gel structure, the amount of DEAE groups can be considerably larger than on cellulose, resulting in a resin with an adsorption capacity about 3.5-5.0 times greater than cellulose. Because of its ease of handling, this resin has found great application in batchwise adsorption as well as column chromatographic application. Now TEAE-, CMC-, etc., derivatives of Sephadex are also sold. Bio-Gels are also available as CMC derivatives (Appendix 11).

### 3.3e  Nucleic Acid Complex Columns

Chromatographic techniques utilizing more specific interactions are increasingly used in separations of complex molecules. The formation of complexes between nucleic acids and proteins or other nucleic acids is often quite specific, and this specificity is taken advantage of in using derivatives of these materials as adsorbents.

#### (i) Histone-Kieselguhr columns

Histone forms complexes with nucleic acids. Brown and Watson (45) adsorbed histone onto Kieselguhr (Celite or diatomaceous earth) and also chemically coupled histone to diazotized PAB cellulose and used these as adsorbents for DNA. The DNA was eluted with increasing concentrations of sodium chloride.

The ratio of adenine to guanine and thymine to cytosine in the eluted nucleic acids increased with the salt concentration. DNA from different bacterial species eluted according to the ratios of these nucleic acid bases. Yeast RNA was not adsorbed, so the column could be used to separate yeast DNA and RNA. Heat denatured DNA was also not adsorbed with this column, showing the specificity of the adsorbents for the structure as well as the composition of the nucleic acids.

### (ii) MAK (Methylated Albumin on Kieselguhr) columns

These are prepared by methylating bovine serum albumin and adsorbing it onto Celite. Elution is again by increasing the ionic strength, generally with sodium chloride. Lerman (46) used this material in a column and was able to increase the specific transforming activity for streptomycin resistance per unit of DNA. Mandell and Hershey (47) gave detailed directions for preparation of this adsorbent. These authors were able to separate $T_2$ and $T_4$ virus DNA, despite their very similar base composition, using columns of this material. They attributed the separation to the fact that all the hydroxymethyl cytosine residues of $T_4$ DNA carry one glucose, while in the case of $T_2$, 25% carry none, 70% carry one, and 5% carry two.

Sueoka and Cheng (48) found by cesium chloride gradient centrifugation (Chapter 6) that the elution of nucleic acids from MAK columns follows their guanine plus cystosine content. They could separate soluble RNA from ribosomal (16 and 23S) RNA.

### (iii) DNA columns

Columns containing DNA have yielded important results in genetics and molecular biology. DNA-cellulose columns (49) are made by acetylating phosphocellulose with acetic anhydride, then coupling the DNA to the phosphocellulose acetate with dicyclohexyl carbodiimide in pyridine. The denatured DNA is presumably linked to phospho-cellulose acetate through the glucosidic hydroxyl.

RNA from E. coli grown in $^{32}$P and $^{3}$H-uracil and after $T_4$ infection was adsorbed on this column by increasing the temperature to 65° and lowering the salt concentration. The RNA adsorbed by the column had eight times the $^{3}$H/$^{32}$P ratio of the starting material. This high ratio RNA had a base composition similar to that obtained by measuring the $^{32}$P distribution in RNA nucleotides after phage infection.

The RNA from rII E. coli was adsorbed on R$^+$ DNA cellulose. This was eluted and passed repeatedly through a column of r 1272 DNA (which lacks the r$^+$ genome). This RNA would still bind to a R$^+$ DNA column. Thus the "message" for the two cistrons of the rII region was presumably isolated.

DNA-Agar gel columns and DNA-cellulose acetate columns have been put to similar uses (50-52). DNA-agar gels are prepared simply by mixing denatured DNA and 3% aqueous agar at a moderately high temperature, then cooling.

### 3.3f  Antibody-Antigen (AB-AG) Columns

Webb and Lapresle (53) diazotized polyamino polystyrene and added human serum albumin. About 10% of the protein was fixed, and the protein-resin was used to make a column. The column adsorbed nearly the theoretical amount of rabbit anti-albumin, which was then eluted. They repeated this using another protein (54). The reverse system, i.e., bound antibody, was studied using polyamino polystyrene and polystyrylmercuro-perchlorate to bind the antibody (55).

### 3.3g  Preparation of Affinity Chromatographic Adsorbents

The biological functions of proteins are based on their ability to adsorb other substances strongly and specifically. Affinity chromatography takes advantage of this by making a resin consisting of a matrix to which are attached either a strong competitive inhibitor of the enzyme to be isolated or a protein which binds strongly to the protein of interest. This technique is in a sense the ultimate extension of adsorption chromatography, since it involves using the very binding properties which are characteristic of a particular substance to isolate it.

The procedures developed recently for preparing various resins have been reviewed by Cautrecasas and Anfinsen (56). We will merely summarize a part of their article.

The best matrix is beaded agarose, especially Sepharose 4B (56). This is porous enough to allow even large molecules within, is chemically and mechanically stable to the reaction, chromatographic and cleaning conditions, has a large number of possible reaction sites and is itself almost inert to proteins. The ligand to be coupled should be bound as strongly and specifically as possible to the protein of interest and must have chemical groups that can form the bond to the matrix without changing its affinity for the protein too much. It must also be stable to coupling, binding, and washing. The best results are obtained when the ligand can be attached to the matrix by a long hydrocarbon chain. Evidently proteins tend to bind ligands in crevices and the chain is necessary to allow insertion of the ligand.

Elution can be done with ionic strength, substrate, the same or another competitive inhibitor, etc. Sometimes the protein adsorbs so strongly that extreme pH or high concentrations of urea or guanidine hydrochloride must be used.

## 3.4 PRACTICAL ASPECTS OF COLUMN CHROMATOGRAPHY

Columns are used in most preparative work because of the advantages of being able to scale up almost indefinitely. Also, effects of contact of the material with any surfaces other than those of the resin are minimized. Consequently this section on column chromatography is included.

A complete column setup is shown in Figure 3.34.

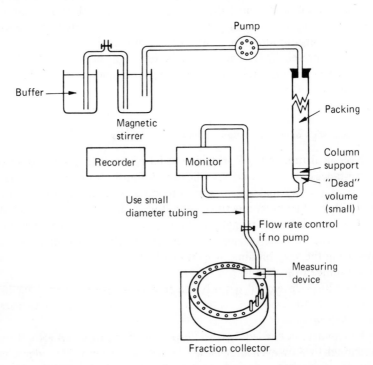

Figure 3.34   A complete column chromatographic setup.

### 3.4a   The Column

Generally, longer columns give better resolution, particularly with "molecular sieve" or "gel" chromatography, and wider columns give greater capacity. To avoid the possibility of zone remixing, any column used should have a minimum amount of dead space at the outlet. The column tube can be coated to avoid "wall effects" (deforming of the zones due to the tendency of the buffer to flow faster in a thin layer near the glass wall), using dichlorodimethylsilane in benzene (see Chapter 1).

*3.4b   Packing the Column*

### (i) Preparation for packing

The column must be vertical before packing, no matter what resin is used. Otherwise column chromatographic procedures tend to be different for ion-exchange celluloses and materials for gel filtration.

The column support can be glass wool, sintered glass, or porous polyethylene (51 micron pore size). It is good practice to check the fritted glass discs mounted in columns for their flow rate. These often become clogged by small resin particles. If this occurs, these particles may be removed by hot 50% $H_2SO_4$. The column is then rinsed thoroughly with distilled water before packing is attempted.

The stopper-tubing connection must be air tight. Tubing can be easily inserted in a rubber stopper to provide such a fit. Push a cork borer up through a one-hole rubber stopper. Then slide tubing which is sufficiently larger than the hole but smaller than the borer down through the cork borer so it extends an appropriate length beyond the stopper. Hold the tubing and stopper with one hand, then pull out the borer with the other, leaving the tubing held in the stopper.

### (ii) Preparation of resins

*1. Ion-exchange celluloses.* Large particles with high internal surfaces equilibrate slowly, necessitating low flow rates, and have poor column packing characteristics. Smaller particles increase flow resistance, forcing lower flow rates than needed for equilibration and reducing protein recovery. Consequently, there is an optimum size. Cellulose ion exchangers from various manufacturers nearly always contain a wide variety of particle sizes. The most useful range is 60-200 mesh. Coarser particles (less than 60 mesh) are, however, very satisfactory for batch type operations and should be set aside for that purpose. Cellulose ion exchangers, when sieved to yield the 60-200 mesh particles, may be packed in columns under 5-10 psi pressure to yield very reproducible performance. This can be extremely important for gradient elution work.

Ion-exchange celluloses must be thoroughly equilibrated with the buffer in which the sample is applied. This can be determined by comparing the pH and conductivity of the eluant buffer with that of the original buffer—they should be the same. DEAE-celluloses supposedly tend to form complexes with carbonate and bicarbonate, which can interfere with chromatography (57). This can be removed by suspending the cellulose initially in the acid component of the buffer and degassing, followed by titration back to the proper pH, or by simply suspending the cellulose in the buffer made up with $CO_2$-free water and titrating to the proper pH with the acid component of the buffer. The titrated cellulose is resuspended in more of the buffer and the pH checked again. If it is still too high, the titration is repeated.

Equilibration can also be done after the column is poured.

*2. Molecular sieving materials.* The dry beads of Bio-Gels should be hydrated by adding them slowly with constant stirring to a solution of the buffer that will be used for the elution. The water regain values in Appendix 11 show the minimum quantity necessary to hydrate each gram of dry gel beads; five to ten times this amount should be used to prepare a slurry suitable for pouring a column. Bio-Gels P-2 through P-10 swell rapidly and are ready for use after two or four hours. The larger pore size gels require up to 48 hours. The hydration time for higher Bio-Gels may be reduced by gently boiling the gels for two hours in distilled water or in buffer (below pH 10).

Sephadex G-100, 150 and 200 must be hydrated by soaking 24-48 hours at room temperature with the buffer they are to be used in. Even then, columns of the higher Sephadexes must be further equilibrated by running their buffer through them for three to five days; otherwise little or no resolution will be obtained. Apparently the penetration of solvent into the smaller channels within the beads is very slow. To shorten these equilibration times, the gels can be hydrated and equilibrated by heating to 80-90° in the appropriate buffer for five hours before pouring the column. The lower Sephadexes and Bio-Gels can be used immediately after the prescribed swelling time, though heating can shorten that.

If swollen Sephadex, Bio-Gel, or cellulose ion-exchanger suspensions have been stored in the cold, dissolved oxygen must be removed before column packing can proceed at room temperature. Degassing can be quickly accomplished by gentle warming while stirring the resin suspension under vacuum. One should never attempt to use at room temperature a column packed or stored in the cold without first repacking, or serious interference will occur from released gas bubbles. It is, however, generally good practice to degas all resin preparations before packing a column, irrespective of what their storage temperature was. This is especially true of the higher Sephadexes (G-75 to 200), and mandatory for substituted Sephadexes.

### (iii) Pouring columns

No column should ever be allowed to run dry.

*1. Cellulose ion exchangers.* In packing columns with cellulose ion exchangers, the more dilute the slurry, the better the packing (and the slower the packing as well, so a balance must be struck). A 5% slurry is generally adequate. The celluloses should be packed in the buffers in which they are to be used.

Column heights of 10-20 cm are usually sufficient for cellulose ion-exchange columns. The width of the column should be chosen according to the amount of protein. About 100 mg protein can be adsorbed per gm of cellulose (dry weight), and 1 gm of cellulose ion exchanger gives 6-9 ml of bed volume.

A convenient setup for packing columns with cellulose adsorbents is shown in Figure 3.35. The column is shut off and the cellulose suspension is poured into it and allowed to settle until about 1 cm of bed volume is established. Then the column is opened to allow a slow flow of buffer. More slurry is added (Figure 3-35) until the proper column size is attained.

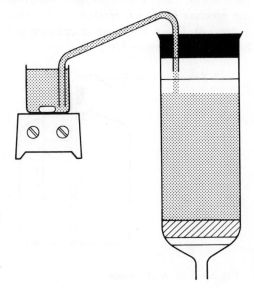

**Figure 3.35**
Procedure for packing a column with
an ion-exchange cellulose.

Flow rates should be 5-25 ml/cm²/hr or higher for cellulose columns. If the flow rates are not that high, you probably have not removed all the fine particles, and should do so.

*2. Molecular sieving materials.* The higher numbered Bio-Gels and Sephadexes should be settled repeatedly to remove "fines." It is also good practice

**TABLE 3.4**

| Sephadex | Maximum Pressure Head (cm) | Maximum Flow Rates ml/cm²/hr |
|---|---|---|
| G-200 | 10 | 4 |
| G-150 | 20 | 8 |
| G-100 | 40 | 16 |
| G-75 | 80 | 32 |
| A-50 and C-50 | 35 | 12–15 |

to use a 1-2 cm layer of washed sand or fine glass beads over the column support. The column is filled $\frac{1}{3}$ full of water, and the sand added as a suspension in water. After settling, the walls of the column are washed down with water and excess removed. Sephadexes and Bio-Gels are then poured as a thick slurry into the column all at once.

Do not pack or elute these columns under more than the recommended pressure head: Sephadex particularly deforms under pressure and a Sephadex column under excess pressure will eventually stop. Also, do not change the pressure head, temperature or flow rate after the slurry is added (see Table 3.4). Use a Marriotte or constant pressure flask to keep the pressure head constant (Figure 3.36).

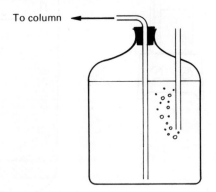

To column

**Figure 3.36**
A Marriotte, or constant pressure reservoir.

#### (iv) Application of sample

*1. Ion exchange celluloses.* It has been found by experiment that it does not matter whether the applied protein is adsorbed in a narrow band at the top of the column or distributed homogeneously throughout the bed of cellulose ion exchangers.

*2. Molecular sieving materials.* To apply the sample to molecular exclusion columns, however, the thickness of the zone of application is of crucial importance to the resolution (see 3-4c(ii) below). The sample zone can be kept sharp by using inverse flow and following the sample with 15% sucrose.

For regular descending flow, the sample is added in buffer with its density increased by addition of sucrose. Softer gels such as G-100 or P-200 tend to be easily disturbed by sample addition. Use of a pump to layer the sample onto the gel can help. Adding a 3 cm layer of a harder gel such as G-25 or P-60 to the top of a Sephadex or Bio-Gel column can also help stabilize the gel bed.

Not only will these precautions result in a more stable bed, but there will be an initial partial separation of proteins from lower molecular weight substances, and the density of the protein zone will be comparatively low when it enters the looser bed, minimizing zone deformation due to convection.

### 3.4c  Elution of Columns

#### (i) Ion-exchange celluloses

It is best to use cationic buffers (such as Tris, or piperazine) with anion exchangers (which are cationic) to reduce interactions between the buffer and the column, and vice versa. The most selective elution with celluloses is obtained with monovalent buffers. Phosphate, though a powerful eluting agent, usually displays poor selectivity. However, trivalent buffers may serve well as supporting buffers in dilute solution, with elution being accomplished by addition of sodium chloride to the supporting buffer.

With ion-exchange resin columns, the gradient used for elution is the chief factor in determining the resolving potentialities of the column (58-60). Many kinds of gradients can be used. The usual apparatus consists of two vessels, connected so that liquid from the second, containing the higher ionic strength or different pH solvent, flows into the first and is mixed there as liquid is removed to the column from the first (Figure 3.37). With vessels of equal shape and size, a linearly increasing ionic strength solvent is obtained, but concave exponential or convex exponential gradients can be produced by varying the relative gradient vessel sizes and shapes. The "varigrad," an apparatus of six vessels (61), provides more flexibility in producing gradients. The gradient produced may be varied in several ways by changing concentration ratios of adjacent vessels, moving the limit concentration buffer to different vessels, or distributing various solutions in different vessels.

Whatever gradient device is used, its effectiveness should be checked by pH or conductivity.

Resolution is also a function of the ratio of gradient volume to column volume since a more slowly changing solvent (the flow rate being equal) will tend to separate different proteins better.

It seems best to elute proteins from ion exchange celluloses as soon as possible; leaving proteins on these columns for long periods—several days—seems to result in greatly reduced recoveries, for unknown reasons.

#### (ii) Molecular sieving materials

With gel columns, resolution is a function of the ratio of the sample size to the column length for a given fraction size (see 3-4h below).

Gel columns usually perform better when run at flow rates lower than their free flow rate. This allows more complete equilibration between the mobile solvent and the stationary solvent within the pores of the beads. At low temperatures it is especially important to keep flow rates low, since the higher viscosity of water at low temperatures tends to restrict solute diffusion into the smaller pores and channels within the beads.

Reverse flow chromatography has some advantages. During development of a normal column the higher Sephadexes especially tend to pack, decreasing the flow rate. This may be minimized by running the column "backwards,"

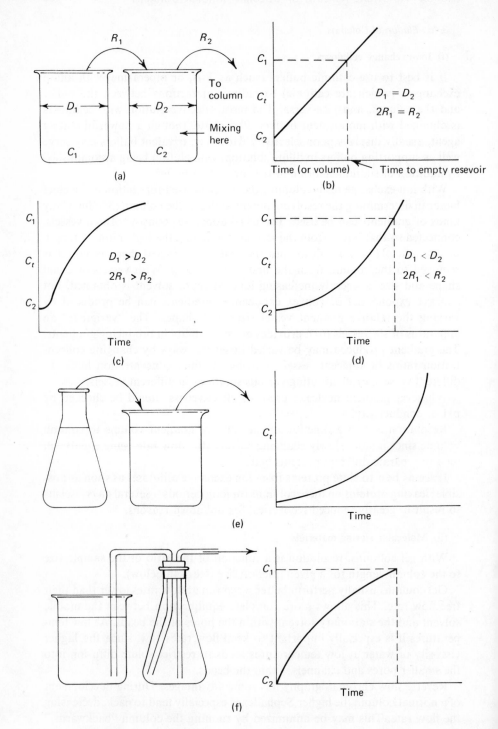

and this has a further advantage of giving sharp zones on sample application. The column should be run at the same flow rate in all experiments.

Recycling chromatography is used when the separations wanted require an inordinately long column (58). The effluent from one column is simply led into another, or pumped back into the first.

Aromatic and heterocyclic substances tend to be adsorbed by Sephadexes. In addition, in columns prepared in distilled water, adsorption of basic substances and exclusion from gel particles of acidic substances is seen. This effect is eliminated when salt is added and has been shown to be entirely due to the degree of esterification of the resin. Methylated dextran, a highly hydrophilic resin known as Sephadex LH-20, exhibits the greatest aromatic or heterocyclic effect.

For gel columns growth of mold can be inhibited by addition of $1/1000\%$ $NaN_3$.

### 3.4d The Pump

Many are available; the Buchler is particularly versatile. It allows a wide range of flow with a continuous choice and a steady rate without contact with the solution.

### 3.4e Flow Rate Control

The flow rate control must always be upstream from the monitor; a Harvard apparatus clamp, Kontes, Owens-Corning or a wire in a clamped tube may be used.

### 3.4f The Monitor

This consists of a light source and filter, generally transmitting 260 nm or 280 nm light, and a photocell. The column effluent flows through a tube bent

---

**Figure 3.37**
(*Opposite page.*) Apparatuses for gradient elution and types of gradient produced: (a) two-chamber type; (b) linear gradient with chambers of equal diameter; (c) convex exponential gradient with rearmost (reservoir) chamber larger (the best separation is achieved near the end of the gradient; more highly charged proteins will be eluted at that point); (d) concave exponential gradient with rearmost chamber smaller (the best separation is at the beginning of the gradient; the less highly charged molecules will be eluted at that point); (e) gradient using an Erlenmayer flask as reservoir; (f) convex exponential gradient produced using an Erlenmayer flask to provide a constant mixing volume (the shape here is unimportant; it is the constant mixing volume which matters). In the figure, the rearmost (reservoir) chamber is 1, the chamber where mixing occurs is 2; $R$ is rate of solvent flow, $D$ is diameter, and $C_1$ and $C_2$ are initial solvent concentrations in vessels 1 and 2.

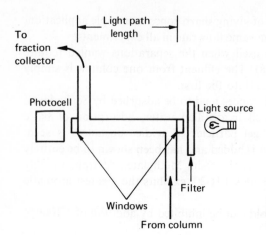

**Figure 3.38**
Schematic representation of a column monitor.

to give a certain path length (Figure 3.38). Watch out for air bubble formation in the tube because of temperature changes or aspiration.

The monitor records percent transmission. The adsorbance (OD) is the $\log_{10}$ of this, and the percent transmission curve sharpens when converted to OD. However, the initial advantage of measuring transmission is that there is greater detail and small peaks are amplified. (see Figure 3.39).

However, the monitor should not be trusted to give an accurate idea of the quantity of eluant. For this you must measure OD. Remember that the extinction coefficients of various proteins as seen by a monitor may not be the same as that in any one fraction; i.e., the monitor looks at each small drop as it moves off the column, where an individual tube is a composite of many drops.

The $OD_{280}$ is not related to weight of protein by the same factor for all proteins. It is a function of aromaticity.

### 3.4g  Measuring Devices

Several devices to regulate eluant fraction size are available. Volumetric siphons generally perform poorly. The Teflon plunger types are better. Fractions are often collected at fixed intervals of time, but the flow rate may change unless it is controlled by a pump. Drop counting is best, but the drop size changes with effluent composition because of changes in density and surface tension. Usually the change is not important, but if it is one can check fraction weights.

### 3.4h  Eluant Fraction Size

For gel columns, the best resolution is obtained when the fraction volumes are the same as that of the original sample. For cellulose columns, the smaller the fraction collected, the better the resolution. In both cases, there are practical lower limits to the volume of the fractions collected.

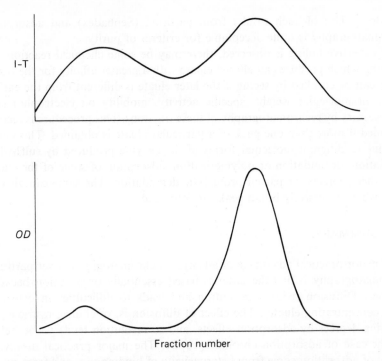

**Figure 3.39**  Comparison of an elution profile as measured by percent transmission (upper curve) or optical density (lower).

## 3.5  ANALYSIS OF DATA

Partition chromatography (paper, Sephadex, GLC) should provide Gaussian
(from diffusion—see Chapter 6) peaks if the substance is pure; the presence
of shoulders suggests contaminants. Tailing or streaking can also result
from adsorption effects. The usual procedure of measuring some specific
property of the eluate across its peak to see if the ratio is constant is often
used in these cases. Sometimes the peak is analyzed by measuring the differential, as is done by schlieren optics (Chapter 6). This is very sensitive to irregularities in the peak or small amounts of other boundaries.

While a substance may appear to be a single component upon chromatography under a specific set of conditions, the rule is that several criteria
of purity should be applied, preferably measuring very different properties
of the protein (chromatography on Sephadex and electrophoresis, for example). At the least, chromatography should be done under several conditions,
if you plan to provide chromatographic evidence for purity.

Peaks from adsorption chromatography are seldom Gaussian; some tailing is almost always observed, even using gradient elution. Here the same
procedure, assaying some specific property of the eluate across its peak, can

be done. Use of such results from partition (Sephadex) and adsorption chromatography is quite acceptable for criteria of purity.

If excessive tailing is observed, there may be some chemical reaction occurring which produces an altered eluate with a greater affinity for the resin. This can be checked by seeing if the later eluate is different from the earlier (e.g., in molecular weight, specific activity, mobility on electrophoresis), if necessary by rechromatography of these fractions. This procedure is recommended if more than one peak of a particular eluate is obtained. This could be due to different molecular forms of the enzyme produced by sulfhydryl oxidation, deamidation or polymerization, adsorption of some of the eluate by other proteins, or partial proteolytic degradation. The same can also be said where excessively broad peaks are obtained.

## 3.6 SUMMARY

The major practical advantage of adsorption chromatography over partition chromatography is that the latter is based essentially on diffusion between phases. Diffusion reduces resolution and leads to difficulties in detecting and concentrating eluates. The effect of diffusion is, of course, restricted to varying degrees by adsorption effects, and so resolution tends to be better in the case of adsorption chromatogrophy. The major practical disadvantages of adsorption come from heterogeneity of binding sites and from excessively strong adsorption of substances so that recoveries are poor. This problem can be severe in the case of affinity chromatography. Still, purifications of material such as proteins are currently largely done by adsorption chromatography, though chromatography on molecular sieving materials is popular because of that feature of their operation.

## REFERENCES

1. Choivin, P. in *Comprehensive Biochemistry*, Vol. 4, M. Florkin and E. H. Stotz, Eds., Elsevier, New York (1962), Ch. 3.
2. Moore, W. J., *Physical Chemistry*, 4th ed., Prentice-Hall, New Jersey (1972).
3. Craig, L. C. in *Comprehensive Biochemistry* (reference 1), Ch. 1.
4. Keller, R. A. and J. C. Giddings in *Chromatography*, 2nd ed., E. Heftmann, Ed., Rheinhold, New York (1967), Ch. 6.
5. Lederer, E. and M. Lederer in *Comprehensive Biochemistry* (reference 1), Ch. 2.
6. Humphrey, J. H. and R. R. Porter, *Biochem. J.*, **62**, 93 (1956).
7. Porter, R. R. and E. M. Press, *Biochem. J.*, **66**, 600 (1957).
8. Flodin, P., *Dextran Gels and Their Application in Gel Filtration*, Uppsala, Sweden (1962).
9. Pederson, K. O., *Arch. Biochem. Biophys.*, Suppl. 1, 157 (1962).
10. Andrews, P., *Biochem. J.*, **91**, 222 (1964).
11. Ackers, G. K., *Biochem. J.*, **3**, 723 (1964).

12. Laurent, T. C. and J. Killander, *J. Chromatogr.* **14**, 317 (1964).
13. Ogston, A. G., *Trans. Faraday Soc.*, **54**, 1754 (1958).
14. Siegel, L. M. and K. J. Monty, *Biochim. Biophys. Acta*, **112**, 346 (1966).
15. Russell, B., T. H. Mead and A. Polson, *Biochim. Biophys. Acta*, **86**, 169 (1964).
16. Haller, W., *J. Chromatogr.* **32**, 676 (1968).
17. Hummel, J. P. and W. J. Dreyer, *Biochim. Biophys. Acta*, **63**, 530 (1962).
18. Fairclough, G. F., Jr., and J. S. Fruton, *Biochem. J.* **5**, 673 (1966).
19. Andrews, P., *Nature*, **196**, 36 (1962).
20. Hjerten, S., *Biochim. Biophys. Acta*, **79**, 393 (1964).
21. Steere, R. L. and G. K. Ackers, *Nature*, **194**, 114 (1962).
22. Porath, J., *Nature*, **196**, 47 (1962).
23. Iwatsubo, M. and A. Curdel, *Comptes Rend. Acad. Sci.* (Paris), **256**, 5224 (1963).
24. Loening, U. E., *Biochem. J.*, **102**, 251 (1967).
25. Polson, A. G., *Biochim. Biophys. Acta*, **50**, 565 (1961).
26. Ackers, G. K. and T. E. Thompson, *Proc. Nat. Acad. Sci. U. S.*, **53**, 342 (1965).
27. Martin, A. J. P. and R. L. M. Synge, *Biochem. J.*, **35**, 1358 (1941).
28. Keulemans, A. I. M., *Gas Chromatography*, Rheinhold, New York (1957).
29. Dal Nogare, S. and R. S. Juvet, Jr., *Gas-Liquid Chromatography*, Interscience, New York (1962).
30. James, A. T. and A. J. P. Martin, *Biochem. J.*, **50**, 679 (1952).
31. Van Deemter, J. J., E. J. Zuiderweg and A. Klinkenberg, *Chem. Eng. Sci.*, **5**, 271 (1957).
32. Thompson, B. and R. D. Cook, *Industrial Research*, **14**, 52 (1972).
33. Lovelock, J. E., *Nature*, **187**, 49 (1960).
34. Special Report of the Fatty Acid GLC Committee, *J. Lipid Res.*, **5**, 20 (1964).
35. Hofstetter, H. H., N. Sen and R. T. Holman, *J. Amer. Oil Chem. Soc.*, **42**, 537 (1965).
36. Schlenk, H., J. L. Gellerman and D. M. Sand, *Anal. Chem.*, **34**, 1529 (1962).
37. Snyder, L. R. in E. Heftmann, *Chromatography*, 2nd ed., (reference 4), Ch. 4 and 5.
38. Hagdahl, L., *Acta Chem. Scand.*, **2**, 574 (1948).
39. Claesson, S., *Arkiv. Kermi. Minerol. Geol.*, **23**, A, No. 1 (1946).
40. Tiselius, A., *Naturwissenschaften*, **37**, 25 (1950).
41. "Dowex: Ion Exchange," Dow Chemical Company, Donnelley and Sons, Chicago (1959).
42. Peterson, E. A. and H. A. Sober, *J. Am. Chem. Soc.*, **78**, 751 (1956).
43. Dalziel, K., *J. Biol. Chem.*, **238**, 1538 (1963).
44. Wolf, F. J., *Separation Methods in Organic Chemistry and Biochemistry*, Academic Press, New York (1969), p. 177.
45. Brown, G. L. and M. Watson, *Nature*, **172**, 339 (1953).
46. Lerman, L. S., *Biochim. Biophys. Acta*, **18**, 132 (1955).
47. Mandell, M. and A. D. Hershey, *Anal. Biochem.*, **1**, 66 (1960).
48. Sueoka, N. and T. Y. Cheng, *J. Mol. Biol.*, **4**, 161 (1962).
49. Bautz, E. K. F. and B. D. Hall, *Proc. Nat. Acad. Sci. U. S.*, **48**, 400 (1962).
50. Bolton, E. T. and B. J. McCarthy, *Proc. Nat. Acad. Sci. U. S.*, **48**, 1390 (1962).
51. McCarthy, B. J. and E. T. Bolton, *Proc. Nat. Acad. Sci. U. S.*, **50**, 156 (1963).

52. Cowie, D. B. and B. J. McCarthy, *Proc. Nat. Acad. Sci. U. S.*, **50**, 537 (1963).
53. Webb, T. and C. Lapresle, *J. Exp. Med.*, **114**, 43 (1961).
54. Webb, T. and C. Lapresle, *Biochem. J.*, **91**, 24 (1964).
55. Kent, L. H. and J. H. R. Slade, *Biochem. J.*, **77**, 12 (1960).
56. Cuatrecasas, P., and C. B. Anfinsen in *Methods in Enzymology*, Vol. XXII, W. B. Jakoby, Ed., Academic Press, New York (1971), p. 345.
57. Whatman Data Sheet (No. 1-15M-1/68), H. Reeve Angel and Co., Clifton, New Jersey (1968).
58. Porath, J. and H. Bennich, *Arch. Biophys.*, Suppl. **1**, 152 (1962).
59. Lakshmanan, T. K. and S. Liberman, *Arch. Biochem. Biophys.*, **53**, 258 (1954).
60. Peterson, E. A. and J. Rowland, *J. Chromatogr.*, **5**, 330 (1961).
61. Walton, H. P., in *Chromatography*, 2nd ed., (reference 4), Ch. 12 and 13.

# 4 | IMMUNOCHEMISTRY

## 4.1 INTRODUCTION: THE "IMMUNE RESPONSE" (1-6)

The term "immune response" originally arose from observations first made in ancient times that those who survived an infectious disease rarely contracted it again. More generally, the term means the introduction of a foreign material (antigen) into the tissues of higher animals leading to the production by the host of proteins (antibodies) which interact with the antigen. This is "active" immunization as opposed to "passive" immunization. Active immunization involves *de novo* protein synthesis from amino acids rather than selection of proteins from the pool of serum globulins.[1] Passive immunization is immunity acquired from colostrum, transplacental transfer of antibodies from mother to fetus, or from the injection of immune serum.

Phylogenetically, the immune response is well-established as a general

---

[1]An immunoglobin is a protein which may have antibody activity. All antibodies are called immunoglobins.

97

phenomenon for warmblooded vertebrates (mammals and birds). At least some aspects of the response occur in cold-blooded vertebrates (reptiles and fish) and possibly in invertebrates. It has not been observed in plants, although some plants have agglutinins[2] against certain human red blood cells.

Normally antibodies can be detected within a few days after "challenge" (injection of an antigen). After an initial rise, the concentration ("titer") of antibody falls in 2-3 weeks, but can be restimulated by a second exposure to antigen (see Figure 4.1). The secondary response is more pronounced than the first even after years have elapsed—there is a "memory."

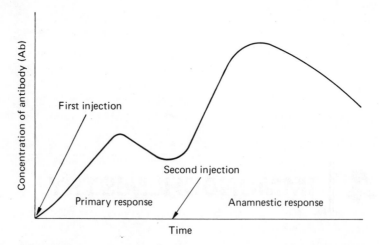

**Figure 4.1**   The "anamnestic response" of an organism to a second introduction of antigen.

These observations give rise to two questions: What is the nature of the "memory", and how did the antibody-forming cells recognize the antigen as foreign? Taken together, these questions summarize one of the major unsolved problems in biochemistry.

The immune response also has a negative side. We are becoming increasingly aware of its adverse effects in autoimmune diseases (such as systemic lupus erythematosus), allergic reactions, and tissue-graft rejections (resulting from kidney transplants, etc.).

Biochemists use immunochemical methods either to study the phenomenon of the immune response or as a technique for the specific detection, identification or preparation of materials such as proteins, polysaccharides and even entire microorganisms. The latter use is somewhat analogous to affinity chromatography in that you are preparing material with a specific affinity

---

[2]Agglutinins are proteins with the ability to clump and precipitate red blood cells.

for a substance. This chapter will stress the use of immunochemical methods as a technique in protein preparation and analysis. We will not cover some of the more "practical" procedures such as those involving complement fixation (the basis of the Wasserman test), passive cutaneous anaphylaxis, etc.

## 4.2 ORIGIN OF THE IMMUNE RESPONSE

### 4.2a The Nature of the Antigen

The antigen is a molecule possessing a configuration which, all or in part, is recognized by the organism as foreign. Clearly, any animal must be capable of distinguishing between its own substance and alien material (7).

The antigen is often (but not necessarily) protein: dextran (a polysaccharide) is antigenic for man, but not for rabbits. The response can be against the configuration of amino acids (in which case there would be many different sites of response); or it could be against an attached group (called a hapten), such as bacterial polysaccharide (cell wall, capsule), or a chemical hapten like a 2,4-dinitrophenol (DNP) group.

There is apparently a lower limit for antigenicity of about 10,000 molecular weight, and it is thought that smaller molecules act only after coupling (*in vitro* or *in vivo*) to a larger molecule.

The antigen is often polyvalent, i.e., it can induce the formation of a variety of antibodies, each able to react against a different antigenic site. Consequently, "antigen" (or "antigenic site") has two possible connotations: it is that which induces antibody formation, or it is that which reacts with the antibody produced. This need not necessarily be the same molecule. For example, the isolation of antibodies against the 2,4-dinitrophenol (DNP) group uses the reaction of the antibody with $\epsilon$-DNP lysine (which has a very strong affinity) or the reaction with DNP. In the latter case, the affinity is not as great, but the DNP can be used in much higher concentration.

An antigen often cannot react with as many antibodies as it can cause the production of, for purely spatial reasons. Although the antigen induces the formation of a variety of antibodies it only reacts with a few of these, since steric hindrance by one antibody already attached prevents the approach of an antibody specific for a nearby site.

### 4.2b The Nature of the Antibody (8, 9)

Blood consists of 45% red blood cells and 1% white blood cells and platelets. The remaining 54% of the volume is "plasma." The "plasma" contains the antibodies. It consists of 0.3% fibrinogen and 6.7% of other proteins by weight. "Serum" is the plasma without the fibrinogen, and its composition is shown in Figure 4.2. The antibodies are all $\gamma$-globulins. The immunoglobins

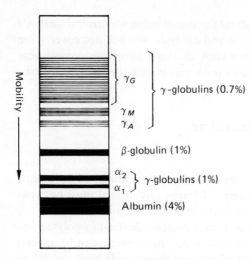

**Figure 4.2**
Representation of serum composition demonstrated by disc electrophoresis (Chapter 5) at an alkaline pH. Note the heterogeneity of the immunoglobulins.

differ from other serum proteins in not being homogeneous ("microheterogeneity") in sequence, number and type of polypeptide chains, number and location of disulfide bonds, and carbohydrate content.

There are three major classes of antibody immunoglobins in the rabbit.[3] The IgG or $\gamma$G class of antibody has a molecular weight of about 158,000–160,000, a sedimentation coefficient of 7S and contains 2–4% complex carbohydrate by weight. It is the classical "antibody". The IgA or $\gamma$A immunoglobin usually has a molecular weight of 160,000, but some range up to a molecular weight of 1,300,000. This type is mostly 7S material and contains approximately 9% carbohydrate. The IgM or $\gamma$M immunoglobin has a molecular weight of 900,000, and its sedimentation coefficient is 19S. It has a high carbohydrate content (about 10% by weight). The molecule can be broken up into 5 subunits of 180,000 MW by reducing disulfide bridges. The fragments are not identical to $\gamma$G, since they have only partial cross-reactivity. $\gamma$M is formed early in the immune response. It has interesting properties, but will not be further discussed.

The $\gamma$G antibody (IgG) is synthesized in cells of the reticuloendothelial system (RES): lymph nodes, spleen, bone marrow, thymus, Kupfer cells of liver, and some cells in the lungs. Plasma cells and some lymphocytes are the major cell types which synthesize immunoglobulins. They are estimated from fluorescent antibody techniques in *in vitro* experiments to release from 9-30 $\times 10^4$ molecules of antibody per minute. The half-life of these antibodies is about 15-20 days.

Their combining sites are each estimated to be less than 700 Å$^2$ in area (10). They are Y- or T-shaped, with the combining sites at the top of the Y or T.

[3]There are now known to be at least five classes of human and several of mouse immunoglobins.

The $\gamma$G molecules should be at least bivalent in reactivity for the buildup of a large three-dimensional lattice which ultimately precipitates (see 4.4a). In fact, the $\gamma$G immunoglobulins appear bivalent by electron microscopy (11). Antibodies are known that do not precipitate with their antigen and cannot form a lattice. Although they once were considered to be monovalent, their nonprecipitating character is more likely a reflection of the nature of the antigen or structure of the antibody.

In experimental determinations of the chemical structure of the antibody, two monovalent fragments were obtained by Porter (12) through papain digestion, and also by thiol reduction. The $\gamma$G may also be split into two types of subunits with different molecular weights ("$\gamma$" and "L" polypeptides) by treatment with mercaptoethanol, followed by Sephadex chromatography at low pH (12). From this and other work, the following description of the $\gamma$G immunoglobulin has been formulated (modified from reference 13). See Figure 4.3.

Some people suffering from myeloma excrete "Bence-Jones proteins"—actually free L chains—which are so homogeneous they can be sequenced. These proteins have been isolated from several people and found to consist of a region of constant sequence (the C-terminal half), and a region whose

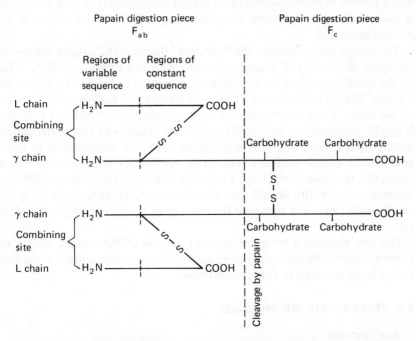

**Figure 4.3** The general chemical structure of rabbit IgG (13). The combining sites seem to be the regions of variable sequence in the four chains.

sequence varies from person to person. There is a similar variable region in the $\gamma$ chains. These variable regions are different for different antibodies.

### 4.2c  Origin of Antibody Specificity

Any acceptable theory of the origin of the combining sites must account for a number of phenomena: the specificity of the antigen–antibody reaction; the fact that for every molecule of antigen introduced 100,000 or more molecules of antibody may be manufactured; the kinetics of antibody production and the difference between primary and secondary responses; the capacity of the body to respond to foreign but not to its own potentially antigenic molecules; immunological tolerance (i.e., the failure to make a normal response, or even a response at all, to the subsequent introduction of a potential antigen which has been introduced into an organism sufficiently early in its development); and the apparently very large number of distinct antigenic groupings against which specific antibodies can be made.

There are currently two types of theories. One group is called the "instructive" theories. The antigen is known to adsorb onto the surface of the membrane of the cell which will make antibodies and apparently acts as a template for synthesizing molecules of immunoglobulin. This causes the newly formed immunoglobulin to be shaped (sequenced?) in such a way that it has a surface configuration complementary to part of the surface of the antigen molecule.

The second group includes the "selection" theories: the antigen induces an increased production of immunoglobulins with preexisting specificity. That is, the specific template for the antibody is not conferred on the cell by the antigen. The "clonal selection" theory of Burnet and Lederberg is a version of the latter. Certain mesenchymal cells (lymphocytes) which are immunologically competent are supposedly genetically so endowed that under certain circumstances (right phase of maturation, etc.) they can respond to molecules with a specific antigenic pattern or template by making antibodies against this template. Burnet and Lederberg think that the variety of different antibodies is actually limited and that their number, though large, is still small enough so as not to exceed the capacity of the genetic apparatus of the cells.

One can visualize a repressor system (reaction of the antigen with a repressor removes the repressor) or an activator system (the antigen is the activator) being involved in either mechanism.

## 4.3  PREPARATION OF ANTIBODY

### 4.3a  Materials

The basic research materials are the antigen and the antibody. The former must be supplied; the latter must be synthesized by an organism and isolated. New Zealand white rabbits (young adults) are suitable; provide sufficient antigen, whether serum, fractions from serum or urine, or pure lyophilized proteins. Solutions necessary are normal saline (0.9%), 70% alcohol, xylene and 1% sodium azide. In addition, Freund's adjuvant (complete or incomplete) must be obtained. Difco Laboratories, Detroit, Michigan, sells "Difco Bacto Adjuvant" (incomplete Freund), "Difco Bacto Adjuvant" (complete Freund) and "Difco Bacto M. butyricum." "Vacutainers" may be obtained from Becton Dickinson & Co. The model which holds 20 ml is good for mixing adjuvant and protein solutions. Also, 10 ml sterile syringes may be obtained from the same company. Get syringes with Luer Lok tips. With these, use 18 and 20 gauge $1\frac{1}{2}$ inch needles, with Luer Lok Hubs, made of stainless steel and disposable. A Swinnex-13 millipore filter, cotton gauze and centrifuge tubes are also necessary.

### 4.3b Preparation of Antigen

The material injected may be in the form of a solution or a suspension. In the latter case, the size of the particles suspended is limited by the route of injection: they cannot occlude blood vessels. The suspension may be particulate, such as bacteria or phage, or alum-precipitated protein (protein adsorbed to potassium aluminum sulfate). Alternatively the material can be injected as an emulsion such as Freund's water-in-oil emulsion.

The alum and Freund's adjuvant increase the antigenicity of substances by converting them into a particulate form, thereby influencing the uptake of antigen by phagocytes and increasing the persistence of the antigen in the body. These are called immunological adjuvants.

It is best to prepare the protein solutions for injection on the day they are to be used. Lyophilized proteins are dissolved in normal saline, and the protein solutions must be sterilized to prevent infection of the animal. This would cause undue suffering and possible loss of the animal. Sterilization is done either with millipore filters or, for nonfilterable solutions, by adding merthiolate to 0.1 mg/ml.

For the millipore filter method, the protein solution, which may be anywhere from 1 to 5 ml, is drawn into a 10 ml syringe and pressed through the Swinnex-13 millipore filter into a 10 ml sterile "vacutainer." The protein solution is now sterile.

To prepare the first injection, complete Freund's adjuvant is used. This consists of 17 ml Shell Ondina 17 oil, 3 ml Arlacel A, and 6 mg dried mycobacterium. Or, incomplete Freund's adjuvant (without the mycobacterium) may be added to Bacto M. butyricum using a sterile syringe and an 18 gauge needle. An equal volume of this oily mixture is then transferred into the same vacutainer as the protein solution. The Freund's adjuvant and the protein solution are thoroughly emulsified with either a Spex shaker, a hand mixer, or with two syringes connected by Teflon tubing. Complete emulsification is required so the emulsion will break down very slowly and release the antigen into the animal over a long period of time. To test for completeness of emulsification, let a drop of the emulsified material fall into cold water in a beaker. If it is complete, the material forms a spherical drop; if it isn't, the material disperses.

### 4.3c Introduction of Antigen

Before injection of the rabbit, a sample of blood should be drawn to serve as a control when the antiserum is examined (see 4.4). Once this is done, several routes of antigen introduction may be used.

Intravenous injection is often used. Frequent injection (four consecutive days on alternate weeks) is required. There is a rapid entry and response. There is also danger of anaphylactic shock: this is a form of hypersensitivity, and it happens when the interaction of antigen is with antibody which is fixed in tissue. What results is a contraction of the smooth muscle (e.g., intense constriction of bronchi, bronchioles, producing asphyxia) caused by the release of pharmacologically active agents such as histamine, bradykinin, serotonin or SRS-A (a "slowly reacting substance," released by anaphylaxis).

The antigen can also be injected into foot pads. It is thought, on the basis of the size of the lymph nodes draining the area, that antibody production is primarily there. This gives a more localized response.

Intraperitoneal injection is good for particulate substances, since many macrophages are present in that region. Another possible location for injection is under the shoulder blade: the pocket formed between the fascia, covering scapular muscles and rhomboideus muscles over the dorsal ribs. Intramuscular injection is probably longer lasting, since there is a slower leakage into the circulatory system.

For general purposes, intramuscular injection is best. The animal should be held firmly between the body and the arm of the investigator. The investigator's hand should be placed firmly under the chin. This leaves the posterior portion of the animal's body free. Stretch one of the hind legs and inject the Freund's adjuvant intramuscularly. The best site of injection is the upper part of the thigh near the lymph node. It is best to attempt to get as much material as possible into the lymph system, for this is where antibodies are generated. Approximately 1.0-1.5 ml is all that should be placed into the animal at one particular site. However, one may use both thigh muscles.

To further enhance the response of the animal, a second injection should be given no earlier than one week after the first injection. The second injection may be made intramuscularly into the thigh, or subcutaneously on the upper back. A good method for subcutaneous injection is to raise the skin on the part of the back near the collar bone. The skin should be held loosely with the fingers and the needle inserted between the skin and muscles. Sometimes, however, subcutaneous injection results in ulcers on the skin. Some workers make the first and second injections in several sites on the animal; this gives a faster response.

The amount of antigen injected is arbitrary. It depends on the availability, antigenicity, animal, and the "prejudice" (physiological conditions). Usually 10 mg is injected, but responses have been obtained with as little as 6 micrograms. The material should be of high purity; in the hyperimmune animal over half of the serum immunoglobulin may be antibody. With so much antibody produced, a considerable amount would be made against an impurity.

If the antigen has been previously administered in a quantity greatly exceeding that required to elicit an immunological response, immune paralysis can occur. There is no effect on the response to unrelated antigens. This response occurs only in immunologically mature animals.

The frequency of injection depends on the route and when the antibody is needed.

### 4.3d  Isolation of Crude Antibody

The first bleeding for testing of antisera should not be made before three weeks after the first injection of antigen. You may withdraw 50 ml of blood a week, but no more than 70 ml.

The animal is placed in a restraining box. The animal must be restrained sufficiently to prevent stretching and possible injury to his back. Rabbits have a peculiar ability to stretch or kick their hind legs in such a manner as to luxate or actually break their backbones. After the animal is placed in the restraining box one ear is selected, shaved, and washed with 70% alcohol. Gently rub a small area of the center ear vein with a gauze pad soaked with xylene. This causes the vein to become very large. Then pressing the vein very gently with one thumb, insert a 20 or 21 gauge needle. Let the blood run through the needle and into a tube until 5-20 ml of blood is collected.

Remove the needle and stop the bleeding by slightly pressing the site of the injection with a gauze pad soaked in 70% alcohol. Since xylene irritates the skin, the ear must be washed with alcohol and soap and water. Dry the ear and place some medicated lotion (such as Jergens skin lotion) on it. This helps the ear to heal faster. If this is not done, some necrosis of the tissue of the ear will occur.

The blood which has been withdrawn must be left at room temperature to clot for several hours. Then, using a Pasteur pipette, transfer the serum into a centrifuge tube, and centrifuge it at 10,000 × g for 20 minutes. For preservation of the antiserum, 0.1 ml of 1% sodium azide in saline is added to each 10 ml of antiserum. Store the antiserum in either a freezer or a refrigerator.

### 4.3e  Preparation of Pure Antibody Using Immunoadsorbents (14)

The antibody in serum is of course contaminated with many proteins. Antibody in whole serum should be adequate for the methods in this chapter, but for some purposes more highly purified antibody is preferable.

### (i) Theory

Immunoadsorbents can be used to prepare highly purified antibody. The crux of the technique is to prepare insoluble antigen in such a manner that the structure of the antigen remains immunologically unchanged. That is, the insolubilized antigen presents the same sites to the antibody as when it is free in solution. The procedure described here uses glutaraldehyde, a bifunctional reagent, for formation of a protein gel from the antigen. This gel is the immunoadsorbent. The antibody is removed from this immunoadsorbent by lowering the pH to approximately 2.8 where the antibody and antigen dissociate.

### (ii) Experimental procedure

The antigen used for immunization should be homogeneous.

The procedure described is scaled for approximately 5 ml of antibody serum, but the amount may be increased or decreased at will.

To 12 ml of antigen at 20-50 mg/ml buffered at pH 7.0 with phosphate buffered saline (PBS) are added drop by drop 3 ml of 2.5% glutaraldehyde. A twofold concentrated stock PBS is made from 12.25 g $KH_2PO_4$, 2.05 g NaOH, 17.55 g NaCl and 0.1 g $NaN_3$ per liter. If there isn't enough antigen to immunize the rabbit and form the immunoadsorbent gel, bovine serum albumin may be used as a carrier protein with whatever antigen is available to help form the gel.

The solution is stirred vigorously during addition and the reaction allowed to continue for 2-3 hours with rapid stirring (but don't allow the solution to froth). At the end of this time, a protein gel should be formed. If it hasn't, add several more drops of glutaraldehyde and allow the reaction to continue for another hour.

Add five times the volume of PBS to the gel. The mixture is stirred and centrifuged. The precipitate is collected and broken up with a homogenizer using five or six times the starting volume of PBS. The resuspended gel is recentrifuged. The supernatant is discarded and the gel is again dispersed by homogenization in the same manner. This process is repeated until no free protein is found in the solution—that is, until the optical density at 280 nm of the supernatant is less than 0.04.

The washed gel is then suspended in five times the initial volume of 0.1 $M$ glycine buffer, pH 2.8, for 5 to 10 minutes. The gel is homogenized as above and centrifuged. The supernatant is discarded. The protein gel is then resuspended in phosphate buffered saline and dispersed, washed and centrifuged as before until no free protein is present. The gel is then ready for antibody adsorption.

To the washed protein gel are added 5 ml of antibody serum or fraction thereof. This is stirred vigorously (without frothing) for 15 minutes and the

suspension centrifuged. The supernatant is carefully removed from the protein gel.

The protein gel with antibody attached is washed with PBS until the optical density of the solution is again less than 0.04. Then add 30 ml of 0.1 $M$ glycine buffer, pH 2.8, to the protein gel. The gel is suspended and stirred five minutes in the buffer. The suspension is centrifuged and the supernatant, which contains the released antibody, is kept. The glycine buffer wash may be repeated until no more protein can be eluted from the gel. After the protein gel has been thoroughly washed at pH 2.8, it may be washed with PBS until no more protein is eluted. After this, protein gel can be used again. The purified antibody is brought to neutral pH and concentrated or used as is convenient.

Once the antibody has been prepared, it should be analyzed for homogeneity by ultracentrifugation and electrophoresis as well as by immunochemical techniques. See Chapters 5 and 6.

## 4.4 MEASUREMENT OF ANTIBODY ACTIVITY: EXPERIMENTAL PROCEDURES

Immunologists use a variety of methods to measure antigen-antibody interaction, ranging from *in vivo* methods such as passive cutaneous anaphalaxis (PCA), to phage inactivation, to methods for measurement of ligand interactions such as fluorescence quenching or polarization of fluorescence (Chapter 7). The ligand is most conveniently a hapten, but can be a protein. The methods we will cover depend on precipitation, the formation of an insoluble antigen-antibody complex, as the assay.

### 4.4a Theoretical Considerations

The formation of a precipitate depends on having the correct proportions of antigen and antibody, or "equivalence." Equivalence is the situation where the solubility of the antigen-antibody complex is minimal. Consider Figure 4.4.

Since techniques involving precipitation are based on the optimal proportion of antigen to antibody, the quantity of precipitate rises with antigen concentration (if antibody concentration is constant and sufficient) to the equivalence point, then falls. The precipitate can be dissolved by too much or too little antigen or antibody.

Furthermore, it is necessary that both antigen and antibody solutions contain sufficient antibody or antigen to give a precipitate; i.e., exceed the solubility of the specific complex in the medium. So if one solution does not have sufficient reactant, it is useless to increase the concentration of the other. More generally, one solution will become controlling by virtue of concentra-

a. Excess antibody:

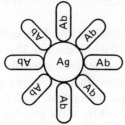

b. excess antigen:

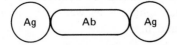

c. Equivalence: a 3-dimensional lattice is formed

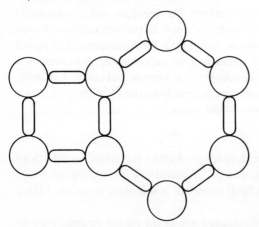

**Figure 4.4**
Antigen–antibody complexes in solution under conditions of: (a) antibody excess; (b) antigen excess; (c) equivalence. The antibodies are assumed oval-shaped instead of T- or Y-shaped for convenience of representation.

tion; i.e., if the ratio of concentrations of *A* to *B* are greater than their ratio at equivalence, then *B* is the controlling solution. Any precipitate will form in the *B* solution, so increasing *B* will both increase the density of the precipitate, and decrease the time necessary for its appearance. This can be expressed diagrammatically (Figure 4.5).

Since one generally does not know what the exact equivalence point will be—if indeed the actual concentrations of antibody or antigen are known—methods have been developed which use concentration gradients of antigen, antibody, or both to increase the chances of providing conditions for equiv-

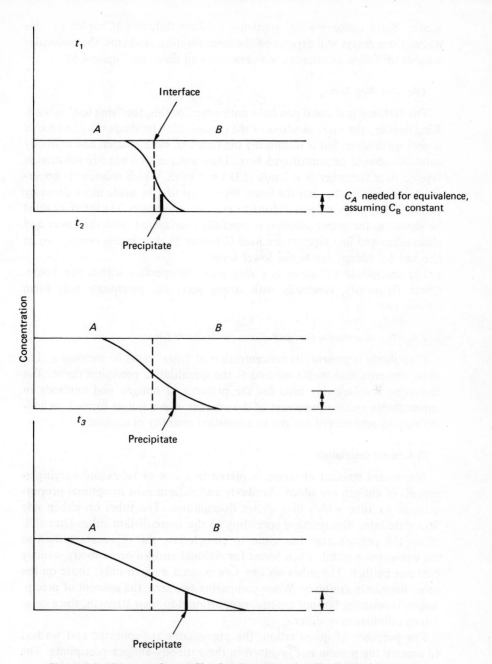

**Figure 4.5** Concentration profile of antigen (*A*) and antibody (*B*) showing effect of diffusion at increasing times ($t_1$ to $t_3$) on the location of the zone of equivalence. The antibody is assumed to be immobilized. $C_A$ and $C_B$ are the concentrations of *A* and *B*.

$C_A$ needed for equivalence, assuming $C_B$ constant

alence. Since concentration gradients produce diffusion (Chapter 6), this means these assays will depend on the concentration gradients, the molecular weights (diffusion constants), temperature, and time (see Figure 4.5).

### 4.4b The Ring Test

The first step is to see if you have antibody. For this, the "ring test" is used. Ring tests are the most sensitive of the precipitation methods. The technique is only qualitative, but is technically the simplest. Both antigen and antibody solutions should be centrifuged first. Then antigen or antibody solution is layered over the other in a 3 mm (I.D.) test tube. Which solution is uppermost does not matter, but the lower layer must first be made more dense by adding 2-4% polyvinylpyrrolidone to prevent convection. The interface must be sharp, so the upper solution is carefully overlayered onto the lower and clean tubes and fine pipettes are used (Chapter 5). Any excess mixing can be checked by adding dye to the lower layer.

The precipitate will form as a ring near the interface within two hours. Check frequently, especially with strong sera; the precipitate may form quickly and fall.

### 4.4c The Quantitative Precipitin Curve for Antibody Titer

If antibody is present, its concentration or "titer" must be measured. The most common and useful method is the quantitative precipitin curve. The technique involves ring tests for the presence of antigen and antibody in supernatants and measurement of the amount of precipitate formed on adding varying amounts of antigen to a constant quantity of antibody.

#### (i) General description

A constant amount of serum is placed in a row of tubes and varying amounts of antigen are added. Antibody and antigen exist in optimal proportions in the tube which first shows flocculation. The tubes on either side flocculate later, the reaction spreading to the more distant tubes later still. When the precipitation has gone to completion, the supernatant fluid at the equivalence point, when tested for residual antigen or antibody, usually contains neither. The tubes on one side contain antigen only; those on the other have only antibody. When comparing two sera, the amount of precipitation is roughly, but not exactly, proportional to their strength, since combining affinities may differ.

For purposes of quantitation, the precipitates are collected and washed to remove the protein not involved in the antibody-antigen precipitate. The greatest sources of error by far are from either not removing all the nonreacting protein or loss of the precipitate through handling.

### (ii) Materials

Antisera—1 ml for an experiment, and 1 ml of a 1 : 10 dilution—and two solutions of antigen, at concentrations of 1 and 5 mg/ml in phosphate-buffered saline, will be sufficient. Both antigen and antibody solutions should be centrifuged to remove aggregated material before use. The buffer is phosphate-buffered saline (PBS) [see 4-3e (ii)]. For the protein measurements, use biuret reagent (Appendix 6); the Folin reagent can be omitted. A standard protein solution for the protein measurements is also needed. Use a 10 mg/ml bovine serum albumin solution. Use $10 \times 75$ mm test tubes.

### (iii) Procedure

Set up the tubes as shown in Table 4.1a. Incubate the tubes at 37°C for 30 minutes, then leave them in the cold room or refrigerator overnight. Centrifuge the tubes at 1500 rpm for 30 minutes, preferably in the cold. Carefully decant the supernatant and save it. Add 0.3 ml of PBS to each precipitate. Break up the precipitate by flicking the tube. Centrifuge again, decant and discard the supernatant. Do this twice more. To the final precipitate add 1 ml

**TABLE 4.1a** PROTOCOL FOR QUANTITATIVE PRECIPITIN TEST

| Tube | 1 | 2 | 3 | 4 | |
|---|---|---|---|---|---|
| Serum (ml) | 0.10 ml | 0.10 | 0.10 | 0.10 | |
| Antigen (1 mg/ml) | 0.05 | 0.10 | 0.15 | 0.20 | |
| Buffer (PBS) | 0.15 | 0.10 | 0.05 | 0.00 | |
| Tube | 5 | 6 | 7 | 8 | 9 |
| Serum (ml) | 0.10 ml | 0.10 | 0.10 | 0.10 | 0.10 |
| Antigen (5 mg/ml) | 0.05 | 0.06 | 0.08 | 0.10 | 0.12 |
| Buffer (PBS) | 0.15 | 0.14 | 0.12 | 0.10 | 0.08 |

**TABLE 4.1b** QUANTITATIVE TEST OF SUPERNATANTS FROM a

| Tube | S1 | S2 | S3 | S4 | etc. |
|---|---|---|---|---|---|
| Supernatant (ml) | 0.05 ml | ⟶ | | | |
| Antigen (1 mg/ml) | 0.05 | ⟶ | | | |
| Tube | S1 | S2 | S3 | S4 | etc. |
| Supernatant (ml) | 0.05 ml | ⟶ | | | |
| Antiserum (1/10) | 0.05 | ⟶ | | | |

of PBS and 2 ml of biuret reagent. For the standard curve, use between 1 and 10 mg of protein. Record the biuret reading and plot the points. The maximum amount of protein precipitate corresponds to the equivalence point. To calculate the amount of antibody precipitated at equivalence, subtract the amount of protein precipitate due to the antigen, assuming it is all precipitated.

To be certain that the point of greatest precipitate is the equivalence point, additions of antigen and antiserum should be made to tubes containing the supernatants. Add 0.05 ml of each supernatant to 10 × 75 mm tubes (see Table 4.1b). Note whether (+) or not (−) a heavy precipitate forms. The equivalence tube should provide essentially no precipitation on addition of either antigen or antiserum.

#### (iv) Interpretations

Ideally, one wants to straddle the equivalence point with several tubes. Therefore, some preliminary testing may be necessary. This can also be done by exhausting the antiserum by adding increasing quantities of antigen until no additional precipitate is formed. Centrifuge after each addition to remove the precipitate.

If the ratio of antibody nitrogen to antigen nitrogen can be calculated, then the molar ratio of antibody to antigen can be obtained. An unknown antigen concentration can also be determined measuring the amount of precipitate obtained and interpolating using the data obtained above. You must, of course, know which side of the equivalence point you are on, so you must do at least two measurements, seeing if the amount of the precipitate increases or decreases with antigen concentration.

#### 4.4d   Oudin Diffusion

This is an extension of the ring test. A precipitation reaction is carried out at an interface, but it avoids the effects of gravity or mechanical agitation by the use of an immobilizing gel layer. The technique gives a solution to the problem of finding the number of antigens to a given serum (15).

Using this technique, it has been found that each antigen is responsible for, or can give rise to, one zone of precipitation by virtue of its reaction with a specific antibody, and that when two distinct antigens are present in the same solution each behaves as though the other were not present.

#### (i) Theory

The theoretical bases for the technique are Fick's Laws of Diffusion (1, 14). If a given solution and the corresponding pure solvent are layered one over the other, without mixing, in a vertical tube, molecular agitation (Brownian motion) will result in penetration of molecules of solute into the lower layer. This is described by

$$\left(\frac{dc}{dt}\right)_x = D\left(\frac{d^2c}{dx^2}\right)_t \qquad (4\text{-}1)$$

The change in concentration with time at any level $x$ is equal to the diffusion coefficient times the rate of change of concentration gradient at that level at a given time, $t$.

In this technique, the antibody or antigen (generally the former) is immobilized in gel, and its concentration remains approximately constant throughout the experiment. The antigen diffuses in one dimension so its concentration changes with time and distance from the interface (see Figure 4.6). Con-

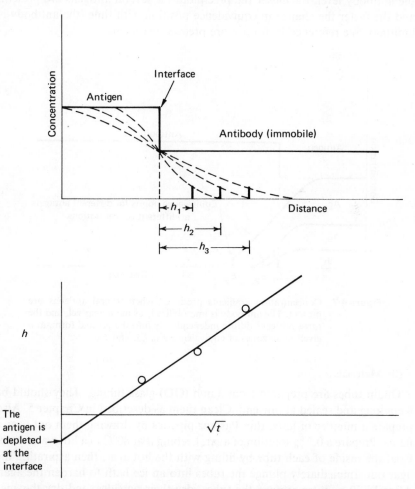

**Figure 4.6**  Representation of diffusion with time of antigen solution into an agar gel containing immobilized antibody. $h$ is the distance of the precipitate from the interface between antigen and antibody at increasing times ($t_1$ to $t_3$).

sequently, the distance from the interface to the point of equivalence (front)—i.e., the penetration, $h$—is proportional to $t^{1/2}$ (see Chapter 6). It has also been found that

$$h = K \log (\text{antigen}) \qquad (4\text{-}2)$$

and

$$h = K' \log \frac{1}{(\text{antibody})} \qquad (4\text{-}3)$$

A concentration gradient is formed for each antigen (Figure 4.7). The higher the antibody level, the closer the precipitates if several antigens are present, and the faster the change in equivalence position with time (the antibody is limiting). See reference 16 for a more precise treatment.

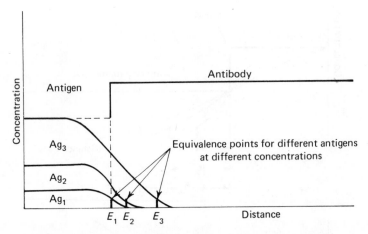

**Figure 4.7**   Concentration gradients produced when several antigens are present. The antibody is immobilized, as in an agar gel, and the three antigens diffuse independently into the gel and form, at a given time, zones of equivalence $E_1$, $E_2$, and $E_3$.

**(ii) Materials**

Oudin tubes are prepared from 3 mm (OD) glass tubing. They should be 8 cm long and sealed at one end. Clean them as described in Chapter 5. Also prepare a number of long, thin Pasteur pipettes by drawing them out over a flame. Prepare a 0.2% solution of agar, keeping it at 80°C in a hot water bath. Coat the inside of each tube by filling with the hot agar, then aspirating the agar out. Immediately plunge the tubes into an ice bath to harden the agar. After 15-30 minutes remove the tubes, dry their outsides, and dry the agar completely by lyophilization.

The reagents include glass ink for marking the tubes, 0.8% Noble agar + 1% $NaN_3$ in saline for the bottom layer, and the solution for the top layer:

antigen (usually 10 mg/ml in saline) $+ 0.25\%$ NaN$_3$ (or antiserum, 9 parts to 1 part $2.5\%$ NaN$_3$).

The fronts are sharper if the lower layer contains the antibody, but the lower layer must be controlling; i.e., the zone must form in the lower layer. With weak antisera, the lower layer should contain the antigen.

### (iii) Procedure

For the experiment, prepare a 48°C water bath.[4] Make agar for the bottom layer. Do not cool it below 48°C. Heat three parts of antiserum or a dilution in plastic sterile test tubes in the water bath just long enough to come to temperature (any longer than necessary may destroy the activity). Add one part agar, mix and fill the Oudin tube to about 4.5 cm with it. To obtain an even interface, fill above 4.5 cm, then draw out the excess with a Pasteur pipette connected to an aspirator. Allow the agar to solidify and come to any temperature below 40°C which can be maintained exactly.

Allow the component for the top layer to come to this constant temperature. Then fill the tubes to 0.5 cm from the top, and cap them with clay. For added temperature stability, place each tube in a screw cap culture tube. Keep the tubes at this constant temperature for one week, *undisturbed*. The slightest change in temperature may cause artifacts.

Find the zones of precipitation. These do not actually move, but rather form and resolve as they pass first through regions of antibody excess and then antigen excess (if antibody is on the bottom). Precipitation occurs at the region of equivalence. The results are best recorded by photographing with P/N 55 film. Lay a centimeter scale along the tubes for converting penetrations into mm.

### (iv) Practical aspects

Temperature artifacts can be very troublesome with this system. A rapid change in temperature provokes the formation of a narrow band whose density is different from that of the surrounding precipitate. This band appears at the level of the leading edge of the precipitate at the time of the temperature change and remains at this level while the zone migrates (see Figure 4.8). The effect produces a sharp change in the progress of the diffusion of the antigen and demonstrates that the rate of diffusion of the antibody is not actually zero, even though this assumption is made.

### (v) Interpretations

To identify particular bands, observe the general appearance of the tubes, comparing knowns with unknowns. The density of precipitates should be the same for any given serum against a given antigen. One can add a large amount

---

[4]Use a calibrated thermometer or calibrate one yourself. Commercial thermometers can be off by as much as 20°.

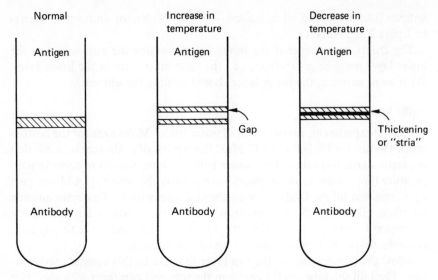

**Figure 4.8** Effect of sudden temperature changes on the precipitation pattern (17).

of a known antigen to the upper layer and see what happens to the pattern—that antigen should penetrate farther into the antibody region than the others. Or one can exhaust the antibody with a known antigen. By eliminating its reaction, one may see others.

### 4.4e  Ouchterlony Diffusion Method (18)

#### (i) General aspects

The Ouchterlony technique is a two-dimensional, qualitative technique; the use of two dimensions makes the technique more informative. Antigen and antibody are placed in separate holes in an agar plate. As the reactants diffuse toward each other, a precipitation line occurs at the equivalence point of each antibody-antigen system.

#### (ii) Materials

Prepare coated 100 × 15 mm petri dishes by placing some vacuum grease on the bottom of the dishes. Spread it evenly with some tissue paper or cotton gauze into a very thin, almost invisible, layer.

Prepare agar by dissolving ionagar in 100 ml of phosphate-buffered saline. Heat until the ionagar is dissolved and the solution is clear and viscous.

Put the petri dish on a level table or tray. Using a 20 ml plastic disposable syringe, place 13 ml of hot ionagar solution into each plate. Allow the plates to set for a while, then cover them. After cooling, the plates are ready to be

Sample well template

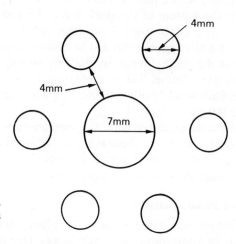

**Figure 4.9**
Sample well template for the Ouchterlony technique. The outer wells
are for antigen.

cut. Cut the desired number of wells (holes) with the template shown in Figure
4.9. A 13-gauge needle, beveled on the inside and connected to a vacuum
line, is used for the smaller wells. A cork borer can be used for the center well.
The cutting is done by firmly inserting the needle into the gel from a vertical
position using one stroke. The vacuum will pull the gel into the needle and
leave a clean hole. If the needle is not held vertically or the holes are cut with
a noncircular shape, the resulting immunochemical patterns will be irregular
and inaccurate. One should also be careful that the distances from the center
well to the peripheral wells are equal (see Figure 4.9). A commercially available template cutter can be helpful.

### (iii) Procedure

Using a disposable pipette, put the antibody into the large center well and
the antigen into the smaller outside wells. Do not overfill the wells. Suggested
amounts of antigen are 2.0, 1.0, 0.5, 0.25, and 0.1 mg, in approximately equal
volumes. One well may be left without antigen as a control. With time, the
antibody and antigen solutions will seep into the agar, so it does not matter
if the wells appear to be dry.

Cover the petri dish. Place it in a metal or plastic container. Put a wet
sponge next to the dish to prevent the agar from drying out. Then cover the
container and leave it at room temperature for 24 to 48 hours. However,
the lines should develop after 18 hours and can be seen then.

To prepare the plates for permanent record, soak the plates in PBS with
0.25% azide for 48 hours, changing the PBS at least twice a day, or for a
longer total time if one does not wish to change the PBS as often.

Remove the plate from the saline and very carefully break the edge of the petri dish with a pair of pliers. Using a flat spatula, very gently lift the agar from the bottom of the petri dish and transfer it to a glass plate. Place the plate on a paper towel. Pour some distilled water over the plate and cover it with a filter paper saturated in distilled water. Let the paper-covered plate dry in the air. Stain the dry plate by placing it for 15 minutes in an Amido Schwartz solution. This is made from 1 g Amido Schwartz ("Amido Black") ("Buffalo Black" NBR), 450 ml 12% acetic acid, 450 ml 16% sodium acetate, and 100 ml of glycerol.

Rinse the plate well with 50% methanol and 1% acetic acid solution. When all the excess dye has been removed, which may take one day, allow the plate to dry. This will give a permanent record of the Ouchterlony plate. The sharp blue lines show where the antibody and antigen precipitated.

### (iv) Practical aspects

The position of the precipitate is a function of the relative antigen and antibody concentrations and other factors. One must get them balanced so that the reaction takes place in the area between the wells and not *in* the wells. The concentration falls off with the square of the distance from the wells, so if you have a very concentrated antiserum, you can increase the distance between center and outside wells, but don't space them too far apart.

### (v) Interpretations

See Figure 4.10.

#### 4.4f   Radial Immunodiffusion (19)

### (i) Theory

This is a quantitative version of the Ouchterlony technique discussed in Section 4.4e. The rate of diffusion can be measured by the movement of the precipitate, and the concentration of antigen (or antibody) estimated.

### (ii) Materials

The "Immuno plates" or "Diffusion plates" sold by Hyland Division of Travenol Laboratories, Los Angeles, California, are satisfactory for this experiment. A cellulose membrane protector is supplied with the plates. A water bath at 56°C, a 10 ml pipette with a wide opening, a 100 ml Erlenmeyer flask, a thermometer, and sufficient twofold concentrated stock phosphate buffered saline are also needed.

A stock agarose solution must be prepared for the plates. Add 3 grams of agarose slowly while stirring to 100 ml of phosphate buffered saline in a 250 ml flask. Stir the solution with a glass rod, and heat it carefully over a low flame or over a hot plate until all the agarose is dissolved and the solution

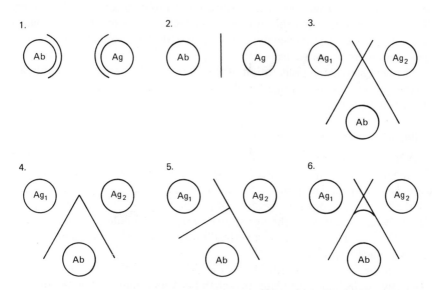

**Figure 4.10**   Some possible patterns from Ouchterlony experiments and their interpretations: (1) two antibody–antigen systems; (2) if the antibody and antigen concentrations are equal and the line is closer to the antibody well, then the molecular weight of the antigen is probably less than that of the antibody; (3) the antibody is against both antigens, and the antigens have no antigenic sites in common; (4) the two antigens have all antigenic sites in common for the antibody or antibodies; (5) all antigenic sites in the first antigen to which antibodies are present are also in the second, but the second antigen has antigenic sites not present in the first (producing the "spur" extending up from the curved line); (6) the two antigens have some sites in common, and some not in common, to which antibodies are present. Note the two "spurs." Whether the lines are bowed or not, and in which direction, depends on the antigen concentration and diffusion constant.

is clear. Don't burn the agarose. Set the solution in the water bath until its temperature is near 56°C. The agarose may be stored in a refrigerator in its gel form and be used again by simply reheating.

The antibody can be a pure antibody or simply serum containing the desired antibody, but it should give a single band against the antigen, as shown by Ouchterlony double diffusion and/or by immunoelectrophoresis (see Sec. 4.4g). A good initial dilution is 1.5 ml of antiserum into 8.5 ml of PBS.

The following directions are for a batch of five plates. Final agarose concentration is always 1.5%. Place 10 ml of the stock agarose solution in a 100 ml Erlenmeyer flask which is in the 56°C water bath. Add 10 ml of the diluted antiserum solution, and mix well with a glass rod or a pipette. Avoid forming bubbles in the agar. Add about 4 ml of warm agarose-antiserum solution to each "Immuno plate." Allow the gel to cool for several minutes. They may be used when the surface is not tacky to the touch. If the plates are not to be used immediately, cover them with the cellophane membranes and store them in plastic bags.

The next step is the preparation of the plates for the antigen. To cut holes in the agarose for the antigen wells, use the 13 gauge needle and vacuum described in 4.4e (ii).

### (iii) Experimental procedure

After the wells are cut, several concentrations of antigen (a range of 0.4 to 2.0 mg/ml is common) are placed in the wells using 10 microliter disposable capillary pipettes. The pipettes will actually deliver about 6 $\mu$l (microliters) of the 10 $\mu$l in the pipettes, because of capillary action. The antigen should fill the well to even with the agarose surface. Care should also be taken to load the antigen from an almost vertical position so that the sides of the well are not made irregular.

When the various concentrations of antigen have each been put into their respective wells, place the plates in a chamber kept humid with a wet sponge. Allow the antigen to diffuse overnight. Distinct rings should form. If they are too diffuse, then a more concentrated solution of antiserum should be used in a new batch of plates.

### (iv) Treatment of data and interpretations

The diameters of the precipitate rings are measured with a ruler or microscope. The concentration of protein is plotted on a logarithmic scale and the ring diameter on a linear one. A straight line should result. If the rings are greater than 7 mm in diameter, discard the data, as this is the region of non-linearity. Unknown concentrations of antigen can be determined from the graph.

### 4.4g Immunoelectrophoresis (5, 18, 20)

### (i) Theory

The techniques mentioned so far rely on diffusion to separate different antigens. If these are present at equal concentrations and have the same molecular weight, only one band of precipitate may be seen. Immunoelectrophoresis is another qualitative technique, but antigen separation is based on both electrophoretic mobility and immunological specificity. The antigens are placed in wells and separated by electrophoresis. Antibody is then added to nearby troughs and any immunological reaction allowed to occur.

### (ii) Materials

An Amido Schwartz solution (see Sec. 4.4e) and barbital buffer are required. The barbital buffer is at pH 8.6, and is made from 19.6 g of sodium acetate (anhydrous), 50.0 g of sodium barbital (diethyl), 34.2 ml 1 $N$ HCl and 3.84 g of calcium lactate. Add water to 10 liters, and then add 50.0 ml of 5 % thymol in isopropanol.

For the immunoelectrophoresis plates, dissolve 1 g of ionagar in 100 ml of barbital buffer and heat carefully, preferably on a hot plate, until clear. Don't burn the agar. If the solution must be heated on a bunsen burner, do not allow the flame to heat only one spot. Excess agar may be stored in the refrigerator and reused by reheating until it liquifies.

Have a clean 3-1/4 × 4 inch glass plate available (a Kodak Projection Slide Cover Glass #B 324 is good). Clean it with soap and water, or acetone, and dry. Apply hot ionagar solution to the plate using a 10 ml disposable plastic syringe with a 20 gauge needle. About 10 ml are necessary to cover the plate. Try not to let the solution drip over the edges of the plate.

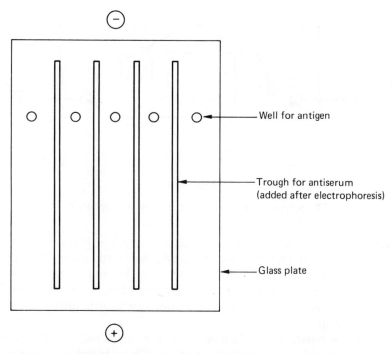

**Figure 4.11**  Pattern for the antigen wells and antibody troughs for immuno-electrophoresis. The troughs are added after the antigen has been electrophoresed.

Allow the plate to gel, and then cut the wells for antigen. The 13 gauge needle mentioned in 4.4f (ii) is very good for this. The gel can be taken from the cut by a smaller needle hooked to a vacuum line. Again, the needle should descend vertically onto the gel—try not to cut the well at a slant. The pattern for the wells and troughs for immunoelectrophoresis is shown in Figure 4.11.

### (iii) Experimental procedure

With a 20 $\mu$l pipette, place antigen into the wells. Let it diffuse into the ion-agar for at least 10 minutes. Cover the wells with a drop of ionagar solution. Do not use a very hot solution as this will destroy the antigen. A solution cooler than 50° is preferable.

Put buffer in both electrode chambers. In most immunoelectrophoresis apparatuses, it is possible to level the buffer in both troughs by tilting them to one side (see Figure 4.12). Do not forget to do this or you will get increased

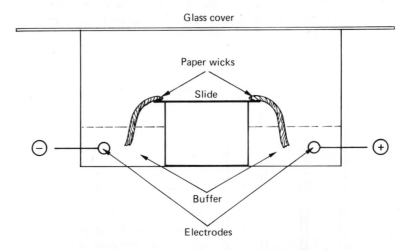

Glass cover

Paper wicks

Slide

⊖ Buffer ⊕

Electrodes

**Figure 4.12** Schematic drawing of an immunoelectrophoresis apparatus.

electroosmosis, which will cause erroneous migration. Place the plate into the apparatus. Saturate the wicks in barbital buffer and place them on both sides of the plate (about $\frac{1}{2}$ inch from the edge). The other edge of the wick must be in the barbital buffer. On one side of the plate, by the wells nearest to the edge, put some bromophenol blue indicator to determine the time of electrophoresis.

Cover the apparatus and turn on the power source and set the voltage to 150 volts. The gel should draw about 15 ma. When the indicator reaches the other edge of the plate, discontinue electrophoresis.

Remove the plates from the apparatus and using the template (Figure 4.11)

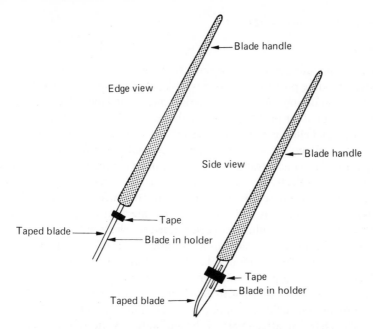

**Figure 4.13** A blade for cutting the troughs for immunoelectrophoresis.

cut the predetermined number of troughs, using the special blade (Figure 4.13): an ordinary scalpel with a disposable blade and a second blade taped to it.

Place antiserum in each trough. Put the plate into a plastic container, and place it in a refrigerator for 48 hours. Be certain the plastic container has some sponges soaked in distilled water or buffer placed inside and that it is air-tight. Failure to do this will result in drying of the agar and poor results. At this point, one may see the developed bands.

For a permanent record, remove the container from the refrigerator. Soak the plate for 48 hours in PBS, changing the saline at least twice a day. Then treat the plate as described in 4.4e (iii).

### (iv) Practical aspects

Since the gel is buffered at an alkaline pH, proteins will tend to be nega-tively charged and move through the gel toward the anode. But some will migrate toward the cathode as a result of electroosmosis (see Chapter 5). This phenomenon occurs because the supporting medium, the agar, is nega-tively charged. All the proteins are affected by this, but the immunoglobulins are affected most: their normal slight movement toward the anode is com-pletely nullified.

*4.4h  Radioimmunoassay (21-24)*

**(i)  Theory**

The radioimmunoassay is an ultramicromethod. It consists of three components: a binding protein, a radioactively labelled compound, and the same, unlabelled compound. A stoichiometric amount of labelled compound is added to the binding protein, then unlabelled compound is added and the amount of unlabelled compound is estimated from the amount of labelled compound displaced. This system has been applied to plasma proteins which bind steroids. The concentration of steroids can be measured by this method.

With protein antigens the interaction is not easily reversible. For example, antibody-insulin complex can be electrophoresed, since the insulin bound migrates with the antibody and does not dissociate to any appreciable extent. Thus, bound and free insulin can be separated by electrophoresis (23).

The major technical consideration is how to easily and quantitatively separate the bound from unbound labelled material. The essentially irreversible binding between protein antigens and antibodies makes this relatively easy. Methods for separation of free antigen from antigen–antibody complex include ammonium sulfate precipitation, electrophoresis, adsorption of small ligands with charcoal, attachment of the antibody to various insoluble materials such as cellulose, and adsorption of the antibody onto polystyrene (24).

**(ii)  Materials**

1. $^{125}$I-labelled human serum albumin will be prepared (21). For this two millicuries of $^{125}$I are necessary (22). The material is shipped in a polyvial container. In addition, prepare 0.1 or 0.2 $M$ potassium phosphate buffer, pH 7.0; 0.1 $M$ potassium phosphate buffer, pH 7.0, containing crystalline bovine serum albumin (BSA) at a final concentration of 10 mg/ml; 2.5 mg of chloramine $T$ dissolved in 1 ml of 0.1 $M$ phosphate buffer (immediately before use); sodium metabisulfite in 0.1 $M$ phosphate buffer at a final concentration of 5 mg/ml; potassium iodide in 0.1 $M$ phosphate buffer at a final concentration of 1 $M$ (0.166 g/ml); 20% trichloroacetic acid (TCA); and human serum albumin in 0.1 $M$ phosphate buffer at a final concentration of 4 mg/ml.

Fifty $\mu$l of 0.2 $M$ phosphate buffer is added to the polyvial. Next, 50 $\mu$l of human albumin is added (200 micrograms). After the solution is mixed, 20 $\mu$l of chloramine $T$ is added. The solution is mixed by drawing the solution up and down the pipette several times (use a bulb), and the reaction is allowed to go to completion (1 minute). Twenty $\mu$l of sodium metabisulfite is added to stop the reaction, the solution is again mixed, and 50 $\mu$l of potassium iodide is added. From this reaction mixture 1 $\mu$l of labelled material is transferred into 0.5 ml of phosphate buffer with BSA. To this is added 0.5 ml of trichloroacetic acid (TCA). The mixture is shaken and the tube containing the flocculent precipitate is centrifuged. The supernatant and precipitate are counted (Chapter 8). The ratio of supernatant counts to precipitate plus supernatant counts gives the amount of free iodine. Ordinarily, 50 to 70% of the $^{125}I$ is bound to the protein.

The remainder of the material is transferred to 0.5 ml of phosphate buffer with BSA. The polyvial is washed with this solution. The resulting 1 ml or so is passed through a G-25 column equilibrated with phosphate buffer with BSA. The initial solution is followed by approximately 20 ml of 0.1 $M$ phosphate buffer. Take 20 drop (0.8 ml or so) fractions, count them, and pool the three fractions with the highest counts.

Five $\mu$l of the labelled material is added to 0.5 ml of phosphate buffer with BSA and 0.5 ml of 20% TCA is added to that. The precipitate is collected by centrifugation, and the supernatant and precipitate are counted to estimate the amount of free iodine remaining. Better than 90% of the $^{125}I$ should be in the precipitate. If there is less, then the 2.5 ml or so of sample from the G-25 column must be lyophilized, redissolved in 1 ml and again passed through the G-25 column. Normally at this point the amount of free iodine is minimal, on the order of 3%. This material will be used for the radioimmunoassay.

2. For the radioimmunoassay itself, obtain 12 × 75 mm polystyrene tubes (Falcon Plastics sells these) (24), the phosphate buffer with BSA made for the labelling (above), and the $^{125}I$ albumin. This should be diluted 1/5000 or 1/500 into phosphate buffer with BSA to give either 1 or 10 nanograms of $^{125}I$ albumin per 0.1 ml of solution.

Antihuman albumin from a rabbit is also necessary. The antibody serum is diluted 1:500 in barbital buffer, pH 8.7. The antiserum is prepared fresh each time for coating the tubes. Sodium barbital (57.72 grams) is adjusted to pH 8.7 with concentrated HCl, then diluted to make 4 liters. A saline wash solution is made of sodium chloride (0.9 g/l) plus sodium azide, 0.01% (10 mg/l).

3. Standards are prepared from the stock albumin solution. 1 mg of human albumin is dissolved in 1 ml of saline wash solution. A 1:100 dilution (10 $\mu$l into 1 ml) is made into phosphate buffer with BSA. Serial dilutions are then made to prepare solutions for the standard curve. Sufficient albumin is added to the tubes from these dilutions for 0.05 nanograms ($10^{-9}$ g or ng), 0.10, 0.20, 0.50, 1.0, 2.0, 5.0, 10.0, 20.0, 50.0, 100.0 and 200.0 ng.

#### (iii) Experimental procedure

0.2 ml samples of the diluted antibody are added to each polystyrene tube and incubated overnight in the cold. The antibody adsorbs to the polystyrene (24). In the morning, the tubes are washed several times with saline, the fluid being removed with a vacuum line. To the empty tubes are added 0.2 ml of

phosphate buffer-BSA. The standards or the unknown albumin solutions
are added next using microliter disposable pipettes, in a final volume of no
more than 20 $\mu$l. After addition of the standards or unknowns, 0.1 ml of
labelled albumin containing either 1 or 10 ng of human albumin is added with
an automatic pipette. The tubes are again incubated in the cold overnight.
Aliquots of the solutions in the tubes are counted. This gives the amount of
unbound albumin.

### 4.5 SUMMARY

For preparing pure antibody with an immunoadsorbent very large amounts
of highly purified antigen are required, unless carrier protein is used. The
ring test is for qualitative detection of antibody. It requires moderate amounts
of material and is fast and easy to do. The quantitative precipitin technique
requires relatively large amounts of material. Neither technique is used to
give information about homogeneity of antigen or antibody. The Oudin
diffusion technique is used to provide information on homogeneity, but
since it gives results in only one dimension at one antigen concentration and
because of its extreme temperature sensitivity, the Ouchterlony is preferred.
In addition, the Oudin requires relatively more material. While the Ouchterlony is somewhat more trouble to set up, it gives more information. The
fact that the pattern can be stained increases its sensitivity. Both the Oudin
and Ouchterlony are qualitative techniques.

The radial immunodiffusion technique is used for measuring antigen concentrations; it is a quantitative variation of the Ouchterlony. Immunoelectrophoresis combines some of the information of an Ouchterlony with the
additional information provided by electrophoresis. Theoretically, it is the
most informative of the techniques. Otherwise, it is a qualitative technique
using similar amounts of material. The extreme sensitivity of the radioimmunoassay is its chief feature, deriving from the use of isotopes (Chapter 8).
It is used for measuring antigen concentrations at the ultramicro level.

The major advantage of immunochemical techniques in general is their
specificity; for example, the Ouchterlony method is frequently used in comparisons of the antigenicity of a series of proteins, to see how closely related
they are. However, while a given protein will have many antigenic sites, the
Ouchterlony technique will give only the minimum number of precipitation
lines. For a given protein, this is usually one.

The sensitivities of the precipitation techniques depend on the concentration of antigen: too little antigen will result in no visible reaction. This limits
the usefulness of these techniques as criteria of protein purity.

Note that for the techniques described, variations in binding affinities
between different batches of antibody and antigen or the presence of heterogeneity in the antibody population will not be detected. Indeed, it has

recently been discovered that even a simple hapten such as DNP elicits a large number of different antibodies, presumably with different sequences, and the distribution of this population changes with time.

## REFERENCES

1. Kabat, E. A. and M. M. Mayer, *Experimental Immunochemistry*, 2nd ed., Thomas, Springfield, Ill. (1961).
2. Kabat, E. A., *Structural Concepts in Immunology and Immunochemistry*, Holt, Rinehart and Winston, New York (1968).
3. Williams, C. A. and M. W. Chase, *Methods in Immunology and Immunochemistry*, Vol. I and II, Academic Press, New York (1967).
4. Topley, W. W. C. and G. S. Wilson, *Principles of Bacteriology and Immunity*, Vol. II, 2nd ed., Williams and Wilkins, Baltimore (1964), pp. 1107–1460.
5. Campbell, D. H., *Methods in Immunology*, Benjamin, New York (1963).
6. Humphrey, J. H. and R. G. White, *Immunology for Students of Medicine*, 3rd ed., F. A. Davis, New York (1970).
7. Jerne, N. K., *Ann. Rev. Microbiol.*, **14**, 341 (1960).
8. Good, R. A. and D. W. Fisher, Eds., "Immunobiology," Sinauer Associates, Inc., Stanford, Conn. (1971).
9. Edelman, G. M. and W. E. Gull, *Ann. Rev. Biochem.*, **38**, 415 (1964).
10. Campbell, D. H. and N. Bulman, *Fortschr. Chem. Org. Naturstoffe*, **9**, 443 (1952).
11. Almeida, J., B. Cinader and A. Howatson, *J. Exp. Med.*, **118**, 327 (1963).
12. Fleischman, J. B., R. H. Pain and R. R. Porter, *Biochem. J.*, **88**, 220 (1963).
13. Edelman, G. M., *Fed. Proc.*, **28**, 245 (1969).
14. Avrameas, S. and T. Ternyck, *Immunochemistry*, **6**, 53 (1969).
15. Oudin, J., *Methods in Med. Res.*, **5**, 335 (1952).
16. Becker, E. L., J. Monoz, C. Lapresle and L. J. LeBeau, *J. Immunol.*, **67**, 501 (1951).
17. Crowle, A. J., *Immunodiffusion*, Academic Press, New York (1961).
18. Ouchterlony, O., *Handbook of Immunodiffusion and Immunoelectrophoresis*, Ann Arbor Science Publishers, Ann Arbor, Mich. (1968).
19. Fahe, J. L. and E. M. McKelvey, *J. Immunol.*, **94**, 83 (1965).
20. Heremans, J., *Les Globulines Seriques die Systeme Gamma*, Paris Masson et C$^{ie}$ (1951), p. 501.
21. Weir, D. M., Ed., *Handbook of Experimental Immunology*, Blackwell Scientific Publications, Edinburgh (1967).
22. Yalow, R. S. and S. A. Berson, *Radioisotopes in Medicine. In Vitro Studies* Hayes *et al.*, Ed., U.S.A.E.C. Conference 6711 (1968).
23. Morgan, C. R. and A. Lazarow, *Diabetes*, **12**, 115 (1963).
24. Catt, K. and G. W. Tregear, *Science*, **158**, 1570 (1967).

# 5 || ELECTROPHORESIS

## 5.1 INTRODUCTION

Many biological macromolecules are charged at physiological pH's, and proteins are amphoteric—i.e., their net charge varies from positive to negative, depending on the protein and the pH. Consequently, these macromolecules can be made to move at most pH's by application of an electric field. In this way, we can analyze or separate charged macromolecules as is done in partition chromatography using solvent flow. Under a specific set of circumstances, a given molecule will migrate at a specific velocity for a particular external electric field. Its velocity divided by the electric field strength is called its mobility, and this concept is central to a discussion of electrophoresis.

In this chapter, we first discuss isoelectric focusing, a method in which proteins are made to migrate in a pH gradient until they reach a pH at which their net charge is zero. At other pH's, the proteins will have some net charge, and electrophoresis can be performed. We give a qualitative description of how the necessity for a system which is discontinuous in protein concentration, together with the charge and size of proteins, affects their electrophoretic behavior, leading to anomalies in observed electrophoretic patterns.

We describe several types of zone electrophoresis, including sodium dodecyl sulfate (SDS) gel electrophoresis, which is analogous to Sephadex chromatography, and a refinement of zone electrophoresis, disc electrophoresis. Then, we give methods for starch gel, SDS-gel, and analytical disc electrophoresis, as well as isoelectric focusing.

## 5.2 THEORY OF ELECTROPHORESIS

### 5.2a   Isoelectric Focusing

In this method, a column is set up with a strong base such as ethylenediamine at one end and a strong acid, such as phosphoric acid, at the other (1). In the middle is a solution of protein and amino acids or aliphatic aminocarboxylic acids which have isoelectric points spread throughout a given pH range (say pH 5-8). When a voltage is applied, a pH gradient forms over a period of time, the proteins and amino acids migrating until they reach a pH at which they are isoelectric,[1] at which point they will accumulate.

The amino acids or aminocarboxylic acids ("ampholytes") are chosen for good conductivity at their isoelectric point, their ability to buffer strongly at their isoelectric point (that is, the p$K$'s of their acidic and basic groups should be as close as possible), their low molecular weight and optical density, and their inertness to proteins. To stabilize the pH gradient and the protein bands that form, the experiment is done either in a sucrose density gradient or in a solid support, as described in (iv) of Sec. 5.2b.

### 5.2b   Electrophoresis (2)

A *charged* particle in an insulating medium will experience a force if an electric field is applied. The particle will move at a velocity ($v$) determined by its charge ($q$), the applied electric field ($E$) and the frictional drag ($f$) produced by the passage of the particle through the medium:

$$Eq = fv \qquad (5\text{-}1)$$

The frictional coefficient, $f$, of the particle is a measure of its hydrodynamic size. The mobility of the particle is the velocity it attains for a given external electric field:

$$\text{mobility} = u = \frac{v}{E} = \frac{q}{f} \qquad (5\text{-}2)$$

This is generally a constant which is characteristic of the particular ionic species.

---

[1] That is, they have no net charge in that solvent. The isoelectric point of a protein may be different from its "isoionic point", which is that pH at which its charge is zero when no ions are present except for protons, hydroxyl ions, and the protein.

While this relation can be experimentally verified, many other factors influence the actual mobilities of charged particles, so that observed mobilities cannot be used to determine actual net charges or frictional coefficients. Hence, electrophoresis is used primarily as a criterion of purity of materials and to a lesser extent as a method of preparation.

### (i) Interaction with supporting electrolyte

A solution which is homogeneous with respect to protein would be of little use for analysis or preparation. So other ions besides the counterions to the protein are needed. This is because the current through all parts of the circuit must be the same, so where protein does not carry the charge smaller ions must. Also, in order to make the solution as electrically homogeneous for protein migration as possible, an excess (often about 0.1 $M$) of low molecular weight electrolyte is added. This greatly reduces, but doesn't eliminate, artifacts caused by discontinuities in concentration of charged species in the solvent. Moreover, the presence of the low molecular weight electrolyte introduces other artifacts.

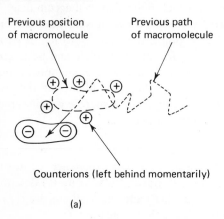

Previous position of macromolecule

Previous path of macromolecule

Counterions (left behind momentarily)

(a)

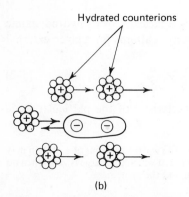

Hydrated counterions

(b)

**Figure 5.1**
(a) "Asymmetry effect" and (b) "electrophoretic effect" in electrophoresis. In both cases the mobility of the macromolecule is reduced.

Two major effects of the supporting electrolyte on the mobilities of proteins are recognized. Binding of ions by a protein directly affects its charge and consequently its mobility, but this is not common. However, the counterions always tend to form a cloud of opposite charge around the protein. Because the protein and ions are hydrated and because in addition there is possibly some short range electrostatic repulsion, the ion cloud is thought to be physically separated from the protein and any ions bound to it. During electrophoresis, the protein moves in an irregular fashion toward the appropriate electrode, and each time a protein molecule moves the counterion cloud is left behind momentarily. Thus the net electric field the protein experiences is lowered because the counterions tend to lag behind the protein and produce a net electrostatic field which opposes the applied field. This is called an "asymmetry effect," and it lowers protein mobilities (3). See Figure 5.1.

The ions of the low molecular weight electrolyte are hydrated, and those moving in the opposite direction to the protein carry considerable liquid along with them. Consequently the protein is in effect moving against a flow of solvent. This factor (called "electrophoretic effect") also reduces its mobility.

These factors can be calculated for spherical molecules, assuming ideal behavior. However, many molecules of interest are not spherical and do not behave ideally. Furthermore, the discontinuous protein concentration and the charge on the protein itself produce artifacts in spite of the excess electrolyte present.

### (ii) Effect of concentration discontinuities

Consider the system shown in Figure 5.2. When voltage is applied, the ions move, producing a current. This current must be uniform, and electrical neutrality must be maintained. Therefore, the number of charges crossing any given cross section must be the same *in both* directions. Since the conductivity in the 0.2 $M$ section is twice what it is in the other, the voltage gradient[2] in the 0.2 $M$ section must be one-half that in the 0.1 $M$ section. Indeed, in any electrophoretic system the current must be constant, so the voltage gradient in any region is inversely proportional to the conductivity ($\sigma$) in that region.

$$\text{voltage} = V = iR = \frac{i}{\sigma} \tag{5-3}$$

$$V_A = \frac{i_A}{\sigma_A} \qquad V_B = \frac{i_B}{\sigma_B} \tag{5-4}$$

$$i_A = i_B, \text{ so } V_A \sigma_A = V_B \sigma_B \tag{5-5}$$

Since the current consists of the movement of both sodium and chloride ions, the position of the boundary between $A$ and $B$ cannot change. If the

[2]The voltage divided by the distance.

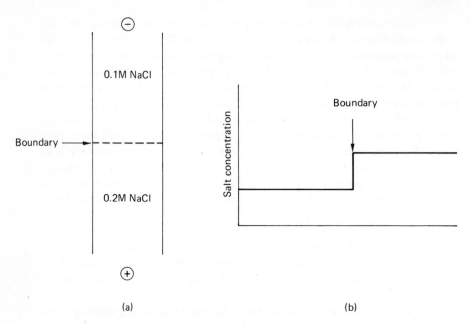

Figure 5.2   An electrically discontinuous solution. (a) The experimental
setup; (b) plot of salt concentration versus distance, neglecting
diffusion.

boundary moved to the left (toward the positive pole), this would mean that
some sodium ions were moving backward. Similarly, if the boundary moved
to the right, a fraction of the chloride ions would have to migrate toward the
negative pole. If this happened, the current would no longer be equal in all
sections, or electrical neutrality would not be maintained. The system described in Figure 5.2 is called a stationary or dilution boundary.

### (iii)  Effect of macromolecular charge and size

Now consider the electrophoresis of a macromolecule, a negatively charged
protein, in the system shown in Figure 5.3. Since the protein solution is
denser than its solvent, and since no supporting medium is used to prevent
convection ("free electrophoresis") the solvent must overlay the protein
solution.

If the protein solution has been dialyzed against its solvent, the concentration of sodium ion will be somewhat higher and the concentration of chloride ion somewhat lower in the protein solution than in the solvent, because
of the Donnan effect. Ideally, the protein concentration should be low enough,
or the salt concentration high enough, so that the Donnan effect and the
protein's contribution to the conductivity of the solution are negligible, but
this is impractical.

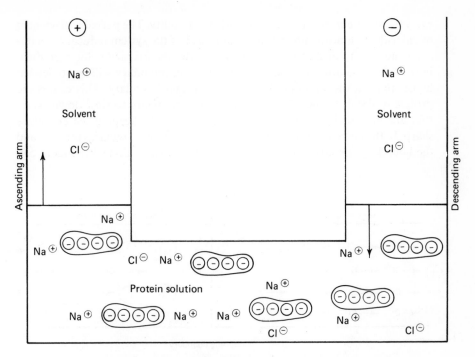

**Figure 5.3** Representation of the experimental setup for free boundary electrophoresis. The arrows indicate the direction of movement of the protein–solvent boundary.

In this case, the protein-solvent boundaries will move in the electric field, since the protein does. The protein's contribution to the conductivity of the solvent together with the Donnan effect introduces anomalies into the electrophoretic pattern.

1. As the protein rises into solution which is "new," i.e., not Donnan-equilibrated, its contribution to the conductivity of the "new" solution depresses the movement of chloride along with it. That is, the transfer of the protein means that much less chloride transfer, since the total charge moving in either direction must be constant. The rate of movement of a protein under standard conditions (its mobility) is generally less than that of chloride ion, since proteins are much larger than chloride ion, and experience much more frictional drag on moving through a solvent. This means that an ion of high mobility (chloride) is being partly replaced by one of lower mobility (the protein).

The conductivity due to a particular ionic species is proportional to the concentration (activity) of the ionic species, and also to the mobility of the species. That is, it takes a higher concentration of an ion of low mobility to carry the same current than an ion of higher mobility. The partial replacement of chloride with protein on the ascending side of the system reduces the average conductivity of that region. Consequently, the average voltage gradient in the "new" solution behind the rising protein boundary will be somewhat higher than in the solution ahead of the protein boundary. This causes the protein molecules which diffuse ahead of the boundary to slow down, since they are in a region of lower voltage gradient, thus keeping the boundary sharp. In the descending arm, the reverse is true: chloride replaces protein, and the boundary spreads faster than it would by diffusion alone. See Figure 5.4.

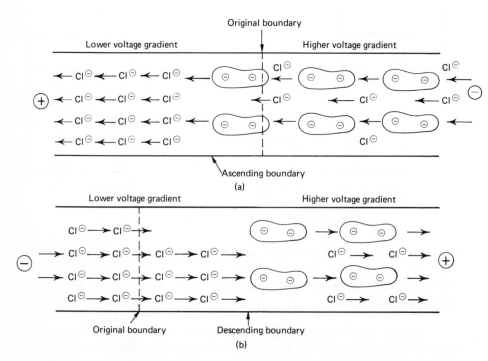

**Figure 5.4** Effect of protein migration on the average mobility in (a) the ascending boundary and (b) the descending boundary.

2. Because of the Donnan effect, sodium and chloride concentration boundaries exist at the original boundaries of the protein solution. That in the descending arm is a sodium chloride boundary, while the one in the ascending arm is a sodium proteinate boundary. If the progress of the electrophoresis is followed using schlieren optics (Chapter 6), these boundaries appear as two extra stationary peaks in the pattern, with the sodium proteinate peak being larger than the sodium chloride peak.

3. The total concentration of ions inside the protein solution is greater than in the dialysate (solvent). Consequently, the conductivity inside the protein solution is greater than that outside. The protein in the rising arm is migrating into a region of lower overall conductivity and higher voltage gradient. So the rising boundary velocity is greater than the velocity of the descending boundary. See Figure 5.5.

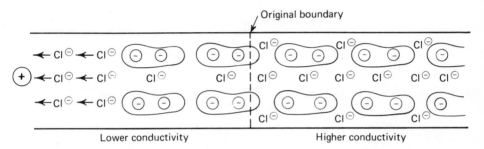

**Figure 5.5**  Production of increased velocity of ascending boundary movement by the Donnan effect.

4. The effective mobility of a buffer ion such as acetate is a function of its net charge, that is, of the pH. This is of course especially true of substances with many ionic forms such as proteins.

When the counterions are used to maintain the pH, the Donnan effect, conductivity of the protein and the relative mobilities of the counterions and protein can produce additional complications. For example, if the pH is such that a protein is positively charged, and the dialysis buffer is acetate-acetic acid, the ratio of acetate to acetic acid will be higher in the protein zone than in the dialyzate. That is, the pH will be higher within the protein zone than without. The net charge on the protein will be lower and consequently the velocity of migration of the protein will be lower within the protein zone (on the descending side) than without (the ascending side). This particular pH effect would tend to enhance the effects mentioned in 3. See Figure 5.6.

If the pH was such that the protein was negatively charged, then the transfer of acetate ion into the "new" solution on the rising boundary side would be decreased by movement of the protein anion into that solution. This would produce a lower pH in the "new" solution and the protein's mobility would decrease. This effect would counteract that described in 3.

In addition, the sharpness of the rising boundary relative to the descending boundary can be enhanced or decreased, depending on the mobility of the protein relative to that of the buffer ions, as described in 1. See Figure 5.7.

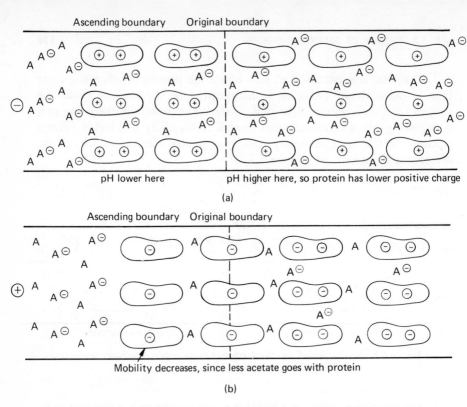

Ascending boundary       Original boundary

pH lower here          pH higher here, so protein has lower positive charge

(a)

Ascending boundary    Original boundary

Mobility decreases, since less acetate goes with protein

(b)

**Figure 5.6**    Effects of pH on ascending boundary velocity of migration and rate of diffusion when (a) the protein has a different charge from the buffer ion and (b) it has the same charge.

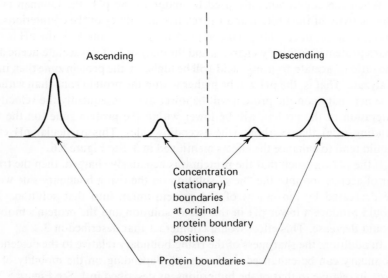

Ascending               Descending

Concentration (stationary) boundaries at original protein boundary positions

Protein boundaries

**Figure 5.7**    Schlieren pattern of a free boundary electrophoresis experiment, showing typical anomalies.

136

### (iv) Zone and boundary electrophoresis

Free boundary electrophoresis is used for analysis of the purity of protein solutions, to see how many moving boundaries are produced at a given pH. Another type of electrophoresis involves the use of thin bands of protein in a medium, and is called "zone electrophoresis." It is used for complete separation of proteins or other substances and is therefore used in preparative as well as analytical work. The thinner the zones, the more sensitive the method is for analytical use (less protein is needed to produce a visible zone). And the more complete the separation in preparative work, just as in partition chromatography. See Figure 5.8.

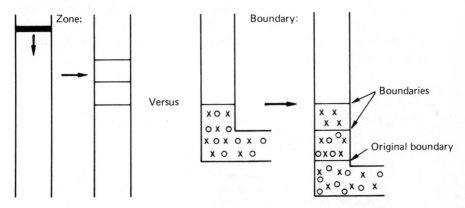

**Figure 5.8** Separations of macromolecules achieved by zone and boundary electrophoresis.

Another important thing to understand about zone electrophoretic methods, aside from the pecularities imposed by the supporting medium, is that the same processes occur as were described for free boundary electrophoresis. For example, if the mobility of the protein is lower than that of the supporting ion of the same charge, the front edge of the protein zone will appear sharp, while the trailing boundary will be diffuse, if the pH effects mentioned above do not counteract this effect. Naturally these anomalies become more obvious at lower ionic strengths and higher protein concentrations.

A supporting medium is generally used with zone electrophoresis to prevent artifacts from convection because of the relatively high density of concentrated macromolecule solutions. Since the flow of a liquid through a capillary is proportional to the fourth power of the radius, and the area of the capillary cross-section is only proportional to the square of the radius, decreasing the pore size of a matrix will effectively prevent more dense regions of a solution from falling into less dense regions—that is, convection.

The most widely used supporting media are cellulose acetate, starch gels, agar gels and polyacrylamide gels. Agar gel electrophoresis (4) is an analytical technique done on microscope slides on which a shallow layer of agar gel has been cast to demonstrate components in serum, enzyme preparations, etc. In immunoelectrophoresis (5-7) (Chapter 4), components of serum are separated electrophoretically (for example, in an agar gel) and then identified immunologically, as in the Ouchterlony plate method. Pevikon block electrophoresis (8, 9) is a preparative method, used for large-scale separation and recovery for enzyme purification, detection and separation of isozymes. The pevikon acts only as a support. Vertical and horizontal starch gel electrophoresis (10.11) are primarily analytical techniques. Separation occurs on the basis of charge and to some extent on molecular size (below). Polyacrylamide gels are made from a mixture of acrylamide and (usually) methylenebisacrylamide, which provides cross-linkages between polyacrylamide strands (see Figure 5.9).

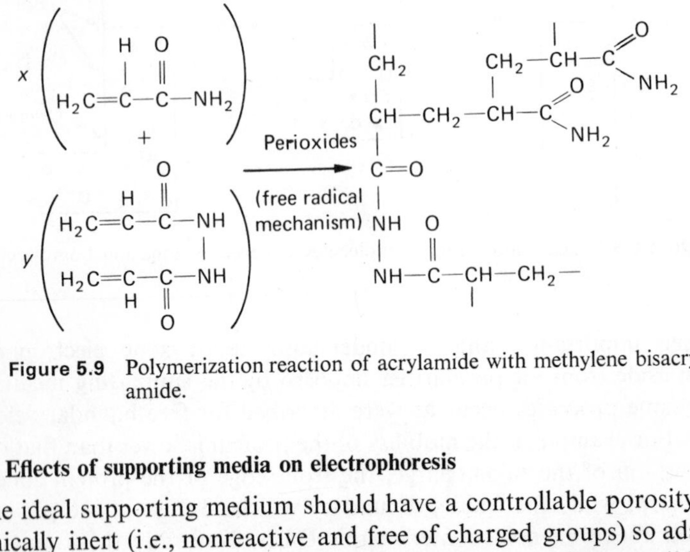

**Figure 5.9**   Polymerization reaction of acrylamide with methylene bisacrylamide.

## (v) Effects of supporting media on electrophoresis

The ideal supporting medium should have a controllable porosity and be chemically inert (i.e., nonreactive and free of charged groups) so adsorption effects cannot occur and electroosmosis is reduced as much as possible. Agar, for example, has a small percentage of carboxyl groups. An equal number of counterions must be present, and only these will move when a voltage is applied, since the agar cannot move. The counterions carry some solvent with them, which produces "electroosmosis," an electrophoretic streaming of the solvent. However, some electroosmosis will occur irrespective of the nature of the suppporting medium, because whenever two different substances or phases are in contact, a difference in chemical or electrical potential occurs—they are experimentally the same thing—at their junction (3). So the solvent at the junction will tend to move in an applied electric field. (Conversely, a moving solvent in contact with stationary phase will produce a potential difference across the stationary phase, called a "streaming potential.") For example, paper at pH 8 has a negative charge due to this phenomenon. So the direction of the electroosmotic flow of the solvent is toward the negative electrode. The extent of electroosmosis depends on the medium; acrylamide gels don't produce much electroosmosis.

There is one kind of interaction with the supporting medium which is often wanted. If the fibers or chains of the medium are close enough relative to the size of the proteins being electrophoresed, the proteins will be literally filtered by the medium. Consequently, larger proteins will be retarded more than smaller ones, and separation can be made on the basis of molecular size as well as electrophoretic mobility.

Of supporting media in common use, only polyacrylamide gel is capable of being easily cast to produce this effect, at least with the bulk of enzyme proteins (20,000 to 200,000 molecular weight) (12). The effective "pore sizes" of "normal" concentrations of agar gel or starch are too large to produce extensive "molecular sieving."[3]

### (vi) Acrylamide gel and SDS-gel electrophoresis

The molecular weight of a globular protein can be estimated by measuring its mobility in a series of gels of various percentages of acrylamide, relative to the mobilities of some proteins of known molecular weight in the same gels (12,13).

In electrophoresis of nucleic acids, the sieving effect of the gels is the important factor in separations, since the charge to mass ratio of nucleic acids is constant, at least at neutral or alkaline pH. Also, because many nucleic acids are large (ribosomal RNA's have molecular weights of from 500,000 on up), they will be concentrated during acrylamide gel electrophoresis because the gel lattice retards the leading nucleic acids long enough for the rest to catch up. See Figure 5.10. The resolution for high molecular weight nucleic acids using acrylamide gel electrophoresis is better than for proteins. And because of the greater choice of solvent composition than in disc electrophoresis (see section vii), acrylamide gel electrophoresis is currently favored for analysis of high molecular weight nucleic acids (14, 15).

At pH 7, in 1% SDS and 0.1 $M$ mercaptoethanol, most multichain proteins apparently bind about 1.4 g of SDS/g of protein and dissociate, and the resulting SDS subunit complexes behave as though they have a uniform negative charge and shape (16). This is not true of all proteins, however (17). The $R_f$ of a reduced protein in an acrylamide gel containing 0.1% or 1.0% SDS

[3]Remember that Sephadex and Bio-Gel "sieve" proteins by offering greater column volumes to smaller proteins, and larger proteins are eluted before smaller (Chapter 3).

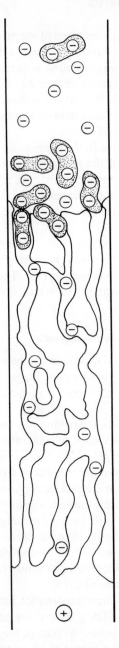

**Figure 5.10**
Concentration of macromolecule solution by molecular filtration in a polyacrylamide gel.

is usually proportional only to its Stokes radius, which in turn is roughly proportional to the logarithm of its molecular weight. This phenomenon has been exploited to obtain subunit molecular weights (13). However, because of inevitable variations between gel preparations, at least one marker protein is often run along with the sample (13). Using marker proteins whose molecular weights bracket that of the unknown will allow the molecular weight of the unknown to be fixed to within 5-10%.

### (vii) Theory of disc electrophoresis (18)[4]

Disc electrophoresis is a refinement of zone electrophoresis. It involves concentrating protein or nucleic acid samples into very thin layers using a discontinuous voltage gradient and then electrophoresing them on a column of polyacrylamide gel. The concentration step allows electrophoresis from starting zones only 1-100 microns thick, which produces the enhanced resolution characteristic of the technique.

The protein concentrates because the solvent is made discontinuous. Consider the "standard" disc electrophoresis system shown in Figure 5.11. When

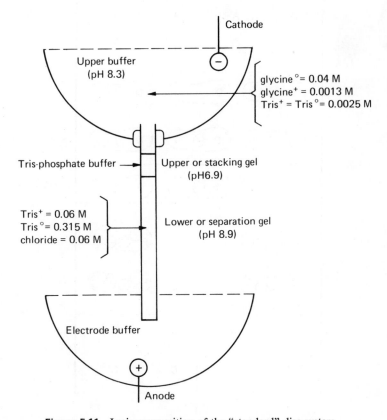

**Figure 5.11**   Ionic composition of the "standard" disc system.

[4]Some of this section is taken from Brewer et al. (19). Reprinted by permission of the publisher.

a voltage is applied across the solution, the glycine and chloride anions move toward the positive electrode. The mobility of glycinate anion is lower than that of chloride anion. In addition, at pH 8.3 the average net charge of glycine is $\frac{1}{30}$ that of chloride (which is one). This reduces the effective mobility of glycine further, to very much lower than that of chloride.

The chloride and glycine solutions are electrically in series, and the current —the movement of ions—must be the same throughout the system. Therefore, these anions must move with the same velocity. For this to happen, the voltage must be considerably higher in the glycine region than in that containing chloride. The discontinuity in solvent produces a discontinuity in voltage gradient which effectively prevents diffusion at the interface between the two solutions. A chloride ion moving into the glycine region is accelerated out again by the higher voltage gradient. A glycine anion attempt-

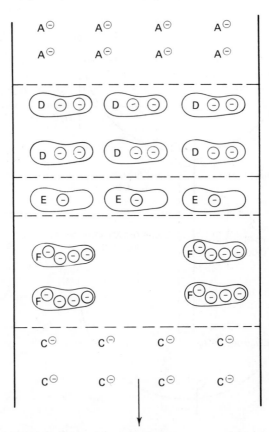

**Figure 5.12** Stacking of several macromolecules of different charge to mass ratios in disc electrophoresis.

ing to diffuse into the chloride solution finds itself in a region of much lower voltage and slows down until the rest of the glycine catches up.

If proteins are also present whose mobilities are between that of glycine at pH 8.3 and chloride, these will "sandwich" themselves between the two solutions to form their own zone of intermediate voltage gradient. Diffusion at both ends of the protein zone and between discs of proteins of different mobilities within the protein zone will be again restricted for the same reason that diffusion between the glycine and chloride solutions is. Note that in this zone the only anions present will be the proteins.

Because of the low net charge to mass ratio of most proteins, they must concentrate until each can carry in its subzone as much current as the chloride and glycine solutions. Thus, the protein zone will consist of thin discs of all the proteins with the right mobilities, arranged in order of increasing mobility. This concentration step ("stacking") occurs in the "upper gel" (see Figure 5.12).

These discs, being immediately adjacent, are of no analytical use. They must now be separated by ordinary electrophoresis. This is done in the "lower gel."

The average net charge, and consequently the effective mobility, of a solution of glycine will vary with pH. Note that when Tris is used as the counterion the conjugate base is uncharged. So when the glycine zone migrates into the lower gel where the concentration of Tris base is high, the pH will rise (to 9.5 in the standard system), and the average charge of the glycine (or concentration of glycinate anion) will increase. This will increase the effective mobility of glycine in the lower gel to the point where it will be higher than the effective mobility of the proteins, though not of course of the chloride which it cannot surpass. The glycine front will then pass through the protein zone, and the proteins will now be electrophoresing in the glycine solution at pH 9.5, called the "running pH." This will occur even though the net charge of the proteins will also increase. See Figure 5.13.

If nucleic acids or very highly charged (acidic) proteins are present, these may still have mobilities exceeding that of glycine at pH 9.5. These will then stay immediately ahead of the glycine solution and will not be visibly separated.

If the lower gel pH is progressively lowered, so that the concentration of Tris base is decreased while maintaining the same chloride concentration, the "running pH" in the glycine region will decrease. Consequently, the concentration of glycinate anion and hence the effective mobility of the glycine will both drop. The voltage gradient in the glycine region will increase, and so the proteins will migrate relatively faster in the lower pH glycine. At some lower gel pH (about 7.5-8.0 in the standard system) all the proteins which "stacked" will have a higher mobility than the glycine, and these proteins will not separate but migrate together just ahead of the glycine front.

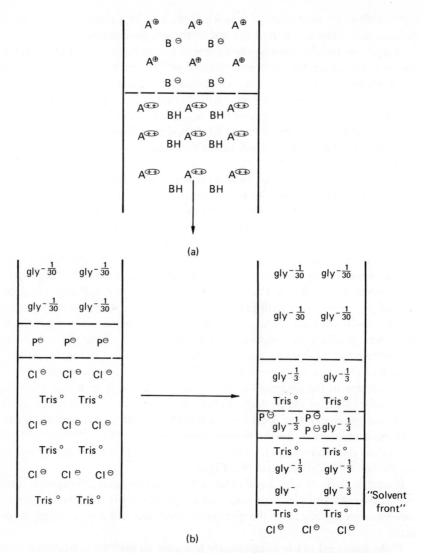

(a)

(b)

**Figure 5.13**  Production of the electrophoresis step in disc electrophoresis by pH changes: (a) cationic system; (b) "standard" anionic system.

The concentration of Tris base is naturally made very low in the upper gel (pH 6.9), so that the protein will "stack" and migrate at the glycine front until it reaches the lower gel. Phosphate has a greater mobility than glycine at pH 8.3, so it is used instead of chloride because of its superior buffering capacity at this pH (originally, a pH 6.9 Tris-chloride buffer was used).

"Tracking dye" (see Sec. 5.4a) is added to the protein or the upper buffer before electrophoresis. It has a mobility which is independent of pH in this region and is intermediate between chloride and glycine. The "tracking eye" remains at the glycine front and marks its progress.

The "electrode buffer" is the Tris-glycine buffer. It simply serves to make electrical contact between the gels and the positive electrode. A Tris-HCl buffer would serve as well.

The mathematical relations between ion concentrations, mobilities, pH's and p$K$'s are given in reference 18.

## 5.3 INSTRUMENTATION

### 5.3a  Power Supplies

A variety of power supplies are available. If you are reasonably knowledgable about electronics, you can save money by building your own power supply, or you can save by having the instrument shop, if one is available, design and build one to specification. The Heath Company, St. Joseph, Michigan, sells power supply kits which are cheap and efficient.

Power supplies for electrophoresis should feature low ac ripple, good stability (for long experiments) and sturdiness. For most electrophoresis experiments, a power supply is convenient which has terminals which supply a constant current to the apparatus. For isoelectric focusing experiments, you will need a device which supplies a constant voltage. The ideal power supply can provide both and has both a voltmeter and an ammeter.

Any power supply you use should be well shielded and grounded.

### 5.3b  Electrophoresis Apparatuses

A variety of these are marketed also, but for starch gel electrophoresis, a suitable apparatus can be made from Lucite (20). A sketch of the apparatus is shown in Figure 5.15 (see Sec. 5.4d). Its design is similar to apparatuses used for paper, cellulose acetate and agar gel electrophoresis. It consists of two buffer reservoirs with electrodes and a detachable cradle, containing the supporting medium, which completes the electrical circuit between the electrodes in the two reservoirs.

For acrylamide gel electrophoresis, whether plain, with SDS or with a disc system, or for isoelectric focusing in acrylamide gels, an apparatus is shown in Figure 5.14. This incorporates the general features of most commercial devices, having two reservoirs connected electrically by glass tubes containing the gels. This particular apparatus can be made for about fifty cents.

## 5.4  EXPERIMENTAL PROCEDURES

The procedure for disc electrophoresis is included first, since it is probably the most popular.

### 5.4a  Procedure for Disc Electrophoresis (21)

The following recipe is for a "standard" analytical system: 7.5% acrylamide lower gel, pH 8.9 (running pH: 9.5); 2.5% acrylamide upper gel, pH 6.9; and a Tris-glycine electrode buffer, pH 8.3. See Figure 5.14 and Appendix 12a.

For analytical columns, glass tubes 8 cm by 0.4 cm ID, marked 2 cm from

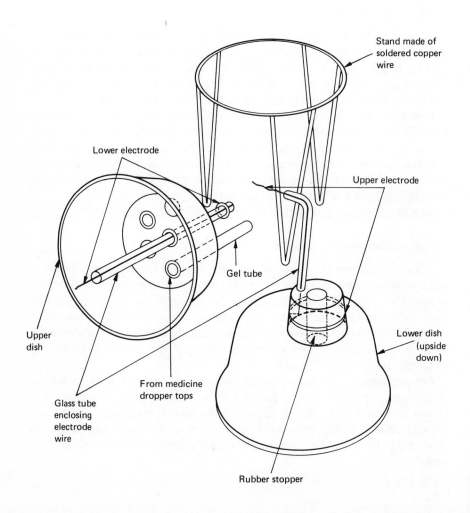

one end (the top) are used. Place the tubes in rubber serological stoppers so that they stand upright. Or, if there is some means of keeping the tubes upright, simply wrap the bottoms of the tubes tightly in parafilm, and clamp them in position. (If the Canalco fluorescent light holder is available, clamp the tubes to the face of the holder, using the metal spring across the front. Leave the light switched off to prevent bubble formation from lamp heat.) Prepare a Pasteur pipette with a drawn out tip for overlayering and fill it with water. This should deliver about one drop per second.

For convenience and to make storage possible, the ingredients for gel preparation are made up in several solutions (Appendix 12a), called A (containing concentrated lower buffer), B (stacking buffer), C (concentrated acrylamide and cross-linking reagent for lower gels), D (acrylamide for upper gels), E (riboflavin, for initiation of polymerization), and F (ammonium persulfate, for polymerization).

Prepare 1 ml of lower gel solution per tube by mixing 1 part $A$ to 1 part $C$ to 2 parts fresh $F$. The solutions should be at room temperature lest bubbles of air form while the gel is polymerizing. Add the $F$ last. Using a disposable pipette, add the gel solution to the 2 cm mark on the tubes, then tap the tubes

---

**Figure 5.14**

(*Opposite page.*) An inexpensive disc electrophoresis apparatus. The upper and lower vessels are "Cool-Whip" containers. The upper one has six equidistant holes bored (from the inside) in the bottom around a central hole. All holes are bored with a sharp number five cork borer. The grommets (seven of them) are made by cutting off the tops of the rubber bulbs of eyedroppers. The electrode material is made from an 18-inch long piece of 20-gauge Inconel number 92 stainless steel welding wire. The anode wire tends to corrode during use and must occasionally be replaced. The corrosion does not affect proteins if they migrate toward the anode (positive electrode). The six gel tubes are $6\frac{1}{2}$ cm long and 7 mm OD. Six stopper tubes are made from 4-inch lengths of the same size tubing and sealed off at one end with a torch.

The stand consists of a 14-inch long piece of number 8 copper wire bent into a tightly fitting circle around the container and soldered. The legs are 6 inches long and are soldered to the circle at equal intervals.

A 7-inch long piece of glass tubing (6 mm OD) is inserted in the center grommet of the upper dish. A 9-inch long piece of Inconel wire is cut and a $\frac{3}{4}$-inch diameter loop is made in one end and bent to a right angle to the stem. This wire goes up the 7-inch piece of tubing and is bent over at the top to hold it in place.

The upper electrode is essentially a loop of electrode wire attached to a rubber stopper. A size 8 rubber stopper has three holes bored in it: a hole through the center made with a number 5 cork borer and two connecting holes made with a number 2 cork borer. The latter two holes are bored: one parallel to the large hole but between it and the edge of the stopper down about three quarters of the distance towards the larger end of the stopper; the other is bored into the side of the stopper to meet the other number 2 hole. The electrode wire is forced down and out the connecting hole, then bent around the diameter of the stopper and its end crimped onto itself as it leaves the connecting hole. This is the negative electrode and slides down the outside of the glass tubing holding the positive electrode. A $5\frac{1}{2}$-inch piece of straight or bent glass tubing can be pushed down the negative electrode and into the stopper to help keep the two electrodes from contact and to make handling easier. Excess electrode wire is trimmed off so that, when electrical connection is made through insulated alligator clips, no exposed wire protrudes.

to insure that no air bubbles have been trapped in them. Carefully overlayer the gels with 4-6 mm of water, using the drawn out Pasteur pipette. The initial sharp boundary line will blur because of diffusion, but a new sharp line will appear ten minutes to one hour later, about 2 mm below the original interface. This indicates that the gel has polymerized. Let the gels stand another 10-20 minutes.

Remove the upper liquid by flicking the tube, and place the tubes near a fluorescent light. An apparatus consisting of three stacked circular 30-watt Cool White fluorescent lamps mounted on an aluminum plate with the ballasts is convenient for polymerization of analytical and preparative gels and can be constructed for about $50 (22).

Upper gel solution consists of 1 part $B$ to 2 parts $D$ to 1 part $E$ to 4 parts water. 0.2 ml of this is pipetted onto the top of each lower gel and is then overlayered with 4-6 mm of water.

Turn on the fluorescent lamp, and let the upper gels photopolymerize for 30 minutes. They will be opalescent. Flick off the liquid. Do not leave the gels exposed to the lamp for excessive periods, as heat from it tends to produce air bubbles in the gels as well as convection in incompletely polymerized regions. The gels must now be used within three hours or problems occur because of diffusion between the solvents in the upper and lower gels.

If stoppers have been used, insert a syringe needle between the stopper and tube to prevent a vacuum from forming when the stopper is pulled off. If there are air bubbles in the bottom of a gel tube, remove them by adding upper buffer with a Pasteur pipette. Dry the outsides of the gel tubes, and push their tops through the rubber grommets (medicine dropper tops) which give a leakproof fit (Figure 5.14).

Add 200 ml of upper buffer to the lower buffer container. Set the upper buffer container with all its holes plugged with gel tubes or stopper tubes onto the metal rack so that the gel tubes are in contact with the buffer. Add 1 ml of tracking dye to 200 ml of upper buffer, and pour this into the upper buffer container. Remove air bubbles in the tops of the gel tubes by directing a stream of buffer onto the gels with a Pasteur or long-tip pipette.

Connect the electrodes to the power supply: positive polarity to the lower chamber electrode, negative to the upper.

If the Canalco apparatus is used, the gel tubes are inserted into silicone grommets in a plastic dish. Don't force the tubes past the lower edge of the dish, as this tends to tear the grommet.

In preparing samples for disc electrophoresis, it is best to dialyze against the Tris-glycine upper buffer. In any event, the presence of large amounts of salts in the samples should be avoided, as they delay stacking for periods proportional to the amount present.

The dialyzed samples should be made 10% in sucrose or glycerol, then layered with a pipette or syringe onto the upper gels of appropriate tubes.

About 50 micrograms of protein is applied per tube. More should be applied if the preparation is very heterogeneous.

Turn on the power supply and run at 2.0 milliamperes per tube for 1.5 to three hours, until the blue tracking dye has migrated to within 2-5 mm of the bottom of the lower gel. Don't run the dye off the gel. Smaller ID tubes should be given less current or they will overheat.

Any tubes may be taken off at any time, but power must be shut off while doing this. The amperage should be cut back after a tube is removed.

Remove the gels from the glass tubes by "rimming." Insert a 2-inch 25 gauge hypodermic needle between the gel and the glass tube. Hold the needle steady and rotate the tube as the needle moves farther into the tube. Rotate the tube when withdrawing the needle also. It will probably by necessary to rim the gel from both ends of the tube. Do not try to use a longer needle as these are hard to control. The needle should be attached to a slow stream of water for lubrication during rimming. Alternatively use an 18 gauge needle insert with a rounded end; gel tubes are rimmed while immersed in water. If the gel is obstinate even after repeated rimming, use a rubber Pasteur pipette bulb filled with water to force the gel out.

Place the gels in the protein stain solution (Appendix 12a) for one hour. Heating the gels in the staining solution at 95°C for 15 minutes provides faster staining. The gels may be destained by allowing them to stand in several changes of 7% acetic acid until clear or a commercial diffusion destainer may be used. Destaining can be done faster by electrophoresis. Place the gels in constricted destaining tubes. These are slightly larger than the tubes the gels were cast in, but pinched off at the lower end to keep the gels in place during destaining. Electrophorese until all the free dye is removed, using 7% acetic acid in both buffer chambers. Alternatively, a transverse destaining apparatus can be used (23).

While the standard system described above is most popular, disc electrophoresis can be set up with other ion systems. See Appendix 12b.

### 5.4b  Procedure for Acrylamide Gel Electrophoresis

The gels can be made up in essentially any buffer, although substances such as sulfhydryl compounds can delay polymerization. The same concentrations of ammonium persulfate (or riboflavin) and TEMED are used for polymerization as are used in disc electrophoresis. The same buffer, without these substances, is used for both the upper and lower solutions. The preparation and overlayering of the gels should be done as carefully as possible, to keep the migration of the protein band(s) as uniform as possible. It is sometimes advantageous to "pre-electrophorese" the gels before electrophoresing the samples. This removes ionic contaminants from the gels before the protein is brought into contact with them.

The protein should be added in as small a volume as possible to increase the resolution. Times of electrophoresis may have to be determined empirically, but inclusion of tracking dye with the protein or upper buffer can help. Rimming, staining, and destaining the gels are done as in disc electrophoresis.

### 5.4c   Procedure for SDS-Gel Electrophoresis (13)

The protein sample concentration should be 1 mg/ml. The protein must be completely reduced before electrophoresis (25) by heating two hours at 37° in 0.01 $M$ sodium phosphate, pH 7.0 containing 8 $M$ urea (freshly deionized) and 1 % mercaptoethanol. The reduced protein solution is then dialyzed 24 hours against three changes of 0.1 % SDS and 0.1 % mercaptoethanol in 0.01 $M$ phosphate buffer, pH 7.0. (If salt is not removed from the protein, you will obtain extra protein bands with spuriously low apparent molecular weights.)

Gels are prepared in buffer-SDS solution as described in Appendix 12c. Generally, larger tubes are used than with disc or acrylamide gel electrophoresis. We use 10 cm by 6 mm (ID) tubes and add 2 ml of gel solution to each tube. After overlayering the gels carefully and letting them polymerize, the tubes are placed in the electrophoresis apparatus and buffer solution is poured into the upper and lower chambers. Then the reduced SDS-protein (10-50 $\mu$l) (microliters) is mixed with 3 $\mu$l of 0.05 % bromphenol blue in water, one drop of glycerol, 5 $\mu$l of mercaptoethanol, and 50 $\mu$l of dialysis buffer, and layered onto the gels. Electrophoresis is done using 8 milliamperes per tube, generally for about four hours.

The gels are rimmed and stained. Staining SDS-gels for protein cannot be done as described in 5-4a. The function of the 7 % acetic acid normally used is to denature the proteins and precipitate them within the gel matrix so they will adsorb the stain. SDS prevents this by keeping the proteins soluble inside the gel. So the SDS must be removed, and this is done by soaking the gels in the staining solution containing 90 % methanol described in Appendix 12c. The gels shrink but return to their original size on destaining in 7 % acetic acid plus 5 % methanol. They are stored in 7 % acetic acid.

### 5.4d   Procedure for Horizontal Starch Gel Electrophoresis

#### (i) Materials

Make up 1 liter of stock 1.0 $M$ sodium borate buffer at pH 8.8 (measured at room temperature). This is done by adding NaOH to a water suspension of enough boric acid such that the concentration will be 1 $M$ when it all dissolves.

Dialyze the protein preparations to be electrophoresed against 0.025 $M$ sodium borate buffer.

The gel bed form should be well greased with vaseline. The toothed plastic sheet that covers the gel bed after adding the gel should also be greased, except for the plastic teeth on the sheet: only their tips should be greased. Assemble the form with the end pieces clamped on, for pouring the gel.

11 g of starch[5] per 100 ml of 0.02 $M$ buffer (the total amount will depend on the particular apparatus used) are mixed in a 4 liter vacuum flask and heated to 85-90°C on a hot plate with constant and thorough stirring (use an overhead stirrer). The heat is then turned off, and the liquid is degassed by vacuum.

### (ii) Procedure

Before the liquid gel cools, it is poured into the greased form (Figure 5.15). There should be a slight excess of liquid. The greased plastic sheet with the teeth is placed, tooth side down, on the liquid gel in the form, then removed and the bubbles which have formed at the base of the teeth quickly removed with a spatula. The sheet is replaced on the gel form and smoothed so that no air pockets are left near the teeth. Long boards, then weights are

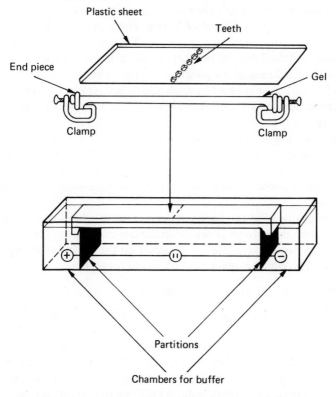

**Figure 5.15**  A horizontal starch gel electrophoresis apparatus (24).

[5]"Connaught Starch—hydrolyzed for gel electrophoresis" is satisfactory.

placed on the sheet to keep it firmly in place. After 3-4 hours at room temperature, the gel has set; remove the sheet carefully, then the end pieces, and place the form in the larger tray. Place the entire apparatus in the cold room. The tray should be filled with 0.1 $M$ sodium borate buffer.

The protein is added in the cold room to the hollow spaces in the gel left by the plastic teeth. There will be 12 spaces. 0.18 ml of protein solution, containing no more than 1 mg of protein, is added with a Pasteur pipette to each space. Electrophoresis is done in the cold room at 250 volts for 15 to 24 hours, no more.

The gel block is divided by plastic partitions into three strips, each with four spaces. In order to locate activity in impure samples, two procedures can be used. Two or three of the strips can be filled with the same protein samples, to be run in duplicate. One of these is carefully removed intact to be stained for protein. An alternative, though more difficult, technique involves stripping off thin layers of the gel strips. First cut off the end sections of the gel. The strips are covered with a strip of parafilm whose width is about 1 cm less than that of the gel, then a plastic strip about $\frac{1}{16}$ to $\frac{1}{8}$ inch thick, and finally with a large stiff sheet of Lucite. The tray is inverted, and the gel strips are carefully removed with a spatula. The gels, on their plastic strips, are replaced, upside down, in the tray. A sharp blade is used to skim off a layer of gel the thickness of the plastic strip. This is stained. Since the stained strips tend to shrink, it is best to punch a matched series of small holes with a hypodermic needle at 1 cm intervals along the sides of the strips before staining.

The gel strips are stained by soaking in a solution of 1 mg Amido Black/ ml in a mixture of methanol, acetic acid and water (5:1:5) for 5-10 minutes. The strips are then rinsed 10 minutes with fresh methanol, acetic acid-water solution without Amido Black and allowed to stand overnight, soaking in fresh solution.

From the positions of the bands appearing on the developed strip(s), one can guess where the protein is on the undeveloped strip(s). These areas can be cut and the protein removed by breaking up the gel in borate buffer and centrifuging. The eluted protein can be assayed for activity.

### 5.4e  Procedure for Isoelectric Focusing in an Acrylamide Gel

A number of people have published directions on this technique (see reference 26). We have successfully used the directions given by Catsimpoulis (27). Sufficient gel solution is made up according to either formula in Appendix 12d. The amount of protein to be included in the gel solution increases with its heterogeneity. For reasonably homogeneous preparations, 15-20 micrograms should be sufficient. Solution for one or two gels should be made up without protein for pH determination after the experiment.

The gel solutions should be degassed under vacuum for about a minute, then added to the gel tubes, and, if riboflavin is used, polymerized with light. No overlayering is necessary. The polymerized gel tubes are placed in the electrophoresis apparatus, and 5% ethylenediamine is added to the lower buffer chamber and 5% phosphoric acid to the upper. The phosphoric acid solution in the upper chamber is connected to the positive side (anode) of the outlets of a power supply, and the negative terminal (cathode) is connected with the lower solution. The electrofocusing is done at constant voltage, with an initial current of 5 ma per tube. After an hour or so, the current should have stabilized, dropping to about 0.5 ma per tube.

The gels are then removed by rimming. If the gels are to be stained with Amido Black, they should be washed for at least 48 hours with several changes of 12% TCA first: the ampholyte must be washed out of the gels before staining, since it takes up the Amido Black stain as well as proteins. A better stain is 0.2% bromophenol blue in ethanol: water: acetic acid (50: 45:5), since the protein will take up the stain but the ampholyte won't (28).

Staining with the bromophenol blue takes two hours. Destaining of the gels is done as described previously, in 7% acetic acid. Alternatively, the absorbencies of the native proteins can be measured with a gel scanner, if one is available, or specific stains for enzymatic activities used.

The pH can be determined as a function of gel length by slicing the "blank" (proteinless) gels and soaking each slice in 1 ml of water for four hours and measuring the pH of the extract.[6] Proteins of known isoelectric point can also be used as internal markers.

Note that, if a protein comes out of solution at its isoelectric point, it cannot settle in the gel. So using a gel to stabilize the protein bands is better than a sucrose gradient. On the other hand, it is almost impossible to prepare significant quantities of proteins with this method.

The pH 3-10 ampholyte should be used for initial determinations, and ampholytes covering narrower pH ranges should be used for more accurate work.

The percentage of solution $C$ in the gel might have to be lowered in working with very large proteins.

## 5.5  PRACTICAL ASPECTS

### 5.5a  Reagents

Most of the solutions are stable in the refrigerator for several months, if good quality chemicals are used. However, the time required for polymerization varies with the age of the solution $F$, so it should be made up fresh every week or so for reproducible polymerization rates. It may be better to use a

---

[6]The pH of ampholytes may vary strongly with temperature so measure the pH of the extract at the same temperature at which the focusing was done.

20- or 50-fold concentrated stock solution of *F* instead. If recrystallized acrylamide is used, the amount of solution *F* used may have to be reduced. Since the polymerization is exothermic, it is better to avoid problems with convection by using minimal amounts of *F* consistent with polymerization in a reasonable time.

The acrylamide polymerization occurs by a free radical mechanism. Consequently, the reaction goes faster at higher temperatures, and is inhibited by a variety of "quenchers," including oxygen, sulfhydryl compounds, reducing agents in general, and some metal ions.

Most commercial preparations of acrylamide contain enough impurities to retard polymerization, and some have enough to prevent it. Acrylamide may be recrystallized by heating in benzene (about 100 g acrylamide per liter of benzene) to a temperature of 60-70°. The impurities will lower the melting point of the undissolved acrylamide and will collect in a molten layer on the bottom of the beaker. Pour off the upper solution, and allow it to cool to room temperature. Do not chill, as benzene freezes at 6°. Collect the crystals of acrylamide by filtration. The impurity-rich material can be re-extracted with more benzene. Loening (14) recommends recrystallizing acrylamide from chloroform. In our experience the benzene method is more effective.

Acrylamide and benzene vapors are poisonous, SO DO THE RECRYSTALLIZATION IN A HOOD. Acrylamide, methylene bisacrylamide, TEMED, and ammonium persulfate are poisonous enough so that you should avoid getting these compounds on external or internal body surfaces.

Methylene bisacrylamide is recrystallized from acetone (14). Glycine can be recrystallized from hot water, and Tris from 95% ethanol.

The glass tubes are washed by soaking in chromic acid cleaning solution overnight, rinsed with glass-distilled water, soaked about a minute in 10% potassium hydroxide in 95% ethanol (w/v), and rinsed thoroughly again in glass-distilled water. Dirty stoppers are suspended in glass-distilled water and potassium hydroxide pellets (5 g per 100 ml) added with stirring. After the solution cools to room temperature, the stoppers are rinsed in glass-distilled water, 0.1 *N* HCl, and twice more in glass-distilled water.

### 5.5b   Variations in Technique

Some workers do not use an upper gel, but simply layer their protein onto the lower gel. We believe use of the upper gel gives better results (see reference 29). For people worried about cross-contamination between protein samples in different gel tubes, the protein may be mixed with a second sample of upper gel solution, and the latter polymerized onto the stacking gel. In earlier literature, this was called a "sample gel." However, we have had no difficulties with cross-contamination using the directions given.

In making up gels containing 8 *M* urea one should either add solid urea to the lower or upper gel solutions, or add concentrated stock solutions to 10 $M$ urea. Either procedure is preferable to making all the solutions 8 $M$ in urea, since urea decomposes with time to ammonium cyanate. The 10 $M$ urea should be deionized by pouring through a short column of mixed-bed ion-exchange resin such as Dowex 501 just before use (Appendix 12e).

With 8 $M$ urea-containing gels, the 7-1/2% standard acrylamide concentration should be cut to $5\frac{1}{2}$% (22). If the acrylamide concentration is cut below 5%, the concentration of methylene bisacrylamide should be increased to 5% of the acrylamide concentration.

The $7\frac{1}{2}$% standard gel is suitable for proteins of molecular weight between 10,000 and 1,000,000, with best resolution from 30,000 to 300,000 Daltons. For electrophoresis of substances of molecular weight less than 10,000 a smaller pore gel is used, with an acrylamide concentration up to 30%. For substances above 1,000,000 Daltons a $3\frac{1}{2}$% gel is used. For heterogeneous mixtures of proteins, such as from serum, better resolution is claimed using gels made with a gradient of pore size (30).

### 5.5c  Staining, Identification and Recovery

Chrambach et al. (31) soak their gels in 1% Coomassie Brilliant Blue in 12.5% trichloroacetic acid. According to those authors, the uptake of stain by the zones of protein can be followed by eye, and the stain is more sensitive.

Proteins containing large amounts of lipid (or carbohydrate) are often difficult to stain using Amido Black. Lipoproteins can be stained with Amido Black (32), but other stains and solvents can be used. Glycoproteins are stained with PAS (Periodic Acid-Schiff) (33).

Staining for nucleic acids is done by soaking the gels overnight in 2% acridine orange, 1% lanthanum acetate, and 15% acetic acid (34). Removal of excess stain is by dialysis against 7% acetic acid.

Instruments for measuring the ultraviolet absorption of unstained nucleic acids in gels as a function of distance of migration have been developed. (Quartz tubes are used.) Some of the constituents of acrylamide gels absorb in the ultraviolet, and the others contain ultraviolet-absorbing impurities, so several techniques have been developed to reduce the "background": recrystallizing some of the reagents (14), electrophoresing the gels before applying any sample (14), and removing, then dialyzing the gels before reinserting in the tubes and running the experiment (15).

Often it is useful to be able to determine whether a particular enzyme activity is associated with a certain protein band. One procedure is to slice the gel and try to elute the enzyme. This can be done with starch or agar gels but it is difficult with acrylamide gels, probably because of the smaller pore size of the lattice enclosing the protein. Good recoveries have been obtained by placing an acrylamide gel slice in a starch gel block, driving the protein into the starch by electrophoresis, and eluting from the starch.

There is a specific stain for dehydrogenases (Appendix 12f) (35). The gels should be soaked in the dehydrogenase staining solution only 5-10 minutes for rings to appear at the site of the dehydrogenases. Too long an incubation gives darker bands which blend into each other, making estimations of the number of bands difficult. Staining solutions have also been developed for leucine amino-peptidase, alkaline phosphatase (36) and many other enzymes (37). Any colorimetric reaction or fluorometric analysis for products can in theory be employed as a stain (37).

For analysis of patterns obtained in electrophoresis of radioactive material, the gels are frozen in dry ice, sliced, the slices dissolved by heating for 1.5 hours at 80° in 30% hydrogen peroxide, and then scintillation fluid is added. If the gels are polymerized using ethylene diacrylate instead of methylene bisacrylamide, concentrated ammonium hydroxide can be used (38). If N, N′-diallyl-tartardiamide is substituted for methylene bisacrylamide, 2% periodic acid can be used instead (39); this does not hydrolyze proteins or quench the scintillators usually employed in counting (see Chapter 8).

## 5.6 ANALYSIS OF RESULTS

The first thing to see in looking at a gel is whether the pattern is anomalous—the bands streaked on one side, cupped, etc. Ritchie et al. (29) describe the causes and cures of some common anomalies in disc patterns. They avoid these by degassing their reagents, using a detergent (Tween 80) for better layering, and generally taking greater pains in preparing the gels. However, the directions in 5.4a give satisfactory results if followed carefully.

Electrophoresis experiments can give several types of information: the number of protein bands, whether smearing has occurred, how heavily the bands stain, and how thick they are.

### 5.6a The Number of Protein Bands

You will obtain one or more bands from a given preparation. If you obtain only one this may mean that the preparation is homogeneous or that more than one species are moving together. For this reason, preparations should be analyzed by electrophoresis at several pH's to be more certain of homogeneity.

Homogeneity of a preparation is of course a relative term. Greater amounts of protein should be applied to establish the purity of a protein, since impurities which were too low in amount to be detected when 50 μg (micrograms) was electrophoresed may be visible when 200 μg is analyzed. The limit of detection of a protein is about 1 μg, depending on the stain, the size, and the "stainability" of the protein. Rough estimates can be made of protein purity from the relative intensities of bands, but this is hazardous.

When all the protein migrates at the solvent front of a disc system, special

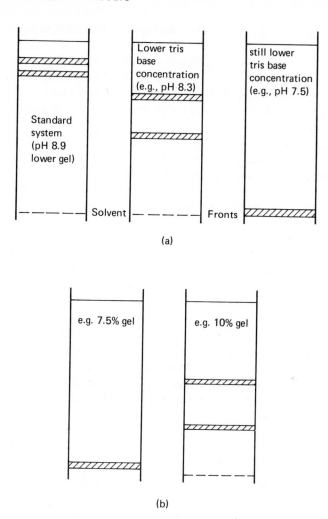

**Figure 5.16** (a) Effect of lower gel pH on electrophoresis patterns; (b) effect of gel concentration.

care should be taken—several components may have high $R_f$'s. Increasing the acrylamide concentration or using another system should be done to lower the $R_f$ of the preparation. See Figure 5.16.

If you obtain several bands, this generally indicates the heterogeneity of the original preparation; very rarely, it can occur because of breakdown of an originally homogeneous preparation into several electrophoretic species. This is especially true of the strongly acid disc systems (Appendix 12b). If you suspect this is occurring, the best procedure is: isolate all the protein species by preparative disc electrophoresis (22), using the same system; see which species are active; and rerun that species using the same analytical system.

Acrylamide gels can oxidize sulfhydryl groups in proteins and under some (rare) circumstances can produce inactivation of enzymes, and even artifactual protein bands (40). This can sometimes be prevented by adding 1-5 micromoles of Tris-thioglycolate to the protein before electrophoresis (stacking may be delayed).

After destaining a "standard" disc gel, there is almost always a thin, sharp band remaining where the solvent front was—sometimes two or three such bands which are very close together. This is protein, not tracking dye, and it can be some fraction of the protein(s) which migrate normally with lower $R_f$'s. It sometimes appears that the higher the $R_f$, the greater the fraction appearing at the solvent front. The reason for this isn't known; it may involve interaction with the chloride ions in the lower gel just as electrophoresis begins (remember, while the protein is "stacked" there are presumably no other anions around, even though the protein will have some positively charged groups).

### 5.6b  Smearing of Bands

If too much protein is added, the heavier bands will probably appear smeared, since the proteins' contribution to the conductivity of the solution has become appreciable (see 5.2b, part iii).

Some smearing is found with some proteins even at low concentrations. This may indicate that a chemical reaction is occurring, affecting the mobility of the protein. Addition of Tris-thioglycolate can be helpful (40).

### 5.6c  Staining of Bands

Proteins will take up characteristic amounts of a given stain even if they contain no lipid or carbohydrate, for unknown reasons.

### 5.6d  Thickness of Bands

When electrophoresis begins (i.e., when the protein enters the lower gel in a disc system), its zone will begin to widen because of diffusion. The extent of broadening will depend on the diffusion constant of the protein, the time of electrophoresis, and the gel porosity.

The diffusion constant of a macromolecule (see Chapter 6) is proportional to the absolute temperature and inversely proportional to the frictional coefficient of the macromolecule. Consequently, proteins of higher asymmetry or molecular weight will give sharper protein bands on electrophoresis. Proteins which aggregate during electrophoresis will show the unaggregated protein to be followed (usually) by much thinner and sharper protein bands. If the thickness of the zone is greater than you would expect from the gel concentration and the molecular weight of the protein, either the protein has dissociated or else two or more proteins are migrating nearly together.

Diffusion is also proportional to the square root of the time. The resolving power of disc electrophoresis therefore depends partly on the running pH (see 5.2b (vii)): in the alkaline (Tris-glycine) system, decreasing the Tris concentration in the lower gel can sometimes lead to separation into more protein components of bands that ran as a single species at pH 9.5, and generally greater separation is obtained in a given time.

## 5.7  LIMITATIONS OF METHODS

The greatest limitation of disc electrophoresis was originally the limited number of systems available, due to the fact that the mobilities of relatively few ions were known. Some systems were not suitable for a given protein. This limitation has been eased since Jovin et al, have recently compiled the characteristics of a large number of disc systems (41).

Acrylamide gel electrophoresis allows essentially any buffer or pH to be used and features the molecular sieving of disc electrophoresis, but except when electrophoresing macromolecules of high molecular weight (see 5.5b) this method does not give the resolution of disc electrophoresis.

SDS-gel electrophoresis provides a fairly reliable method of determining subunit molecular weights, at least for most proteins. Some proteins, however, do not react "normally" with SDS (17) and these can be expected to behave anomalously in SDS-gel electrophoresis.

Isoelectric focusing shows promise as an analytical technique and as a criterion of purity. In theory, this method is the best potential criterion of purity, since the pH range in the gels can be narrowed, increasing the resolving power. Note, however, that the resolving power of the method is also limited by the diffusion constant of the protein. This can perhaps be counteracted by increasing the percentage of acrylamide in the gel.

There is a possibility that the protein(s) to be focused will be altered or inactivated during polymerization, especially if it has exposed sulfhydryl groups.

## REFERENCES _____

An enormous number of papers have been written, particularly about disc electrophoresis. This list is not at all comprehensive, but it should provide any additional information necessary.

1. Haglund, H., "Science Tools", **14**, 17 (1967).
2. Bier, M., Ed., *Electrophoresis*, Academic Press, New York (1959).
3. Moore, W. J., *Physical Chemistry*, 4th ed., Prentice-Hall, New Jersey (1972).

4. Wieme, R. J., *Clin. Chim. Acta*, **4**, 317 (1959).

5. Campbell, D. H., *Methods in Immunology*, Benjamin, New York (1963).

6. Scheidegger, J. J., *International Archives Allergy Appl. Immunol.*, **7**, 103 (1955).

7. Heremans, J., *Les Globulines Seriques du Système Gamma*, Masson et C$^{ie}$, Paris (1961).

8. Muller-Eberhard, H. J., *Scand. J. Clin. Lab. Invest.*, **12**, 33 (1962).

9. Fahey, J. L. and C. McLaughlin, *J. Immunol.*, **91**, 484 (1963).

10. Smithies, O., *Biochem. J.*, **71**, 585 (1959); *Arch. Biochem. Biophys.*, Suppl. **1**, 125 (1962).

11. Poulik, M. D. and O. Smithies, *Biochem. J.*, **68**, 636 (1958).

12. Hjertén, S., *J. Chromatogr.*, **11**, 66 (1963).

13. Weber, K. and M. Osborn, *J. Biol. Chem.,* **244**, 4406 (1969).

14. Loening, U. E., *Biochem. J.*, **102**, 251 (1967).

15. Bishop, D. H. L., J. R. Claybrook and S. Spiegelman, *J. Mol. Biol.*, **26**, 373 (1967).

16. Reynolds, J. A. and C. Tanford, *Proc. Nat. Acad. Sci. U. S.*, **66**, 1002 (1970).

17. Nelson, C. A., *J. Biol. Chem.*, **246**, 3895 (1971).

18. Ornstein, L., *N. Y. Acad. Annals*, **121**, 321 (1964).

19. Brewer, J. M. and R. B. Ashworth, *J. Chem. Educ.*, **46**, 41 (1969).

20. Tsuyuki, H., *Anal. Biochem.*, **6**, 205 (1963).

21. Davis, B. J., *N. Y. Acad. Annals*, **121**, 404 (1964).

22. Jovin, T., A. Chrambach and M. A. Naughton, *Anal. Biochem.*, **9**, 351 (1964).

23. Ward, S., *Anal. Biochem*, **33**, 259 (1970).

24. Tsuyuki, H., *Biochim. Biophys. Acta,* **71**, 219 (1963).

25. Reynolds, J. A. and C. Tanford, *J. Biol. Chem.*, **245**, 516 (1970).

26. Wrigley, C. W., "Science Tools," **15**, 17 (1968).

27. Catsimpoolas, N., *Anal. Biochem.*, **26**, 480 (1969).

28. Awdek, Z. L., "Science Tools," **16**, 42 (1969).

29. Ritchie, R. F., G. Harter and T. B. Bayles, *J. Lab. Clin. Med.*, **68**, 842 (1966).

30. Slater, G. G., *Fed. Proc.*, **24**, 225 (1965).

31. Chrambach, A., R. A. Reisfeld, M. Wyckoff and J. Zaccari, *Anal. Biochem.*, **20**, 150 (1967).

32. Narayan, K. A. and F. A. Kummerow, *Clin. Chim. Acta*, **13**, 532 (1966).

33. Kratoski, W. A. and H. E. Weimer, *Can. J. Biochem.*, **45**, 1577 (1967).

34. Richards, E. G., J. A. Coll and M. A. Gratzer, *Anal. Biochem.*, **12**, 452 (1965).

35. Goldberg, E., *Science*, **139**, 602 (1963).

36. Law, G. R. J., *Science*, **156**, 1106 (1967).

37. Gabriel, O. in *Methods in Enzymology*, Vol. 22, W. B. Jakoby, Ed., Academic Press, New York (1971), p. 578.

38. Weinberg, R. A., U. E. Loening, M. Willems and S. Penman, *Proc. Nat. Acad. Sci. U. S.*, **58**, 1088 (1967).

39. Anker, H. S., *FEBS letters*, **7**, 293 (1970).

40. Dirksen, L. M. and A. Chrambach, *Separation Science*, **7**, 747 (1972).

41. Jovin, T. M., L. M. Dante and A. Chrambach, *Multiphase Buffer Systems Output*, National Technical Information Service, Springfield, Virginia (1970) PB #196085 to 196091 and 203016.

# 6 ‖ ULTRACENTRIFUGATION

6.1 INTRODUCTION

## 6.1 INTRODUCTION

The ultracentrifuge is a device for separating molecules on the basis of their size, using a high gravitational force field (1). For a given force field, a particular molecule will move at a particular velocity, and the ratio of this velocity to the applied centrifugal field is called the sedimentation constant of the molecule. It is analogous to the mobility of a molecule in electrophoresis. A mixture of molecules with different sedimentation constants will be partially separated by the force field, producing concentration gradients of the molecules. This leads to diffusion, which opposes the process of sedimentation.

Analytical ultracentrifugation is used for measurement of purity, molecular weight, and other properties of macromolecules, using very small amounts of material.

In this chapter we discuss the theory of centrifugation, and what particular properties of macromolecules affect their behavior in high centrifugal fields

(2). We describe the analytical ultracentrifuge in terms of its constituent operating systems. Finally, we provide detailed procedures for measuring the sedimentation constant, molecular weight, diffusion constant, and extinction coefficient of a macromolecule.

## 6.2 THEORY OF ULTRACENTRIFUGATION

### 6.2a  Necessity for the Ultracentrifuge

If a mixture of sand and water is stirred to a uniform consistency, then placed on a table, the sand will commence to settle out. It is sedimenting under an acceleration of $1 \times g$. Similarly, protein molecules in solution will experience the same acceleration under similar conditions. However, no appreciable sedimentation will be observed. There are two reasons for this.

The first reason is that the effective weight (particle weight minus weight of liquid displaced) of the protein molecules is much less than that of the sand grains. Consequently, the rate of sedimentation is much lower.

The second reason can be found in the phenomenon of Brownian motion (of visible particles). This random motion is of thermal origin, and can result from collisions of thermally activated solvent molecules with the larger solute particles. Because the motion imparted to the solute particles is random, a uniform suspension of these particles results, even when their particle weights are as high as several millions. This thermal motion is of course the basis of the phenomenon of diffusion. For protein molecules, which have molecular weights of $10^4$ to $10^6$, the thermal motion acquired at temperatures around $300°$ K is quite sufficient to maintain a uniform suspension at $1 \times g$. In order to produce appreciable sedimentation of proteins, much higher forces are needed to overcome the thermal motion, or, in other words, to increase the velocity of sedimentation above the velocity of diffusion.

For particles the size of proteins, the accelerations needed are $10^5$-$10^6 \times g$, and the instrument developed which is capable of exerting such is the ultracentrifuge (1).

### 6.2b  Sedimentation Velocity

Solute molecules are placed in a high centrifugal field, and the rate at which the solute moves as a function of time depends on its effective particle weight

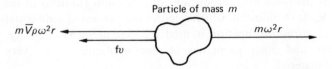

$$m\overline{V}\rho\omega^2 r \longleftarrow \qquad m\omega^2 r \longrightarrow$$

Particle of mass $m$

$f\upsilon$

**Figure 6.1**   Representation of the forces on a particle in a centrifugal field.

and frictional resistance. This experiment can also provide information about purity and concentration of the macromolecule (2).

### (i) Definition of the sedimentation constant

The forces on a particle in a centrifugal field are schematically represented in Figure 6.1. Mathematically the forces are

$$F_{\text{sedimentation}} = ma = (V_p d_p - V_s d_s)\omega^2 r$$

$$= \left(\frac{M}{N} - V_s d_s\right)\omega^2 r = \frac{M}{N}(1 - \bar{V}d_s)\omega^2 r = \frac{M}{N}(1 - \bar{V}\rho)\omega^2 r \qquad (6\text{-}1)$$

$M$ is the anhydrous molecular weight (see 6-2b (iii). 1) of the particles, $N$ is Avogadro's number and $\rho$ is the density of the solvent. $\bar{V}$ is called the partial specific volume of the protein. The force of sedimentation equals the effective mass times the acceleration. The effective mass, due to the buoyancy term, is $m(1 - \bar{V}\rho)$. The angular acceleration equals the square of the angular velocity, $\omega$, times the radius, $\omega^2 r$.

$$F_{\text{friction}} = vf = F_{\text{sedimentation}} = M(1 - \bar{V}\rho)\omega^2 r \qquad (6\text{-}2)$$

The frictional force is the product of the velocity, $v$, and the molar frictional coefficient of the sedimenting particle, $f$. The latter depends upon the parameters of size, shape and hydration of the particle, and viscosity of the solvent.

The particle will quickly reach a constant velocity since the frictional resistance ($F_{\text{friction}}$) will increase with the velocity of the particle. At such a velocity, the net force on the particle is zero, and it continues to move but no longer accelerates:

$$m(1 - \bar{V}\rho)\omega^2 r = fv = \frac{M}{N}(1 - \bar{V}\rho)\omega^2 r \qquad (6\text{-}3)$$

or

$$\frac{v}{\omega^2 r} = \frac{M(1 - \bar{V}\rho)}{Nf} \qquad (6\text{-}4)$$

The velocity attained by the particle for a given angular acceleration is called its sedimentation constant, $S$:

$$S = \frac{v}{\omega^2 r} = \frac{M(1 - \bar{V}\rho)}{Nf} \qquad (6\text{-}5)$$

### (ii) Measurement of the sedimentation constant

1. *The boundary method.* The sedimentation of particles in a solvent creates a boundary between the sedimenting particles and pure solvent

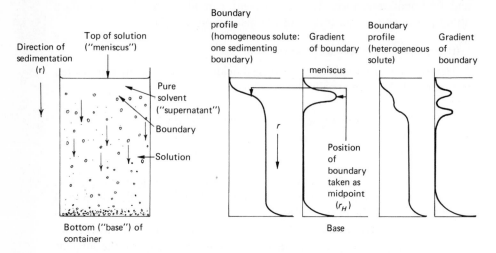

**Figure 6.2**    Experimental determination of sedimentation velocity by measurement of boundary movement. Note how the use of the gradient of solute concentration makes the determination of the boundary position and solute homogeneity easier.

("supernatant")(see Figure 6.2). It is experimentally convenient to measure the rate of sedimentation of the particles by following the movement of the boundary as a function of time. It is also convenient, for purposes of measurement and analysis of the homogeneity of the particles, to use the "schlieren" optical system to follow the boundary movement, as this gives the differential of the boundary.

The velocity of boundary movement is $dr_H/dt$;[1] and

$$ S = \frac{1}{\omega^2 r_H}\frac{dr_H}{dt} = \frac{1}{\omega^2}\frac{dr_H}{r_H}\cdot\frac{1}{dt} = \frac{2.303}{\omega^2}\cdot\frac{d(\log r_H)}{dt} \qquad (6\text{-}6) $$

where $\omega$ is the angular velocity, in radians/second. This is $2\pi N/60$, where $N$ is in revolutions per minute. The distance of the boundary from its axis of rotation (in centimeters) is $r_H$, and $t$ is the time (in seconds).

*2. The zone method.* Another way of measuring sedimentation constants was developed by Martin and Ames (4). Their procedure is a zone method and so can be scaled up for preparative purposes. The protein solution is layered onto a sucrose density gradient and then centrifuged into the density gradient, which serves to stabilize the protein zones and prevent convection. The protein zone slows as it moves farther into the gradient, since the sucrose retards the movement of the protein because of its relatively high density and

---

[1] If the boundary is not symmetrical, a more complex calculation must be made (3).

viscosity. To help measure the velocity of sedimentation in the gradient, two or more proteins of known sedimentation constant ("markers" or standards) are sedimented along with the protein of unknown sedimentation constant, and its migration is measured relative to theirs after a given length of centrifugation. (Another zone technique for measuring sedimentation constants is described in reference 5.)

Martin and Ames (4) showed that, given the proper sucrose gradient, tube size and rotor speed, the ratio of the sedimentation constants of a "marker" (a protein of known sedimentation constant) and an unknown protein is equal to the ratio of the distances each has traveled into the gradient. This is determined by punching a hole in the bottom of the tube with the gradient and collecting fractions of equal volume.

$$\frac{S_{unknown}}{S_{marker}} = \frac{\text{distance (unknown)}}{\text{distance (marker)}} = \frac{\text{fraction number (unknown)}}{\text{fraction number (marker)}} \quad (6\text{--}7)$$

The positions the sedimenting proteins have reached are measured by absorbance, or, more reliably, by measurements of enzyme activity.

### (iii) Molecular parameters affecting the sedimentation constant (2, 3)

The sedimentation constant is characteristic for a particular macromolecule, since it is a function of four macromolecular parameters.

*1. Molecular weight.* Most proteins are hydrated to some extent, and this might be expected to increase their molecular weight. However, the water of hydration will have approximately the same density as the solvent, unless the latter contains high concentrations of salts, sucrose, urea, etc., which interact strongly with water or the protein. Thus, the effect of bound water on the molecular weight of the protein will be approximately cancelled out by the effect of the bound water on the partial specific volume of the protein (see 3. below). Consequently, the sedimentation constant, like other parameters measured by sedimentation methods, is a measure of the anhydrous molecular weight of the protein.

However, the sedimentation constant is not a linear function of molecular weight, since the frictional coefficient of a particle increases with its size. For an unhydrated spherical particle undergoing translational motion:

$$f = 6\pi\eta r \quad (6\text{--}8)$$

where $\eta$ is the viscosity of the solvent and $r$ is the radius of the particle. Since the molecular weight of a rigid spherical particle is proportional to $r^3$,

$$m = \frac{M}{N} = \frac{4\pi r^3}{3\bar{V}} \quad (6\text{--}9)$$

the sedimentation constant of such a particle will be proportional to $M^{2/3}$. This relation is also approximately true for particles which are hydrated and nonspherical but of low asymmetry, such as the "globular" proteins.

2. *The frictional coefficient.* Frictional coefficients of macromolecules are functions of the molecular weight and density of the macromolecules, their asymmetry and the extent to which they are hydrated; that is, of their effective sizes. For asymmetric and/or hydrated particles, the actual radius in Equation 6-8 is replaced by the effective, or Stokes radius, $r_{eff}$:

$$f = 6\pi\eta r_{eff} \tag{6--10}$$

The Stokes radius can be determined by chromatography on Sephadex or Bio-Gel (Chapter 3) relative to those of other proteins, whose Stokes radii were in turn calculated from their diffusion constants:

$$D = \frac{kT}{f} = \frac{kT}{6\pi\eta r_{eff}} \tag{6--11}$$

Values for $f$ and $r_{eff}$ can also be calculated from the sedimentation constant, molecular weight and partial specific volume of a macromolecule. The calculated frictional coefficient includes the experimental errors in all three determinations. It is often assumed that $f$ from sedimentation and diffusion methods is the same quantity (2).

In order to eliminate the effect of the partial specific volume and molecular weight of a macromolecule, the theoretical frictional coefficient, $f_0$, of an unhydrated sphere of the same molecular weight and density is calculated:

$$f_0 = 6\pi\eta r_0; \quad r_0 = \sqrt[3]{\frac{3M\bar{V}}{4\pi N}} \tag{6--12}$$

And the ratio $f/f_0$ (called the "frictional ratio") is then a function only of the shape and hydration of the macromolecule.

The reported frictional ratios for native "globular" proteins range from slightly below 1.0 to 1.4. Some of these are known from X-ray diffraction or electron microscope studies to be relatively compact and symmetrical. Larger values of $f/f_0$ are found for molecules like DNA or myosin which are very asymmetric.

It is easy to see why the asymmetry of a particle should affect its frictional ratio. The velocities of most macromolecules in an ultracentrifuge cell are very low—of the order of 0.2 to 1.0 cm/hr—and the rotational diffusion times for these molecules are in milliseconds or less. Asymmetric molecules consequently cannot be significantly oriented by their flow in an ultracentrifuge cell. This is not the case for these molecules in a capillary viscometer, since the high shear gradients in these instruments can produce considerable orientation. An asymmetric molecule will therefore spend an appreciable fraction of the time sedimenting with its broadest side to the direction of movement, experiencing a greater frictional resistance by the solvent to its movement. On an average, a sphere experiences the least frictional resistance, and has the lowest frictional ratio.

The asymmetry of a molecule affects its behavior in other ways, notably in the concentration dependence of its sedimentation constant and other properties (see 4., below).

X-ray diffraction studies have shown that the shapes and surfaces of proteins are quite irregular. However, measured frictional coefficients are rather insensitive to the detailed conformations of proteins. One reason for this is that when a particle moves through a solvent, the layer of solvent next to the particle moves with the particle. The next layer of solvent moves at a slightly slower speed, and the next at a slightly slower speed than the inner layer. At distances remote from the particle, the solvent isn't moving. This "layering" phenomenon helps spread frictional effects caused by the motion of particles in a solvent, throughout the solvent. However, it also obscures surface detail on the particle, and helps make all but the most irregularly shaped molecules behave like spheres or ellipsoids of revolution. This phenomenon, incidentally, is independent of the macromolecule and the solvent and has nothing to do with water of hydration.

Frictional coefficients of globular proteins also generally increase substantially (by $100\%$ in some cases) when the protein is converted to a flexible, random coil by dissolving the protein in concentrated urea or guanidine hydrochloride solutions. Frictional drag increases since all parts of a random coil are in contact with the solvent, while a native globular protein has only its surface exposed to the solvent.

The fact that several proteins apparently have frictional ratios at or even below 1.0 may be at least partly due to dissociation or other experimental error. It seems unlikely that these proteins are completely unhydrated, even if spherical.

The water of hydration of macromolecules apparently gives a distinctive nuclear magnetic resonance (NMR) signal that can be quantified (6). The "water of hydration" of several proteins was found to be 0.3-0.5 g/g, depending on the protein. These values were in reasonable agreement with previous estimates.

3. *Partial specific volume.* The partial specific volume is defined as the volume increment produced in a solution per unit mass of solute added:

$$\bar{V} = \frac{dV_{\text{solution}}}{dm_{\text{solute}}} \tag{6-13}$$

For proteins which do not contain large amounts of carbohydrate or lipid,

$\bar{V}$ is usually about 0.73 cc/g, or a density of about 1.37 g/cc. If a protein contains enough lipid, its partial specific volume will be greater than 1.0, in which case it will float in an aqueous solution in the ultracentrifuge instead of sedimenting.

Partial specific volumes are obtained by measuring the density of a solution of known protein concentration (directly or from sedimentation measurements) or by calculation from the composition of a protein [see 6-5a (ii)].

Deuterium oxide can be used to measure partial specific volumes of macromolecules by sedimentation methods (7). Deuterium oxide has a greater density and viscosity than water and as a result macromolecules will sediment more slowly in its presence. Macromolecules with high partial specific volumes (low densities) will be more strongly retarded, or buoyed more, than substances with greater density (see 6-5a, ii, 5 below).

Deuterium oxide-water mixtures are used because they behave thermodynamically like a one-component system. Something like sucrose or glycerol cannot be used because of the possibility of preferential interactions with the macromolecule. In sucrose solutions, bovine serum albumin interacts preferentially with water (8), and behaves like a protein-water complex in the sucrose solution. This changes its effective partial specific volume and possibly other properties.

Note that problems arise only when extensive *preferential* interactions occur. If the partial specific volume of a preferentially absorbed substance is the same as that of a macromolecule, the partial specific volume of the complex may be unchanged if the binding of that substance doesn't produce preferential binding of other substances. If preferential interactions are negligible (which is usually the case in 0.1 $M$ solutions of most neutral salts), the hydrodynamic partial specific volume generally agrees with the anhydrous value.

*4. Concentration effects.* For most macromolecules, the sedimentation coefficient is dependent on concentration. One possible reason for this is because the sedimenting macromolecules displace solvent, which must flow in the opposite direction. This "backward flow" of solvent effectively subtracts its velocity from that of the macromolecule (2). Consequently $S_{20,w}$ is measured at several concentrations and extrapolated to infinite dilution to give $S_{20,w}^{\circ}$ (Figure 6.3).

$$S = S^0(1 + kc) \qquad (6-14)$$

The constant $k$ is equal to the slope and is always positive for nondissociating systems.

For dissociating systems, the slope is often negative, since lower concentrations of protein would tend to be more dissociated and consequently have a lower average sedimentation constant. This depends on the type of reaction and whether it is rapidly reversible.

The normal (positive slope) concentration dependence is responsible for the

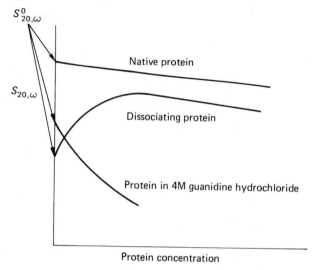

**Figure 6.3** Concentration dependence of sedimentation constants. Note the effect of the guanidine hydrochloride, which converts proteins to random coils, and of rapid reversible dissociation.

"Johnston-Ogston effect" (the apparent change in ratios of concentrations of a mixture of sedimenting proteins upon dilution) and the "self-sharpening" of solute boundaries. Macromolecules at the tailing edge of a sedimenting boundary are more dilute, so they sediment faster and tend to catch up to the less diluted macromolecule solution.

The concentration dependence is higher for macromolecules which are more asymmetric. It almost seems to be a function of the ratio of surface area to molecular volume.

### 6.2c Sedimentation Equilibrium

#### (i) Introduction

The measurement of the molecular weight of a macromolecule is fundamental to its characterization. The discovery of subunit structure of proteins has depended on accurate molecular weight measurement. Sedimentation equilibrium methods are the most reliable and most versatile methods available. The versatility of the method comes from the range of centrifugal force that can be applied with the analytical ultracentrifuge, about $1 \times 10^2$ to $2.5 \times 10^5 \times g$, permitting accurate molecular weight determinations over the range from several million to several hundred daltons.

#### (ii) Theory of sedimentation equilibrium

Macromolecules in a solvent are placed in a centrifugal field until an equilibrium concentration gradient is established when the transport of solute

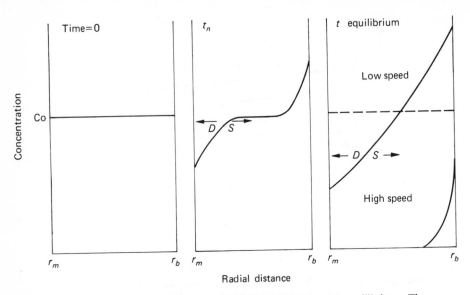

**Figure 6.4** Schematic representation of attainment of equilibrium. The arrows represent the predominant solute flows due to sedimentation and diffusion. From *Fractions*, No. 1 (1967), by K. E. Van Holde and reproduced by permission of Beckman Instruments Co.

due to sedimentation is in equilibrium with the reverse transport due to diffusion (see Figure 6.4). The flow of solute due to sedimentation, $J_s$, in a centrifugal field in terms of grams of solute crossing a unit area in unit time, would be the velocity of the molecules, $V$, times their concentration, $c(r)$:

$$J_s = Vc(r) = S\omega^2 rc(r) \tag{6–15}$$

where $S$ is the sedimentation coefficient, $\omega$ the angular velocity (rpm·$2\pi/60$), $r$ the radial distance, and $c(r)$ is the solute concentration at any point $r$.

Once sedimentation begins, concentration gradients are set up which will be opposed by diffusion. The flow due to diffusion, $J_D$, is given by Fick's first law:

$$J_D = -D\frac{dc(r)}{dr} \tag{6–16}$$

where $D$ is the diffusion coefficient. These two opposing processes will continue to change the distribution of the solute molecules until the back flow due to diffusion balances the flow due to sedimentation at every point in the solution column, and the net flow is zero:

$$J_s + J_D = S\omega^2 rc(r) - D\frac{dc(r)}{dr} = 0 \tag{6–17}$$

Svedberg's and Einstein's laws are, respectively:

$$S = \frac{M(1 - \bar{V}\rho)}{Nf}; \qquad D = \frac{RT}{Nf} \qquad (6\text{--}5;\ 6\text{--}18)$$

where $R$ is the gas constant ($8.314 \times 10^7$ ergs $^\circ C^{-1}$ mole$^{-1}$), and $T$ is the absolute temperature. If Equations 6-5 and 6-18 are substituted on the right side of Equation 6-17, and the terms rearranged, one has an expression that describes the concentration gradient at any point in the centrifuge cell.

$$\frac{M(1 - \bar{V}\rho)\omega^2 rc(r)}{f} - \frac{RT}{f}\frac{dc(r)}{dr} = 0 \qquad (6\text{--}19)$$

Or

$$\frac{1}{c(r)}\frac{dc(r)}{dr} = \frac{M(1 - \bar{V}\rho)\omega^2 r}{RT} \qquad (6\text{--}20)$$

In order to have the equation in terms of concentration instead of the concentration gradient, it is necessary to integrate from $r_m$, the radial position of the top (inside) of the solution column (the "meniscus"), to any point $r$. If the solute is homogeneous, $M$ is not a function of $r$, and we have:

$$\int_{rm}^{r}\frac{1}{c(r)}dc(r) = \frac{M(1 - \bar{V}\rho)\omega^2}{RT}\int_{rm}^{r} r\,dr \qquad (6\text{--}21)$$

$$\ln\frac{c(r)}{c(r_m)} = \frac{M(1 - \bar{V}\rho)\omega^2}{RT} \cdot \frac{r^2 - r_m^2}{2} \qquad (6\text{--}22)$$

Thus the concentration distribution at sedimentation equilibrium is an exponential function and the molecular weight can be obtained from the slope of a graph of $\ln c$ versus $r^2$ (9).

$$M_w = \frac{2RT}{(1 - \bar{V}\rho)\omega^2}\frac{d(\ln c)}{d(r^2)} = \frac{4.606\ RT}{(1 - \bar{V}\rho)\omega^2}\frac{d(\log_{10}C)}{d(r^2)} \qquad (6\text{--}23)$$

Another useful expression is:

$$M_w = \frac{2RT}{(1 - \bar{V}\rho)\omega^2} \cdot \frac{C_b - C_m}{C_0(r_b^2 - r_m^2)} \qquad (6\text{--}24)$$

where $C_0$ is the initial concentration, $C_b$ the concentration at the bottom or base of the solution, and $C_m$ the concentration at the top or meniscus. If $dc/dr$ and $c(r)$ or some value proportional to $c(r)$ or $C_0$ is known at any point in the cell, then the molecular weight can be obtained.

This is the main technical problem in molecular weight determination using schlieren and interference optics. Absorption optics using the monochromator-photoelectric scanner gives concentrations—as absorbencies—

directly throughout the cell. The method of determining an absolute concentration at some point in the cell depends to an extent on whether schlieren or interference optics are used.

Schlieren optics measures concentration gradients and requires initial concentrations in the range of 5 to 25 mg/ml. Numerical integration must be done to convert concentration gradients to concentrations. Determination of $C_0$ is usually done by a second experiment using a synthetic boundary cell, or in a sedimentation velocity experiment (see 6-4c).

Interference optics measures relative concentrations with higher precision, and lower concentrations of protein are used (2 mg/ml or less). After equilibrium has been reached at a relatively low rotor speed, the concentration at the meniscus can be measured by overspeeding the centrifuge so that the protein begins sedimenting. The drop in protein concentration ("fringe drop") at the meniscus is followed by taking pictures until the concentration is zero (10). The technique developed by D. A. Yphantis (9) involves running the centrifuge at a high speed to begin with, so that the protein concentration at the meniscus is zero and all the protein is packed into a narrow layer at the bottom of the cell. The concentrations throughout the cell are then proportional to the refractive index difference between the meniscus region and the other points in the cell.

One technique, called the Archibald or "approach to equilibrium" method, depends on the fact that the net transport of solute across the base and meniscus of the solution is zero at all times (11,12). This particular method is not now used as much as the conventional equilibrium methods, since it is less accurate and requires more material and more work, and will not be covered further.

Measurements of heterogeneous protein mixtures involving refractive indices generally give weight average molecular weights (3). The weight average molecular weight is symbolized by $M_w$:

$$M_w = \frac{\sum c_i M_i}{\sum c_i} \qquad (6-25)$$

where $c_i$ and $M_i$ are the concentration in grams/ml and molecular weight of the $i$th species of a mixture of proteins. For a homogeneous protein, $M_w = M_i = M$.

### 6.2d   Measuring Diffusion Constants (13)

Of the parameters of a macromolecule, the frictional coefficient, $f$, is of interest since it gives a direct estimate of actual effective size of the macromolecule. $f$ can be calculated from the diffusion constant of the macromolecule (Equation 6-18). This can also be measured in the ultracentrifuge.

A synthetic boundary cell and low rotor speeds are used, the latter so that the velocity of diffusion is much greater than the velocity of sedimentation. The cell is used to form the solute-solvent boundary, which spreads with time because of diffusion.

The theory of diffusion is based on Fick's equation:

$$\left(\frac{\partial c}{\partial t}\right)_x = D\left(\frac{\partial^2 c}{\partial x^2}\right)_t \tag{6-26}$$

When solved, Fick's equation yields:

$$C = \frac{C_0}{2}\left[1 - \frac{2}{\sqrt{\pi}}\int_0^{x/2\sqrt{Dt}} e^{-y^2}dy\right] \tag{6-27}$$

This is essentially a probability ("Gaussian") integral. If we use schlieren optics, we obtain the derivative:

$$\left(\frac{dc}{dx}\right)_t = \frac{C_0}{\sqrt{4\pi Dt}}\exp\left[\frac{-x^2}{4Dt}\right] \tag{6-28}$$

At $x = 0$

$$\left(\frac{dc}{dx}\right)_0 = \frac{C_0}{2(\pi Dt)^{1/2}} = Y \tag{6-29}$$

which is proportional to the height of the boundary peak ($Y$).

Experimentally, we call the area under the boundary (schlieren) peak $A_s$:

$$A_s = \int \left(\frac{dc}{dx}\right)dx = C_0 \tag{6-30}$$

In that case,

$$\frac{A_s}{2Y} = (\pi Dt)^{1/2};$$

$$\left(\frac{A_s}{Y}\right)^2 = 4\pi D(t - t_0)\left(\frac{1}{mx}\right)^2 \tag{6-31}$$

Here $mx$ is the magnification of the optical system in the $x$ ($r$) direction.

The time ($t - t_0$) is calculated from the moment the boundary has formed in the synthetic boundary cell ($t_0$). Pictures of the spreading boundary are taken at known times thereafter, as soon as the boundary has widened enough to allow calculation of the area under the schlieren peak.

### 6.2e Measurement of Extinction Coefficients (14)

The extinction coefficient of a macromolecule is invaluable experimentally. The best way to determine concentrations of macromolecules is from the dry weight, but sometimes this is not practicable. An easy and fairly reliable method of measuring the extinction coefficient involves using the ultracentrifuge as a differential refractometer. A protein solution is dialyzed thoroughly against the solvent, the solvent is layered onto the protein solution with a synthetic boundary cell using interference optics, and pictures are taken of the diffusing boundary. The protein concentration is calculated from its refractive index increment—the number of fringes counted across the boundary—and compared with the absorption of the protein solution at some wavelength, usually 280 nm. Babul and Stellwagen (14) claim an accuracy of $\pm 3\%$ in determinations of extinction coefficients.

This method can be employed because the refractive indices of "simple" proteins—those containing no large amounts of lipid, carbohydrate or nucleic acid or any chromophore absorbing in the visible region—are essentially the same (14). See Chapter 7.

### 6.2f  Sedimentation Equilibrium in a Density Gradient (15-17):
### "Isopycnic Gradient Centrifugation"

This method allows a separation and characterization of materials on the basis of their buoyant density. It was applied with great success to the problem of DNA replication by Meselson et al. (15), and to the identification of species-specific DNA (see reference 16). Applications of the method to proteins have not been extensive.

Consider the situation where a high concentration of low molecular weight salt and a low concentration of macromolecule are placed in a centrifuge cell and spun at a high angular velocity until the mixture is at equilibrium. The equilibrium distribution of the salt produces a concentration gradient in the solution, resulting in a continuously increasing density from the meniscus to the base of the cell. The initial concentration of the low molecular weight solute, the centrifugal field strength, and the length of the liquid column may be chosen such that at equilibrium the density gradient $(d\rho/dr)$ encompasses the effective density of the macromolecular species. The centrifugal force will also redistribute the macromolecules into a region of the density gradient where the sum of the forces acting on them is zero, where the effective density of the molecule equals the density of the solution in that region.

With homogeneous macromolecules a nearly Gaussian concentration distribution is formed about the band center whose thickness is determined by the steepness of the density gradient and the molecular weight of the macromolecule. The standard deviation, $\sigma^2$, of the Gaussian distribution (ignoring preferential interactions with the concentrated salt) is:

$$\sigma^2 = \frac{RT}{M_0(d\rho/dr)_{\text{eff},0}\bar{V}_0\omega^2 r_0} \qquad (6\text{-}32)$$

In the equation, $(d\rho/dr)_{eff}$ is the effective density gradient, $M_0$ is the molecular weight and $r_0$ is the radial position of the band center. The subscript zero indicates that the quantities are those at band center, and $\bar{V}_0$ is equal to $1/\rho_0$, the reciprocal of the buoyant density at band center.

## 6.3  INSTRUMENTATION

### 6.3a  The Analytical Ultracentrifuge (18)

An analytical ultracentrifuge must have a system for evacuating the rotor chamber, apparatus for keeping the rotor at a constant temperature (the observed sedimentation rate of most proteins in aqueous solvents changes 2% per degree), apparatus for keeping the rotor speed at a constant pre-selected speed (sedimentation is proportional to the square of the rotor speed) and, for analysis of sedimentation patterns, an optical system and device for recording the patterns. Because of the enormous pressures exerted (up to 400,000 times gravity), the cells holding the solutions to be analyzed and the rotors holding the cells must be strongly constructed.

While several companies sell analytical ultracentrifuges, the discussion in this Chapter will center around the Beckman-Spinco "Model E," as the authors are most familiar with it.

### (i)  Vacuum pump and diffusion pump

Low air pressure is necessary to prevent frictional heating of the rotor at high speeds. A conventional mechanical vacuum pump is used to remove most of the air. The diffusion pump contains butylpthalate, which traps the remaining air when heated and vaporized. Water-cooled baffles condense and cool the oil, releasing the air, and the vacuum pump pulls the air out of the system.

### (ii)  Temperature control: refrigeration system

The refrigeration unit is a conventional compressor, using Freon 12 as the refrigerant. A back pressure gauge is used to control the release of liquid Freon into the coils in the chamber and provide a constant rate of cooling. A heating system is run in opposition to the cooling system, giving a resulting temperature range of 0-40°C. A special accessory is needed for higher temperatures.

### (iii)  RTIC (Rotor Temperature Indicating Control)

A 10,000 ohm thermistor is imbedded in the base of the rotor. A needle in the base of the rotor makes contact with a cup containing mercury. The resistance of the thermistor is held constant by heating against the cooling of the refrigeration. The thermistor resistance is calibrated against known temperatures. The temperature can be read to 0.02°C and controlled to 0.1°C.

### (iv) Speed control: the drive unit

The drive unit is the heart of the ultracentrifuge. Until its development analytical ultracentrifugation did not flourish. A 2 h.p. electric motor, air and water cooled, with a $3\frac{1}{2}$:1 gear ratio is used. It is designed to minimize vibration, precession, and heat transfer, any of which could cause convection in the cell in the rotor and lead to erroneous results. More wear and tear occurs in short high-speed runs than long continuous runs.

Drive units must be replaced periodically: the seal at the bottom of the unit wears out, letting oil into the chamber and coating the lenses.

### (v) The speed control unit

Extreme accuracy in speed control is essential since the force is proportional to the velocity squared. The unit consists of a direct current tach generator attached to the drive (the voltage produced is linear with rotor speed); a control unit which puts out a constant voltage representing a preselected speed; and a microammeter, which detects a difference in current between the tach generator and the reference voltage and increases or decreases the drive current accordingly. Accuracy and sensitivity is $\pm 0.02\%$ (instantaneous) and $\pm 0.01\%$ (continuous).

### (vi) Optical and recording system: the light source

The light source for schlieren and interference optical systems is a General Electric A-H6 1000 watt mercury arc lamp, water-cooled. A Wratten 77A band-pass filter is used to isolate the "green" (546 nm) emission line. A mercury arc lamp is used because such lamps are cheap and their emission energy is concentrated into a few wavelengths ("lines") and so give high intensity monochromatic light (see Chapter 7).

Low energy of the available light sources limits the sensitivity and accuracy of both optical systems. The schlieren slit is 0.008 inches wide. It is apparent that the finer the slit the more accurate and sensitive the schlieren pattern would be, but methods of recording such low energy levels are not available.

The width of the interference source slit (0.002 inches) determines the fringe visibility; the finer the slit, the greater the resolution. Also, as the slits in the upper condensing lens mask (see 6-3a (viii)) are narrowed, the number of fringes in the diffraction pattern increases, and therefore a greater range of concentrations can be measured. However, cutting the width of either lowers the light level and makes it difficult to record the pattern.

The light source for absorption optics is a 100 watt mercury-Xenon arc lamp, which has a fairly continuous output over a wide spectral range, collimated by a DU prism monochromator. The low light level produced in the ultraviolet region causes an increase in the voltage used on the photomultiplier, which produces increased noise and lower accuracy. A lamp that puts out high energy at 230-240 nm is needed, since extinction coefficients are high in this range for most macromolecules. Recently, power supplies have been developed which greatly reduce the instability of the Xenon arc (19). This may make more accurate measurements possible using the scanning absorption optical system.

### (vii) Optics

The three optical systems used in the ultracentrifuge are absorption, schlieren, and interference optics.

*1. Absorption.* Measurement of light transmission can be made by photographing the cell, which records the percent transmitted light. The plates are read with a densitometer. Unfortunately, extensive calibration using standards must be done, and the precision is poor.

Measurement can also be made with a photocell, by scanning the cell with a small slit and recording the percent transmission directly, or converting it to optical density electronically (20). This system is potentially the best, since it offers the advantages of higher sensitivity and selectivity and gives data which can be directly converted to concentrations.

*2. Schlieren optics.* This optical system is probably the one most commonly used. If you were to place a beaker, half filled with say 50% sucrose which is overlayered with water, between yourself and the light, the zone between the sucrose and water would appear dark. This dark line is caused by the deviation of light passing through a region of changing refractive index (see Figure 6.5).

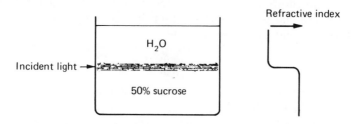

**Figure 6.5** "Schlieren" produced by a region of changing refractive index.

This can be rationalized using Huygen's principle (see Figure 6.6). This says that the propagation of light by matter can be treated as production of many wave fronts which are continually reemitted. The common fronts of these are at right angles to the direction of propagation.

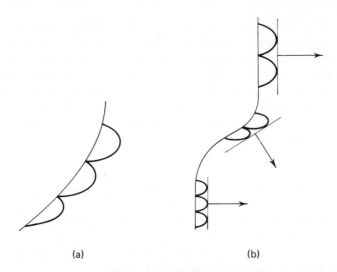

(a)                                                                (b)

**Figure 6.6**   Rationalization of the deviation of light produced by a refractive index gradient, using Huygen's principle. (a) The upper
part of the wave front is represented as advancing faster because
of the lower refractive index, while the lower part of the wave
front is retarded because of the higher refractive index. (b) In
regions of constant refractive index, all parts of each wave front
are represented as advancing at the same velocity, though the
velocity is lower for wavefronts in the sucrose. Light is deflected
(refracted) by the concentration gradient. The angle of deflection
is proportional to the steepness of the change in refractive index,
that is, the concentration gradient, since higher concentration
gradients will deform the wave fronts more.

The ultracentrifuge is designed to be able to produce concentration gradients (refractive index gradients) even in solutions of substances of low molecular weight, as shown in Figure 6.7. Since the degree of deflection ($\theta$) of the
light is proportional to the concentration gradient (the refractive index is
usually proportional to concentration),

$$\theta = k\frac{dc}{dr} \tag{6–33}$$

where $r$ is the distance out from the center of rotation. The schlieren optical
system is described in Figure 6.8.

Undeflected light (i.e., light from areas in the cell where there is no concentration gradient) is focused on and passes through the center of the slit. This
undeflected light is then focused on the photographic plate as a vertical line.
Light deflected by a concentration gradient in the cell strikes the diaphragm
with slit (a phaseplate with a thin line across it is now used) as a horizontal
plane of light. The steeper the concentration gradient, the farther off the
center (vertical) line of the diaphragm that the deflected light strikes. Note

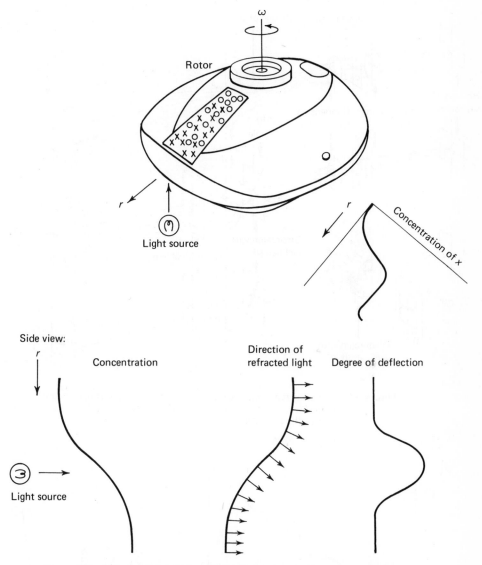

**Figure 6.7**  Showing how the angular acceleration of the ultracentrifuge rotor (Type AN-D) produces concentration gradients in the cell in the rotor. (Adapted from an original published by the Beckman Instruments Co. and used by permission of the company.)

that the slit angle ("bar angle") selects the angle of deviation of the transmitted deflected light hitting the plate. In other words, it amplifies the size of the schlieren peaks.

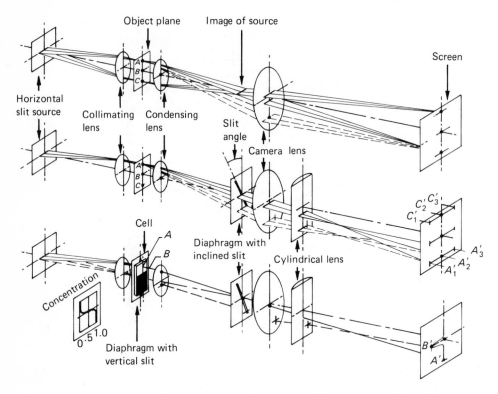

**Figure 6.8** The schlieren optical system used in the ultracentrifuge. From E. G. Pickels, *Chem. Revs.*, **30**, 341 (1942). Reprinted by permission of the publisher.

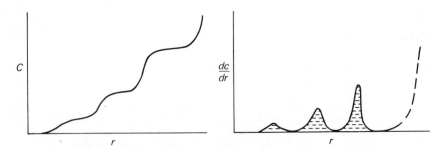

**Figure 6.9** Concentration gradients and schlieren pattern produced by three substances at different concentrations sedimenting at different rates. The effects of different rates of diffusion and the Johnston–Ogston effect are ignored.

180

If three substances are sedimenting at different rates, three concentration gradients and three schlieren peaks would be seen as shown in Figure 6.9.

*3. Interference optics.* This optical system is also quite generally used. In order to have interference, the two light beams must be coherent. In practice, this means that they must come from the same source, and be monochromatic.

The ultracentrifuge uses the Rayleigh system. The interference slit is set parallel to the solution column and to the axis of the cylindrical lens. This insures that one gets a reliable point to point relation from the cell to the photographic plate. See Figure 6.10.

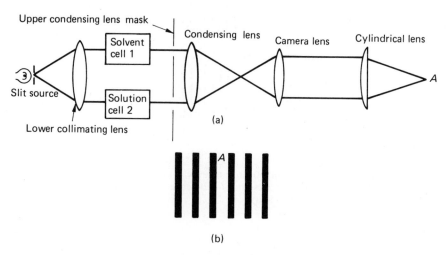

(a)

(b)

**Figure 6.10**  (a) Schematic representation of the Rayleigh interference optical system. (b) Optical pattern formed by interference of light from twin cells, assuming no refractive index gradient.

The optical path from the slit source to photographic plate is equal whether one goes through either cell, providing the refractive index (apparent speed of light) is the same in both cells. The emerging waves reinforce each other and a bright line is formed perpendicular to the paper at point $A$. To the left and right are interference fringes caused by differences in light path length.

Now consider the case shown in Figure 6.11 in which the concentration (refractive index) is constant in cell 1 but changes with the distance $r$ in cell 2. The path length (time of traverse) to band $A$ through cell 2 is increased by an amount proportional to $\Delta n$ over region $\Delta r$:

$$p(r_1) = hn_1 + C \qquad (6\text{--}34)$$

where $p$ is the path length, $h$ the cell width, and $C$ the constant for the path length outside of the cell.

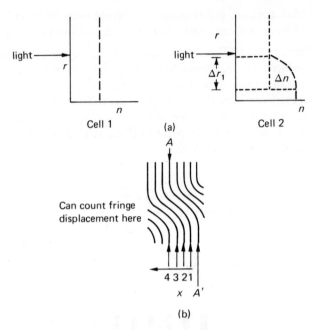

light⟶

$r$

$n$

Cell 1

(a)

light⟶

$r$

$\Delta r_1$

$\Delta n$

$n$

Cell 2

$A$

Can count fringe
displacement here

4 3 2 1

$x$   $A'$

(b)

**Figure 6.11**    (a) Change in refractive index with radial position in one of two
twin cells. (b) Resulting interference pattern produced by solute-
solution boundary.

Similarly

$$p(r_2) = hn_2 + C$$

and

$$p(r_1) - p(r_2) = h(n_2 - n_1)$$

Or

$$p = h\,\Delta n \qquad\qquad (6\text{--}35)$$

The actual magnitude of $C$ is not important.

The band $A$ will be displaced by an amount which will maintain the path
length, the condition for constructive interference (see Figure 6.11).

$$x = kh\,\Delta n \qquad\qquad (6\text{--}36)$$

The distance between fringes is also a function of the wavelength of the
light:

$$x = kj\lambda \qquad\qquad (6\text{--}37)$$

where $\lambda$ is the wavelength and $j$ the number of fringes.

$$h \, \Delta n = j\lambda$$

$$\Delta n = \frac{j\lambda}{h} \tag{6-38}$$

Interference optics yields the difference in refractive index $\Delta n(r)$ between the solution and the reference solvent, which is measured by the fringe displacement. This is proportional to concentration, not concentration gradient. The fringe difference between the solution and solvent at point $r$ relative to that of some other point is

$$j(r) = \frac{h \, \Delta n(r)}{\lambda} \tag{6-39}$$

where $h$ equals the centerpiece thickness (usually 12 mm) and $\lambda$ the wavelength of light used (546-550 nm). For a fringe displacement of one $(j - 1)$, $\Delta n$ corresponds to $4.6 \times 10^{-5}$. The specific refractive increment for most proteins, $dn/dc$, is $1.86 \times 10^{-3}/\text{mg/ml}$, or one fringe for a concentration of 0.25 mg/ml. One fringe from Model E interference optics is about 275 microns wide, and a novice can read the photographic plate to 0.05 of a fringe; that is, to a minimum concentration difference of about $10\mu g/\text{ml}$. This is considerably better sensitivity than can be obtained using schlieren optics.

### (viii) The camera

This is presently the most reliable way available to record the optical patterns. The chief difficulty in the optical alignment is that the three-dimensional cell must be focused on a flat surface.

The camera is automatic sequencing, with pictures taken at intervals of from 2 minutes to 12 hours and 48 minutes. Exposure times are from 0–40 seconds, or with the extended exposure timer, for 1–20 minutes.

The image can be recorded on $2 \times 10$ inch glass plates. These are used because of their durability and dimensional stability, which is essential if accurate measurements are to be made from the plate.

The schlieren exposure time must be short or the boundary will move during exposure. A contrast as high as possible is used so the plates can be read accurately and easily. Interference plates must also be of high contrast, since interference conditions cut the light by half, and of high resolution, since they must be read to $10\mu$. Long exposures using interference optics may lead to inaccuracy if precession[2] or vibrations are occurring, in which case the drive should be replaced or source of the mechanical vibrations eliminated.

---

[2]The period of precession is usually long and not easily observed, but for accurate work should be taken into consideration.

### (ix) Rotors and cells: rotors

These hold the cell during the run. Several types are available. The AN-D is most commonly used. It has two holes and a top speed of 60,000 rpm (260,000 × g). It can take 1.5-12 mm centerpieces. The AN-H is the same as the AN-D, but is made of titanium. It can go to 68,000 rpm (340,000 × g). The AN-J is used for low speed equilibrium runs. It weighs about 22 pounds and has a top speed of 18,000 rpm. It takes 1.5-30 mm centerpieces. Other rotors available are used with longer (18–30 mm) optical path length cells (AN-E) and for use with three or four cells (AN-F).

All rotors save the AN-H are made of an aluminum alloy, because of its high strength to density ratio.

### (x) Rotors and cells: cells

The cell components other than windows, gaskets, etc., are made of aluminum, but centerpieces are also made of Kel-F or filled Epon.

Aluminum components are durable, but Kel-F centerpieces have a higher resistance to solvents (particularly bases). Epon is of high strength and is

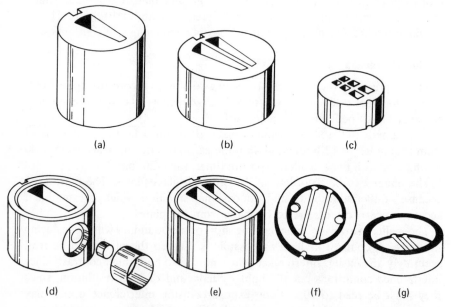

**Figure 6.12**   (a) Single sector centerpiece. (b) Double sector centerpiece. (c) Yphantis multichannel short column equilibrium centerpiece. (d) Valve-type synthetic boundary centerpiece. (e) Capillary-type double sector synthetic boundary centerpiece. (f) Upper window holder for interference double sector cell. (g) Upper (and lower) window holder for schlieren double sector cell. Note the narrow slits on the former. (Redrawn by permission of Beckman Instruments Co.)

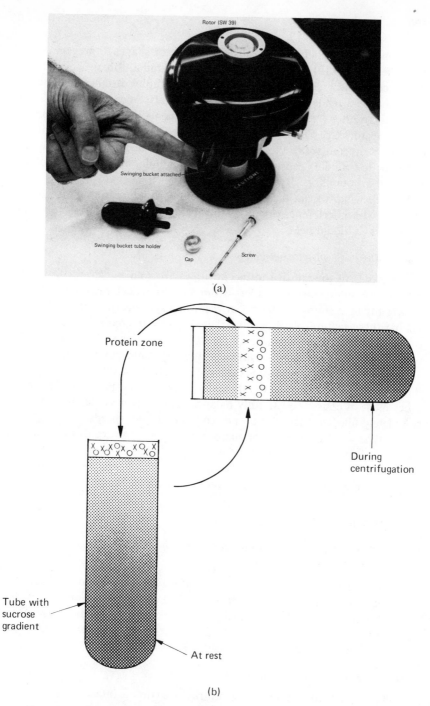

Figure 6.13 (a) A "swinging bucket" rotor. (b) Representation of a tube containing a sucrose density gradient at rest and under a centrifugal force, in a swinging bucket rotor.

easily machined. Single sector cells have 4° sector shapes to prevent convection (see Figure 6.12) (the sedimenting force is a radial force). Since the area of the centerpiece continuously increases with radial distance, sedimenting solute has less tendency to strike the walls. It also means the sedimenting material is diluted ("radial dilution") as it sediments. Double sector cells have two $2\frac{1}{2}°$ sectors (see Figure 6.12). These are used in experiments with interference or absorption optics and with schlieren optics when a baseline is required.

Synthetic boundary cells provide for layering buffer onto a protein solution to create a boundary.

The cell windows are of quartz or synthetic sapphire. Sapphire windows are used with interference optics, as they have less tendency to distort under a high gravitational field.

### 6.3b   The Preparative Ultracentrifuge

A number of preparative ultracentrifuges are sold by several firms. The preparative ultracentrifuge differs from the analytical instrument by not having any light source or optical or recording systems (though models containing these features have been marketed). Since the sedimentation process is not observed, more conventional rotors and centrifuge tubes are used. The preparative ultracentrifuge is consequently more compact and simpler to run.

However, the high rotor speeds still require a high vacuum, and this means the centrifuge tubes must be tightly sealed to prevent evaporation. The sealed tubes must of course be carefully balanced and in some cases they must be completely filled so that they do not collapse at high speeds.

One type of rotor is used for sucrose density gradient centrifugation: the "swinging bucket" rotor, whose tube holders ("buckets") can swing out to a horizontal position as the rotor speed increases. This keeps the sucrose gradient from mixing during centrifugation, as shown in Figure 6.13.

## 6.4   PRACTICAL APPLICATIONS

### 6.4a   General Operating Instructions for the Analytical Ultracentrifuge ("Model E")

These are intended as a general check-off list of the operations involved in analytical ultracentrifugation experiments. Detailed information is provided in the Model E instruction manual (18), supplementing bulletins, and from the literature. No experiments should be attempted without an adequate knowledge of the methods and procedures involved.

#### (i) Preparations prior to centrifugation

1. Reserve adequate time on the centrifuge to do the intended experiment.

2. Have the solutions ready, in the amounts and concentrations needed (see 6.4b). In most cases the solution should be well dialyzed, especially when

interference or absorption optics are used, since the difference in refractive indices (or absorbancies) of solution and solvent are being measured.

3. The rotor and chamber temperature should be equilibrated at the operating temperature in advance. Cooling the rotor and chamber from 25° to 4°C requires about three hours. The An-D rotor itself will cool in about one hour from room temperature to less than 4°C if placed in a −12°C room or a freezer. Generally for a rotor operating temperature of 20°C, the refrigeration back pressure gauge should be set to about 50°F; for an operating temperature of 4°C, the setting should be 15°F.

4. Selection and assembly of cell.

a. Twelve mm centerpieces provide optimum optical conditions and the least amount of parallax, although centerpieces of 1.5 to 30 mm are available for specialized purposes.

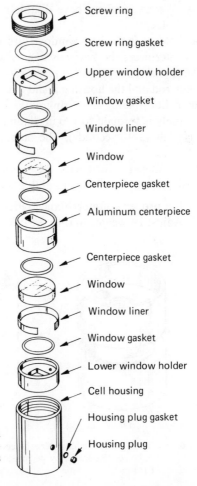

Screw ring

Screw ring gasket

Upper window holder

Window gasket

Window liner

Window

Centerpiece gasket

Aluminum centerpiece

Centerpiece gasket

Window

Window liner

Window gasket

Lower window holder

Cell housing

Housing plug gasket

Housing plug

**Figure 6.14**
Components used in assembling a single sector cell. (Redrawn by permission of the Beckman Instruments Co.)

b. Cell assembly is described in Figure 6.14.

The cell assembly components must be absolutely clean (this is especially true of the windows and centerpieces), or else leakage, optical distortion, or convection can occur. Fingerprints, lint, wrinkled gaskets, etc., can also produce these effects.

The red polyethylene centerpiece gaskets are used only on the aluminum centerpieces.

Place a window gasket in the bottom of each window holder, then place the liner in the window holder. With schlieren optics, Bakelite liners are used, and the gap between the ends of the liner must be away from the arrow or notch on the window and away from the keyway in the holder. Black Hycar liners are used for interference or absorption work; these fit snugly around the windows. Place the window in the holder so the arrow points down in the holder and is toward the keyway. Windows with notches go up and toward the keyway. Wedge windows have double notches and must be used in the *red* window holders.

The two holders with their windows are then stacked with the windows facing the centerpiece, the keyways *aligned* vertically, and the *part numbers on the holders and centerpieces all inverted.* The stack is held in place with a linen Bakelite rod and the housing is slid over the top of the stack. The lightly greased (with Lubriplate) screw ring gasket and screw ring are inserted and the cell assembly is torqued to 120 in.-lb and held there until no perceivable tightening is felt (3–10 seconds). The outside of the housing can be lightly greased.

The cell can then be filled with the solutions, using syringes, and sealed with two red polyethylene housing plug gaskets and a brass housing plug.

c. Select the proper counterbalance (Figure 6.15) for the type of optics being used. Adjust its weight to 0 to 0.5 g *lighter* than the completely assem-

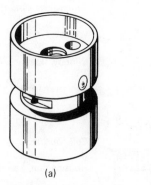

(a)                                      (b)

**Figure 6.15** (a) A schlieren counterbalance. (b) An interference counterbalance. (Redrawn by permission of the Beckman Instruments Co.)

bled and filled cell by changing the weight of the counterbalance with the brass and aluminum weights provided. Two single sector cells and no counterbalance can be used, but the reference hole plugs must be removed from the rotor.

5. Aligning cell and counterbalance in the rotor.

a. Place the cell in the number one cell hole of the rotor with the screw ring pointed up and the housing plug to the *inside*. The cell should be inserted to the bottom of the rotor holes with thumb only. If thumb pressure is not sufficient, there could be a temperature difference between cell and rotor, a burr, dirt, or an unlubricated rotor hole or cell, or an out-of-round cell, due to overtightening the housing plug or warping of the cell housing. Conversely, if the cell and rotor are temperature equilibrated, a loose cell can be tightened by cautiously adjusting the tension on the housing plug.

b. Place the counterbalance in the number two hole with the locking plug to the left (the "2" is upright). Those counterbalances with an arrow have the arrow pointing centrifugally (outward).

c. To align the cell and counterbalance, place the cell aligning tool in the top of the cell, invert the rotor and rotate the cell (or counterbalance) until the scribe lines of the rotor and cell are coincident. Be sure the housing plug still points to the inside of the rotor; that is, do not rotate the cell 180°.

**Note:** When handling the rotor, be careful: the thermistor needle projecting out from the bottom of the rotor is both sharp and easily damaged (Figure 6.16).

### (ii) Steps prior to acceleration

1. Turn off the vacuum pump and open the air valve completely, wait one minute or so for the chamber to come to atmospheric pressure, then place the vacuum chamber switch in the "open" position. The chamber will stop automatically when completely open.

2. Adjust the light source according to the optics to be used (see reference 18, section 3-11, numbers 3-5).

In the center of the top of the rotor chamber is a metal fork under a metal socket (see Figure 6.17). The knob at the top of the rotor has a projecting lip. This lip fits over the fork, and the socket can be pulled down part way towards the top of the rotor. The rotor is lifted up to meet the socket and is screwed on: the knurled projection of the socket is turned outward (using the right hand) until it is finger tight and the rotor is hanging suspended.

The U-shaped metal plate with the edges turned down is a wrench. It is inserted over the socket above the rotor to keep it from turning while the rotor is tightened further. This tightening is done by grasping the ends of the

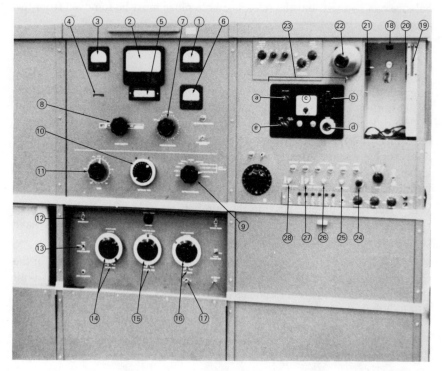

**Figure 6.16** Operating controls of the Model E ultracentrifuge. The numbers refer to: (1) drive current meter; (2) tachometer; (3) drive voltage meter; (4) odometer; (5) microammeter; (6) vacuum gauge; (7) voltage control; (8) speed selector; (9) exposure interval; (10) exposure time; (11) centrifugation time; (12) refrigeration switch; (13) chamber switch; (14) air valve with vacuum pump switch; (15) diffusion pump water valve with diffusion pump switch; (16) water valve for light source with light source switch; (17) Hi-Lo switch for light source; (18) phaseplate dial; (19) plate insertion slot (camera); (20) frame counter; (21) phase-plate adjust; (22) viewing ocular; (23) RTIC system with (a) zero adjust, (b) range switch, (c) control meter, (d) balance potentiometer, and (e) control switch; (24) view mirror knob; (25) view shutter button; (26) plate drive; (27) plate drive switch; and (28) automatic photo switch.

rotor with both hands and turning it (inward with the right hand) until the rotor seems *fairly* tight. Do not overtighten or you will have trouble getting the rotor off after the experiment.

The chamber is closed using the vacuum chamber switch. The vacuum pump is started by throwing the switch after the notched knob has been turned fully clockwise, so that the switch can be flipped into the notch.

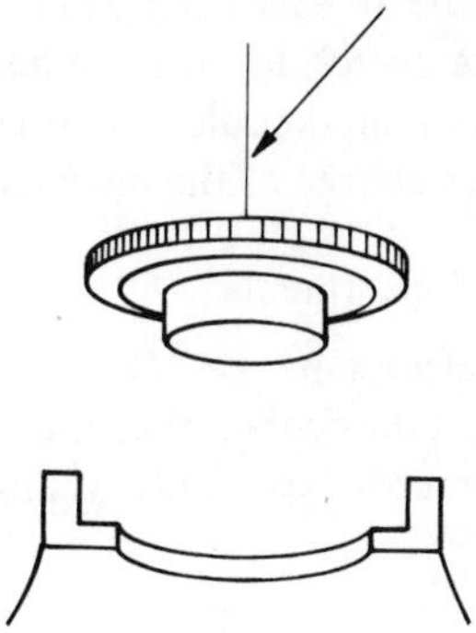

Strong wire (not visible) connecting
the drive motor to the rotor

**Figure 6.17**
Attachment of rotor to drive shaft.

3. RTIC (rotor temperature indicating control), (18, section 3.13, number 7).

Temperature calibrations are made for the individual rotors, since the thermistor in the base of each rotor is different. All of the thermistors, however, give about 50 RTIC units per degree centigrade. A graph showing what a given RTIC value is in terms of degrees centigrade should be available. If the thermistor has not been calibrated recently, the reader should do it, as its characteristics change with time.

There are four positions for the RTIC switch: "off," "indicate," "zero adjust," and "regulate." After the unit is turned on and has stabilized, set the switch to "zero adjust" and turn the "zero adjust" potentiometer (black knob) until the pointer is in the middle of the scale. Then turn the RTIC switch to "indicate." Turn the "balance" potentiometer until the pointer is centered. That is the rotor temperature, in RTIC units. If the potentiometer cannot center the pointer, either the mercury cup is mis-spaced (the pointer always goes to the left) and no contact is being made, or you are in the wrong range. The range switch setting should be changed if the latter is the case.

To set to a desired operating temperature, get the RTIC setting from the temperature calibration plot and turn to this setting with the range switch and balance potentiometer. Set the switch to "regulate"—this activates the temperature control circuitry.

Accelerating from 0 to 60,000 rpm will lower the temperature of the rotor about $0.8°$ due to the adiabatic cooling, so if an operating temperature of $20°$ is desired, the rotor can be accelerated when it is about $20.8°$.

4. Diffusion pump and light source.

a. When the vacuum pump has lowered the chamber pressure to 250 $\mu$ (microns), the diffusion pump may be started. Open the diffusion pump valve (flow rate about 0.5 gallon/min) and put the switch to "on" (18, section 3.14, number 8).

b. Open the water valve to the light source (flow rate 1 gal/min). (The total flow rate must be at least 1.5 gal/min.) Turn the "Hi-Lo" switch to "Hi" and the light source switch to "on." When the lamp fires, put the "Hi-Lo" switch to "Lo." The lamp should fire in less than 15 seconds. If it does not, consult the person in charge of the instrument.

5. Speed control adjustments.

Place the drive selector in the "Hi" or "Lo" position, depending on the desired speed. Adjust the coarse, then fine speed selector to the desired speed. More recent instruments have only a single speed selector.

6. Photography.

a. Metalographic plates are used for schlieren optics, and II-G spectroscopic plates for interference optics. Use a Wratten series 2 safelight in the darkroom.

The plates must be loaded with the emulsion side up, which can be determined by touching a corner of the plate with a moistened finger or lips (the emulsion side is sticky). When the darkslide is closed over the plate, the top of the plate holder (the toothed side is the bottom) must be even with the bar on the end of the darkslide to get into the centrifuge camera. If it is not, the darkslide must be pulled out and flipped over so the metal bar at its end points downward.

b. Place the plate holder in the camera (reference 18, section 3.16, number 11, paragraph 3).

The plate chamber is in the upper right-hand corner of the compartment adjacent to the phaseplate angle control. It is opened by pulling the metal slide down with the small metal knob. The plate holder is inserted with the toothed side down. The middle and index fingers are wrapped around the outer end of the holder while this is done so the darkslide cannot be accidentally pulled or pushed back to prematurely expose the plate. When the plate holder meets resistance, push it in until the metal slide can be just closed over the back end of the plate holder. When it is closed, the camera is loaded.

Place the plate holder drive toggle switch in the "in" (A) position, advance the plate to "0" on the frame counter with the plate holder drive button. The frame counter is located in the camera compartment; the white numbers indicate the frame number when using schlieren or interference optics.

c. Set the exposure time by unscrewing the turned knob on the exposure time dial and turning the outer ring of the dial so that the line engraved on it is set at the proper exposure time. The latter is written on the inner part of the dial and is in seconds (reference 18, section 3.16, number 11, paragraph 4). For schlieren optics, using metalographic plates, an $8 \times 10^{-3}$ inch slit, one single sector cell with non-550 nm absorbing material in the cell and 3 minutes development in D-19, the exposure time is about 5 seconds; using one double sector cell, about 4 seconds; with two single sector cells, about 3 seconds; with two double sector cells, about 2 seconds. With interference optics, II-G plates and offset mask, use about 10-20 seconds; with the symmetrical mask, 20-30 seconds.

### (iii) Acceleration

Acceleration can begin when the vacuum is below 5 $\mu$ and the desired temperature is attained.

The voltage control knob should be at its counterclockwise limit. Set the shut down timer to "hold" or the desired length of operating time. Turn the voltage control until the applied current is 4-5 amps. Check that the rotor is turning, by observing the flashes of green light in the viewer. This can be done by pulling the view mirror knob out and pressing the view shutter button.

When the rotor speed reaches 2000-3000 rpm, the current is increased to 6-8 amps until about 10,000 rpm is reached. The current is then increased to and maintained at 12 amp until a speed of about 1000 rpm less than the desired speed is reached. The voltage control is then decreased until the applied current is only 4.5 amp. The "accelerate" pilot light[3] should go off when the selected speed is attained.

Watch the vacuum indicator while accelerating. If the pressure suddenly rises, end the run by setting the braking rate to "rapid" and returning the timer to zero. This sudden rise in pressure usually indicates a cell leak and can be observed in the viewer.

It may be necessary to increase or decrease the voltage by $\pm 2$ or $\pm 3$ volts if the "accelerate" or "overspeed" lights come on. When the speed control has stabilized, record a simultaneous odometer and time reading.

If using schlieren optics, adjust the phaseplate angle until the desired optical pattern is observed. Note that as the phaseplate angle is decreased from 90°, sensitivity is increased, but the resolution is decreased, so compromise values, usually between 50° and 80°, are used. For reproducible settings, approach the angle from the same direction.

The automatic photo sequence is initiated by setting the exposure interval switch to the desired time between pictures, flipping the plate holder drive switch to "in," turning the seconds exposure dial clockwise past the position at which the camera pilot light comes on, turning the automatic photo switch on and releasing the seconds exposure dial to rotate counterclockwise. The plate will advance and be exposed, repeating at the set time interval.

**Note:** There are only five frames to a plate, so if the plate is not changed after the fifth photo it will be doubly exposed.

---

[3] With the electronic speed control, the microammeter will be at zero.

### (iv) Shut Down

Take a simultaneous odometer and time reading, and set the braking rate to "rapid." Return the timer to zero, turn the voltage control to its counterclockwise limit, turn off the diffusion pump switch and the light source switch. *Do not* turn off the water supply to the light source or diffusion pump at this time. If an excess of air reaches the oil in the diffusion pump while it is still hot, it will oxidize. Put the RTIC on "zero adjust."

After ten minutes, and after the rotor comes to rest (the "accelerate" light comes on), turn off the water valves to the diffusion pump and light source, and open the vacuum chamber air valve. Wait about one minute to ensure the chamber is at atmospheric pressure, then open it and remove the rotor.

Close the chamber and the air valve, and turn the vacuum pump on. The centrifuge can be left in this state for short times. If it is to stand an extended length of time, the vacuum is left "on" until the chamber has warmed to room temperature, then the chamber is brought to atmospheric pressure and the entire instrument is shut down. This prevents condensate from forming on the lens.

The rotor may be placed in the refrigerator if it is to be kept at 4°C, or it may be placed in the chamber where the temperature can be regulated.

Remove the plate holders from the ultracentrifuge and develop the plates.

### (v) Developing procedure

Metalographic plates are soaked in the developer (Kodak D-19) for three minutes, II-G plates for four minutes. This reduces the silver salts on the plates. The plates are then dipped in "stop" solution (32 ml of 28% acetic acid/liter) for a few seconds, then are put into "fix" or "hypo" solution. This contains sodium thiosulfate, which dissolves the exposed silver and hardens the gelatin. Four minutes for fixing is sufficient, and the last two minutes can be done in room light.

The fixed plates are washed several hours with water to remove excess thiosulfate. Hypo clearing agent ("perma-wash") can also be used. It increases the solubility of the thiosulfate, decreasing the washing time to about five minutes.

A wetting agent can also be used. It prevents water spotting of the plates.

### (vi) Clean up procedures

Withdraw the sample from the cell, using a syringe fitted with a blunted 25 gauge, 1 inch needle. Be careful not to scratch or damage the windows or the centerpiece.

Dismantle the cell, being careful that components remain as a matched set; i.e., each window is matched to its original holder, the centerpiece remains with that particular cell assembly, etc. Thoroughly clean all components with dilute Alconox solution and a pipe cleaner. Rinse them with distilled water, acetone, ethanol, more disti'led water and blot dry. Save all gaskets and liners except the red center piece gaskets from the aluminum centerpieces. Return the cell assembly to its proper storage tray.

Clean and dry the syringes needles used, and wipe off any smudges on the rotor. If the rotor holes are dirty, clean them with a moist tissue. Police the area in which you have been working.

### 6.4b Sedimentation Velocity Experiments

#### (i) Solutions and solvents

The solvent should have an ionic strength of 0.05-0.1, and be buffered at a pH near neutrality.[4] It is preferable that it be of known density and viscosity. The protein concentration depends on the optical system to be used. For schlieren optics, use 2–15 mg/ml—the lower the molecular weight, the greater the diffusion constant and the higher the protein concentration you must use. For interference optics, use 0.05-5.0 mg/ml, and for absorption optics, about 0.2-2.5 absorbency/ml.

It is generally best to dialyze the solute thoroughly against the solvent; for experiments using a double sector cell and schlieren, interference, or absorption optics, it is mandatory.

If a single sector cell is used, about 1 ml of solute should be prepared. Double sector cells require 0.5 ml each of both solute and solvent.

#### (ii) During the run

The rotor speed should be as high as is consistent with being able to obtain several boundary positions; generally this means 60,000 rpm. After the schlieren peak pulls away from the meniscus, take five to ten pictures at suitable time intervals, and develop the plate. The phaseplate angle may have to be lowered during the run, as the schlieren peak flattens because of diffusion and radial dilution.

#### (iii) Calculations

The distance between the tip of the schlieren peak and a reference mark must be measured in each picture.[5] See Figure 6.18.

Using the comparator, align the plate on the meniscus or on the undisplaced image of the phaseplate by rotating the comparator stage. Determine

---

[4]Ideally, the pH should be no more than one unit from the isoelectric point of the protein (2, 3).

[5]We use an instrument made by Nikon, but the directions are similar for most commerical comparators.

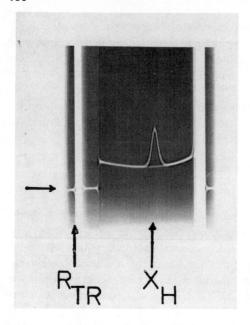

**Figure 6.18**
Measurements of a photograph of a schlieren pattern from a sedimentation velocity experiment. $X_H$ is the horizontal position of the maximum ordinate of $dn/dx$. $R_{TR}$ is the position of the inside reference hole edge. The horizontal arrow indicates the "undisplaced image of the phaseplate." The baseline is curved because of solute redistribution, but note how the use of a double sector cell compensates for this.

$X_H$ and $R_{TR}$, in comparator units (mm). Now the distances on the plate must be converted to actual physical distances from the center of rotation. Using the inside reference hole,

$$r_H = \frac{(X_H - R_{TR})}{mx} + 5.72 \text{ cm} = \text{actual distance between the protein-solvent boundary and center of rotation[6]} \qquad (6\text{-}40)$$

$mx$ is the camera lens magnification factor. It can easily be measured, but is generally 1.8–2.2, and is usually posted on the bulletin board or in the log book.[7] The outer reference hole must be used if two cells and no counterbalance are used.

Plot log $r_H$ (cm) versus time (in seconds) and determine the slope of the line.

$$S = \frac{\frac{d \ln r}{dt}}{\omega^2} = \frac{2.303}{\omega^2} \cdot \frac{d \log r_H}{dt} \qquad (6\text{-}6)$$

[6]The inside reference edge is originally 5.70 cm from the axis of rotation, and the outside edge is 7.30 cm, but the AN-D rotors will stretch 0.02 cm at 60,000 rpm. They exhibit no stretch at 30,000 rpm and below. These distances should be measured for greatest accuracy.

[7]To determine the horizontal camera lens magnification factor, a quartz disc, with scribe lines exactly 1 mm apart, is put in the rotor so the lines are at the focal plane of the camera lens, and rotated so the lines are perpendicular to the axis of the cell. The image is photographed and measured on the microcomparator.

$\omega$ is determined from your odometer and time readings. The odometer reading is in a 1000:1 ratio when the drive selector is on "Lo," and 6400:1 on "Hi" for centrifuges equipped with the mechanical speed control. The ratio is always 1000:1 on ultracentrifuges with the electronic speed control.

Sedimentation constants are expressed in units of Svedbergs ($S$):

$$1 \, S = 1 \text{ Svedberg unit} = 1 \times 10^{-13} \text{ second (cm/sec/dyne/g).}$$

**(iv) Corrections**

The above calculations give $S_{obs}$. To allow comparisons of results obtained in different solvents or at different temperatures, all sedimentation constants by convention must be corrected to the sedimentation constant theoretically obtainable in water at 20°C:

$$S_{20,w} = \frac{\eta_t}{\eta_{20}} \times \frac{\eta_c}{\eta_0} \times \frac{(1 - \bar{V}\rho_{20,w})}{(1 - \bar{V}\rho_t)} \times S_{obs} \qquad (6\text{-}41)$$

$\eta_t$ is the viscosity of water at temperature $t$; $\eta_{20}$ is the viscosity of water at 20°; $\eta_c$ is the viscosity of the buffer at a given temperature; $\eta_0$ is the viscosity of water at that temperature; $\rho_{20,w}$ is the density of water at 20°; and $\rho_t$ is the density of the buffer at temperature $t$. In terms of the expressions used in Appendix 14:

$$S_{20,w} = \frac{\eta_t}{\eta_{20}} \times \left(\frac{\eta_c}{\eta_0}\right)_t \times \frac{(1 - \bar{V}\rho_{20,w})}{(1 - \bar{V}[\rho_{t,w} + \Delta\rho_t])} \times S_{obs} \qquad (6\text{-}42)$$

*6.4c Sedimentation Velocity Measurements Using Sucrose Density Gradient Centrifugation*

**(i) Solvent systems**

There are at least two sets of conditions you can use. The original system was developed by Martin and Ames (4), and utilizes a 5% to 20% sucrose gradient, the SW-39L rotor, and 5 ml tubes ($\frac{1}{2}$ by 2 inch size) at a rotor speed of 39,000 rpm for 10–15 hours. It is usually easier if larger volumes can be used, so another system uses a 3-15% gradient, an SW-25.1 rotor, and three 34 ml tubes (1 by 3 inches) at 25,000 rpm for 24 hours (21). Different times can be used, depending on what is under study.

**(ii) Solution**

The first step is to make up the two solutions for the gradient—one with buffer plus 3% (or 5%) sucrose and the other with buffer plus 15% (or 20%) sucrose. The buffer should be 0.05–0.1 ionic strength at pH 7 to 8.

Alternatively, you can make up 12 solutions of 2.8 ml each, containing 3, 4, 5, 6, etc., percent sucrose by mixing varying proportions of buffer with buffer in 15 or 20% sucrose. Overlayer these, in order of decreasing concentration, upon the next higher concentration. The resulting gradient should be allowed to stand a few hours in the cold before the protein samples are added.

It is better to use a sucrose gradient maker. This resembles the linear gradient makers used in chromatography, except that the *more* concentrated sucrose solution (15% or 20%) is nearest the tube to be filled. The flow rate into the centrifuge tube should be kept at less than 1 ml/minute. Keep the tube cold. The linearity of the gradient can be checked by adding some dye such as fluorescein to either the 3% (5%) or 15% (20%) solution and checking the fractions collected after the run for absorbence or fluorescence. Put the centrifuge tubes carefully into the centrifuge buckets. See Figure 6.19.

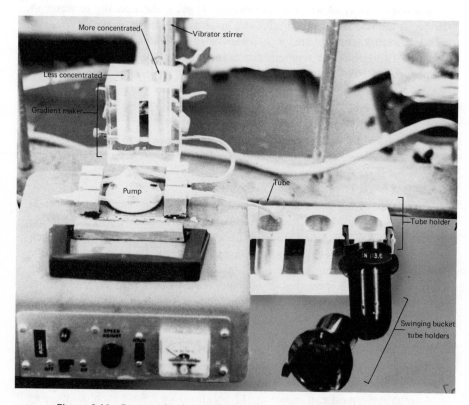

**Figure 6.19** Sucrose density gradient maker, tube holders, and swinging bucket centrifuge tube holders.

For markers, use enzymes which are easy to assay or strongly colored, don't dissociate or aggregate in the gradient or bind to the others, whose sedimentation constants are known and as close to that of the "unknown" enzyme as possible, and whose partial specific volumes are as close to that of the "unknown" enzyme as possible (see Appendix 13). It is best to use markers whose sedimentation constants "bracket" that of the protein you are studying.

The proteins should be in less than 1 ml total volume at no more than 1 mg/ml of any protein. The solution should not contain high concentrations of salt, or the solution will sink into the gradient when layered onto it.

The protein solution is carefully layered onto the gradients.

### (iii) Making the run

Naturally, considerable care should be taken in moving or handling the gradients. However, the major source of possible errors is from evaporation of the samples during the run. So be sure the rubber rings inside the buckets are present and lightly greased (with vacuum grease) before inserting the tubes with the gradients. Screw the tops in to finger tightness before attaching the buckets to the rotor with the long screws.

Directions for the actual centrifugation are given in the instrument manual. After the run, the buckets are detached one by one, and the tubes carefully removed. If any tubes appear crushed or collapsed, or if a great deal of evaporation has occurred, discard the sample (the tubes will expand somewhat during the run, so the liquid levels will drop perhaps 2 mm). Store the tubes in a tube holder and keep them cold.

### (iv) Fractionating the gradients

There are two procedures for removing the gradient solution from the tubes. With the 5 ml tubes, it is convenient to puncture the bottom of a tube with a heated syringe needle and collect one or two drop fractions as the gradient drips out. Using the 34 ml tubes, a density gradient fractionator such as the Isco Model D is more convenient; 0.5 ml fractions should be collected.

The isolated fractions should be assayed for the enzymatic activities (markers and unknown) present. These activities are plotted versus fraction number, and the positions of each of the maximum activities determined from the graph. See Figure 6.20.

The sedimentation constant is calculated using Equation 6-7. Average the values obtained with the different markers and different tubes.

### 6.4d   Yphantis Sedimentation Equilibrium (9)

Once you know the material is homogeneous or at least that it is not contaminated with material lighter than that of the major component, you can calculate the approximate speed and equilibrium time for the experiment.

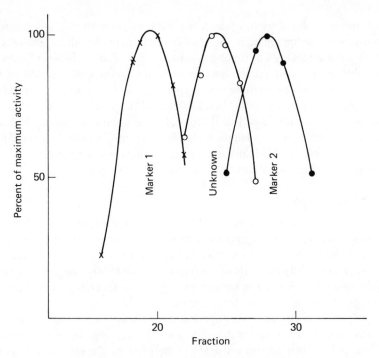

**Figure 6.20** Data from sucrose density gradient centrifugation. All three proteins are assumed to sediment as single active species, so the data are not plotted over the entire tube.

## (i) Rotor speed and equilibrium time calculations

The speed necessary is estimated from:

$$\text{rpm} = 9.54\sqrt{\frac{4.16 \cdot 10^8\ T}{M(1 - \bar{V}\rho)}} \tag{6-43}$$

This is obtained from:

$$\sigma = \frac{\omega^2 M(1 - \bar{V}\rho)}{RT} = 5 = \text{desired slope of } \frac{d(\ln c)}{d(r^2)} \text{ plot} \tag{6-44}$$

You must have some idea of the molecular weight, if necessary from the elution volume of the material from Sephadex or from the sedimentation constant. $\bar{V}$ is usually about 0.730 ml/g, and $\rho$ can be obtained from tables (Appendix 14).

The classical estimate of equilibrium time is twice the time required to sediment across the cell. Therefore, Equation 6-45 should provide a generous estimate of equilibrium time in hours for the high speed method. With short column heights, equilibrium times will generally be less than 24 hours.

$$t_{eq}(\text{hours}) = \frac{2.3\left(\frac{(r_b - r_m)}{\omega^2 S \bar{r}}\right)}{3.6 \times 10^3} \qquad (6\text{--}45)$$

The column height $(r_b - r_m)$ will be about 0.3 cm, $\bar{r}$ is about 6.8 cm, and the sedimentation constant $S$ must be known or measured beforehand. However, the attainment of equilibrium should be checked experimentally during the terminal portion of the run, not assumed from the estimated time given by the equation.

### (ii) Preparation of the ultracentrifuge

Converting the ultracentrifuge from schlieren optics to Raleigh interference optics involves three steps:

1. The push-pull slit assembly on the light source is moved to the forward position so the finer interference slit is now parallel to the sectors of the cell, or the schlieren slit can be rotated 90° and narrowed and shortened with the metal mask.
2. The upper condensing lens mask is replaced with the interference symmetrical mask.
3. The phaseplate (bar) angle is set to 90°.

### (iii) Preparation of the sample and loading the cell

Often the speed of the run is sufficient to redistribute the buffer as well as the protein, so it is necessary to dialyze the solution against the appropriate buffer. Proper loading of the cell is necessary for the same reason in order that the solvent and solute columns are of identical length.

A conventional double sector cell with sapphire windows and the narrow slit lower interference window holder is used (Figure 6.12). Protein concentrations of 0.1-0.3 mg/ml are recommended.

The cell can be first loaded with 10 $\mu$l of base fluid (a heavy oil such as FC-43) in each sector. This can help one to read closer to the bottom of the column, and also insures that the bottom of the column will assume a radial shape. On the other hand, sometimes the oil interacts with the protein. For the fringes to curve upward, the protein solution must be loaded in the sector which passes through the light beam first as the rotor rotates counterclockwise—this is the right sector as the cell is viewed from the screw-ring end. Therefore, 100 $\mu$l (microliters) of solvent or dialyzate is placed in the left sector and 100 $\mu$l of solution is syringed into the right sector.

### (iv) Recording the experiment

Once the run has commenced and the rotor has reached speed, a photograph can be taken on a II-G plate to serve as a reference for fringe straightness or to provide corrections for fringe displacements due to anomalies.

During the 2-4 hours prior to the calculated equilibrium time, several photographs can be taken to check the approach to equilibrium. This is done by checking the position of several fringes or the total number of fringes across the cell (they should be constant within experimental error), and the final photograph can be used for the equilibrium measurements. Also, odometer readings should be recorded during this period to provide an accurate estimate of rotor speed, and temperature readings should be made occasionally to observe if there are any significant fluctuations.

### (v) Calculations

The plate from the high speed equilibrium run will resemble Figure 6.21 and the measurements are made as follows.

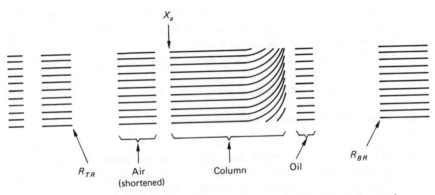

**Figure 6.21**   Rayleigh interferogram from a high-speed sedimentation equilibrium experiment.

The comparator's $x$ and $y$ micrometers are set to "0." The plate is held normally with the emulsion side up, then rotated 180° and placed on the microcomparator stage so the outside reference hole, $R_{BR}$, is centered on the stage. In this way the interference pattern will be easier to align, since it will be a matter of rotating about a center. The $x$ zero set is turned to position the inside reference edge, $R_{TR}$, on the vertical crosshair. The outside reference hole can also be used, but the $x$ micrometer must be set to some higher value.

The plate is aligned along one of the long air fringes by rotating the comparator stage. The $x$ micrometer is moved to a point just inside the meniscus $(X_a)$, and the $y$ zero set is moved to the center of a light fringe that can be followed throughout the column.

Corresponding $X$ and $Y$ readings are recorded for this fringe at preselected intervals down the cell. Readings in the solvent portion of the column (that is, the upper portion of the column where fringes are straight) can be taken at

200-300 $\mu$(micron) intervals. These readings will serve as the zero concentration level. When the vertical fringe displacement is about 100 $\mu$, it is reasonable to take $Y$ readings at every 50 $\mu$ of $X$ displacement to the bottom of the cell or until the fringe density becomes too high to be legible.

The reference plate is read in the same way, and if there are any deviations these are subtracted from the equilibrium plate measurements.

The data are plotted according to Equation 6-23, from which the weight average molecular weight can be obtained.

### 6.4e Measurement of Diffusion Constants

This is done by measuring the spreading of a solute-solvent boundary as a function of time. The diffusion constants obtained should be extrapolated to zero protein concentration, so several experiments must be done.

#### (i) Preparation of the sample and loading the cell

The solutes should be 5–15 mg/ml in protein and approximately 0.1 in ionic strength, well dialyzed against the solvent. The calculations are easiest when schlieren optics are used, and the bar angle must be the same throughout the picture taking.

A valve-type single sector synthetic boundary cell can be used. It will not provide a baseline, but this will not be troublesome if the solutions have been well dialyzed. 0.4 ml of solute is injected into the chamber, and 0.2 ml of solvent is placed in the cup. The cell is sealed, placed in the rotor, and aligned, and the rotor is placed in the rotor chamber. The rotor and its contents must be allowed to come to *complete thermal equilibrium* (0.5–1 hour) before the run begins.

#### (ii) Performing the experiment

The rotor is accelerated at 5 amp until the boundary is formed; then at 8 amp to 10,000 rpm. The diffusion pump is shut off after acceleration starts. It may be better to use the AN-J rotor at a lower speed, but the AN-D at 10,000 rpm can be used with proteins which have low sedimentation constants.

The formation of the boundary is observed constantly, and timing is begun the instant a distinct solvent layer can be seen above the solute. Pictures are taken at intervals once the schlieren peak of the boundary can be visually resolved, but their times should be recorded relative to the time the boundary formed.

#### (iii) Calculations

The calculations involve measurement of the peak areas and heights. The comparator is aligned on a baseline formed by the flat parts of the solute and solvent regions (Fig. 6.22):

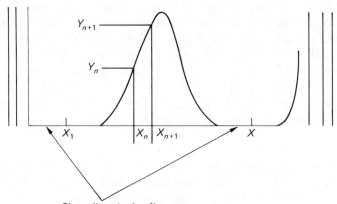

Plate aligned using flat areas

**Figure 6.22** Pattern from a synthetic boundary experiment using schlieren optics. Calculation of the area under the schlieren peak.

The area under the peak is calculated from a series of $x$- and $y$-comparator readings taken throughout the peak area. The $X(r)$ readings should be made at constant increments, if you use a computer to calculate the area (Appendix 15). Otherwise, use the following format:

| reading | $x_n$ | $\dfrac{x_n}{m_x}$ | $r_n$ | $r_n - r_{n-1}$ | $y_n$ | $\dfrac{y_n + y_{n-1}}{2}$ | $(r_n - r_{n-1})\left(\dfrac{y_n + y_{n-1}}{2}\right)$ |
|---|---|---|---|---|---|---|---|
| 1 | $x_1$ | — | $r_1$ | — | $y_1$ | — | — |
| 2 | $x_2$ | — | $r_2$ | — | $y_2$ | — | — |
| 3 | $x_3$ | — | $r_3$ | — | $y_3$ | — | — |
| 4 | $x_4$ | — | $r_4$ | — | $y_4$ | — | — |
| etc. | | | | | | | |

The area $A_s$ is the sum of the terms in the last column. The quantity $(A_s/Y)^2$ is plotted versus the time in seconds to obtain the apparent diffusion constant. Once that is obtained, it must be corrected for the temperature and viscosity of the solvent (22):

$$D_{20,w} = D_{obs} \cdot \frac{\eta_{T,w}}{\eta_{20,w}} \cdot \left(\frac{\eta}{\eta_0}\right)_T \cdot \left(\frac{293}{273 + T}\right) \tag{6-46}$$

$\eta_{T,w}$ is the viscosity of water at temperature $T$, $\eta_{20,w}$ is the viscosity of water at 20°, $(\eta/\eta_0)_T$ is the viscosity of the solvent relative to water at temperature $T$, and $T$ is the temperature in degrees Centigrade.

### 6.4f Determination of Extinction Coefficients

#### (i) Preparation of the sample and loading the cell

The solute must be well dialyzed and be optically clear—no turbidity. The protein concentration should be between 1 and 8 mg/ml to insure enough fringes for an accurate measurement but not more than can be easily resolved. The ionic strength should be 0.05-0.1, preferably in some neutral salt like KCl.

For the synthetic boundary measurement, an interference synthetic boundary cell is used. Add 0.14 ml of solute to the right hand chamber (the screw ring facing you), and 0.40-0.42 ml of solvent to the left chamber of the cell. The cell is sealed; then put in the rotor and aligned. The rotor is placed in the rotor chamber and allowed to come to equilibrium (0.5 to 1 hour after the rotor is at the proper temperature). This thermal equilibrium is important, since temperature gradients can ruin the experiment, especially if working at low temperatures.

#### (ii) Performing the experiment

The rotor is accelerated at 5 amp until the boundary is formed, then at 8 amp to 10,000 rpm,[8] following the formation of the boundary through the viewer. The diffusion pump should be shut off once acceleration is started, to avoid temperature gradient formation.

One picture should be taken when the boundary has formed, to get the fractional fringe number. More pictures are taken after the boundary has diffused enough to allow counting of the fringes in the comparator. It may be helpful to take one or two pictures, develop the plate and examine it in the comparator to see if the fringes in the boundary are resolved. See Figure 6.23.

#### (iii) Calculations

The plate is aligned in the comparator using the air fringes. The comparator cross hairs are then centered on a solvent fringe halfway down the pattern.[9]

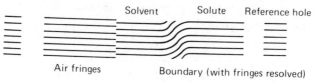

**Figure 6.23** A synthetic boundary using interference optics.

---

[8] The boundary begins to form at about 2,000 rpm, but the AN-D rotor precesses below 10,000 rpm. The AN-J should be used if the protein sediments appreciably at 10,000 rpm.
[9] The fringes are considered to be the light areas.

The stage is then moved horizontally through the boundary, counting the integer number of fringes crossed. At this point the fringes should be level again. There will usually be some fraction of a fringe remaining: the center of the cross hairs will not rest exactly on a fringe. This can be estimated by seeing how far the vertical (Y) stage must be moved to reach the nearest higher fringe. The early picture should be used to measure the fractional fringe, however, since curvature of the fringes due to the diffusion or sedimentation will be minimal then.

The Y comparator distance is converted to fractional fringes. This is added to the integral fringe number, and the result is converted to protein concentration using 1 fringe per 275 $\mu$ (the exact value should be measured) and 4.04 fringes as 1 mg/ml.

The absorption of the protein at some wavelength (usually 280 nm) is divided by the protein concentration to give the extinction coefficient.

### 6.4g  Determination of the Base Composition of DNA (17)

The buoyant density of DNA in cesium chloride is directly proportional to the guanine and cytosine (G-C) content. In a typical equilibrium density gradient experiment, 7.7 $M$ CsCl at equilibrium has a density range of about 1.64-1.76 g/ml, which is sufficient to band most deoxyribonucleates from various sources.

#### (i) Preparation of reagents

A concentrated stock solution is prepared from 130 g CsCl in 170 ml of 0.02 $M$ Tris-HCl, pH 8.0. The solution is filtered through a Millipore filter to remove any insoluble material. The density of the stock solution may be determined pycnometrically (see 6-5a(ii)) or by using the relation between refractive index and density:

$$\rho_p^{25} = 10.8601 \; n_D^{25°} - 13.4974 \tag{6-47}$$

The refractive index of a solution can be determined using a synthetic boundary cell and Rayleigh interference optics, or an Abbe refractometer.

The stock solution is diluted according to Equation 6-48 or 6-49 to the density of the reference DNA:

$$v_{1,w} = \frac{\rho_c - \rho^0}{\rho_c - 0.997} \tag{6-48}$$

$$v_w = \frac{v_c(\rho_c - \rho^0)}{\rho^0 - 0.997} \tag{6-49}$$

The subscripts $w$ and $c$ refer to water and the concentrated stock solution; $v$ is volume; $v_1$ is the volume required to prepare one ml; and $\rho^0$ is the density desired.

### (ii) Making the run

A reference or marker DNA which is well characterized, such as that from *E. coli* ($\rho = 1.710$ g/ml), is used to facilitate calculations. Approximately 0.7 ml of the solution containing 1-2 $\mu$g of the sample DNA and 0.5 $\mu$g of the marker DNA are placed in a standard 12 mm cell assembly with a Kel-F centerpiece and centrifuged at 25°C for 20 hours at 44,000 rpm.

### (iii) Calculations

Photographs are taken with ultraviolet absorption optics on Kodak commercial film and traced with a densitometer as shown in Figure 6.24.

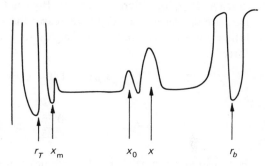

**Figure 6.24**
Densitometer tracing of DNA bands in a cesium chloride gradient. $X_0$ is the position of the marker DNA, and $X$ is the position of the sample DNA.

The density of the sample, $\rho$, can be calculated by using the radial position of the marker, $r_0$, as a reference.

$$\rho = \rho_0 + 4.2\omega^2(r^2 - r_0^2) + 10^{-10} \text{ g/ml} \qquad (6\text{-}50)$$

The density of the marker is $\rho_0$, $\omega$ is the angular velocity in radians per second, and $r$ is the radial position of the band center of the sample.

The density of the sample, $\rho$, can be used in Equation 6-51 to estimate the mole fraction of quanine plus cytosine (G-C):

$$\rho = 0.0098(\text{G–C}) + 1.660 \text{ g/ml} \qquad (6\text{-}51)$$

This method has been used to separate $^{15}$N DNA from $^{14}$N DNA. The density difference between the two types is only 0.014 g/ml, which is an indication of the resolving power of the technique. Determination of densities to within $\pm 0.001$ g/ml is possible.

## 6.5  PRACTICAL ASPECTS

### 6.5a  Sedimentation Velocity

#### (i) External factors affecting sedimentation constants

The sedimentation properties of any macromolecule are affected by the viscosity and density of the solvent:

$$S = \frac{M}{Nf}(1 - \bar{V}\rho) = \frac{M}{6\pi r_{\text{eff}}\eta}\frac{(1 - \bar{V}\rho)}{N} \tag{6-52}$$

The viscosity will directly affect the observed sedimentation constant. However, the density of the solvent and the partial specific volume of the macromolecule also can have a great effect. The latter value seldom differs by as much as 10% for different proteins. But the fact that sedimentation methods measure displacement weights increases the sensitivity of observed sedimentation constants or molecular weights to $\bar{V}$ and to $\rho$ by a factor of about three:

$$(1 - \bar{V}\rho) = (1.0 - 0.75 \cdot 1.0) = 0.25$$

and

$$\frac{0.75}{0.25} = 3$$

For this reason, partial specific volumes and solvent densities must be measured as accurately as possible. For example to determine $(1 - \bar{V}\rho)$ for protein to 1%, the error in the determination of $\bar{V}$ must not exceed 0.3%, which is an accuracy of about 1 part in 50,000 in the density determination of a 10 mg/ml solution.

The temperature also strongly affects observed sedimentation constants, but this is largely because the viscosity of water (and many other liquids) is strongly temperature dependent. A change in sedimentation constant of about 2% per degree can be expected.

Larger variations in temperature can affect protein structure—by denaturation if the temperature is high enough—and possibly affect the partial specific volume of a protein as well (1). The latter can be corrected for. The former phenomenon is usually rather obvious when it occurs.

#### (ii) Determination of partial specific volumes of proteins

*1. Calculation from the amino acid composition.* If the amino acid composition of the protein is known, a value for $\bar{V}$ can be calculated from the partial specific volumes of the individual amino acids:

$$\frac{\sum p_i \bar{V}_i}{\sum p_i} = \bar{V}_{\text{ave}} = \frac{\sum n_i M_i \bar{V}_i}{\sum n_i M_i} \qquad (6\text{-}53)$$

$p_i$ is the percent by weight of the $i$th amino acid, $n_i$ is the number of residues per mole of that amino acid in the protein, $M_i$ is the residue molecular weight (the molecular weight of the amino acid minus the weight of one mole of water) and $\bar{V}_i$ is the partial specific volume of the residue. $\bar{V}_i$ varies from 0.60 cc/g to 0.90 cc/g (23).

This procedure gives accurate results (20). If lipid, flavin, carbohydrate, etc., are also present, their relative amounts and partial specific volumes must be included. The density of biological materials can range from 0.9 g/ml for lipoproteins to 2.2 g/ml for nucleic acids.

*2. Cahn electrobalance to measure specific gravity.* This is accurate to $10^{-7}$ g (0.1 $\mu$g), and uses 1 ml of solution at 10 mg/ml. Its cost is \$3000 (less recorder). This instrument can naturally also be used to measure solvent densities (see iii below).

*3. Linderstrom-Lang and Lanz density gradient column (24).* This is set up by layering bromobenzene and kerosene and mixing gently to form a linear density gradient. Alternatively, a standard gradient-making device can be used. The density range is $\pm 0.02$ g/ml of the material being studied in a 25 cm column. The column is calibrated with standards containing known concentrations of potassium chloride, using 1 $\mu$l drops. The method requires a constant temperature bath accurate to $\pm 0.01°$C. This can also be used to obtain solvent densities.

*4. Pycnometer.* This is essentially a volumetric flask.

A solution of accurately known concentration is used to fill the pycnometer and this is weighed. The volume of the pycnometer can be measured by filling it with water and weighing. Of course, the temperature of the water and the solution must also be accurately known. It is the most direct way to get partial specific volumes of proteins and densities of solvents, but requires considerable material.

*5. Sedimentation velocity or equilibrium in water-deuterium oxide mixtures (7, 21).* This is done like a conventional sedimentation velocity or equilibrium experiment, except the solvent contains a high concentration of deuterium oxide. Since $D_2O$ is only about 11% denser than $H_2O$, concentrations as high as practical are used.[10] To cut down on experimental error, it is gen-

---

[10] Deuterium oxide solutions have a higher dielectric constant and a higher pH than water solutions. This can affect the conformation of proteins. In addition some exchange occurs between dissociable protons in the protein and deuterium ions, which produces small but significant increases in the molecular weight of the protein (6). Such effects may be considered negligible over the times involved in sedimentation velocity experiments.

erally best to run two experiments, simultaneously if possible, one with $D_2O$ and the other without.

The sedimentation rate of the protein in the two solutions is compared and corrected for the viscosity increase produced by the $D_2O$. The difference remaining in observed sedimentation rate is due to the greater density of the solvent containing $D_2O$. The corrected sedimentation constants, without and with $D_2O$ ($S-$ and $S+$) are simply related to the partial specific volume:

$$\bar{V} = \frac{(S-) - (S+)}{(S-\rho+) - (S+\rho-)} \tag{6-54}$$

where $\rho+$ and $\rho-$ are the densities of the solvents with and without $D_2O$, respectively.

Similarly, in sedimentation equilibrium experiments, the slopes of plots of $\ln c$ versus $r^2$ (equation 6-23) without and with $D_2O$ ($m-$ and $m+$) are used to calculate $\bar{V}$:

$$\bar{V} = \frac{(m-) - (m+)}{(m-\rho+) - (m+\rho-)} \tag{6-55}$$

6. *The precision density meter.* This device determines densities of liquids (and gases) from changes in the resonance frequency of a mechanical oscillator containing the sample. Extremely high precision is claimed for this instrument, which in addition uses sample volumes of less than 1 ml.

### (iii) Determinations of solvent densities and viscosities (22)

It is best to use a solvent whose density and viscosity are known (see Appendix 14). However, if this cannot be done, you must measure the density and viscosity of the solvent.

Densities of solvents are normally measured with a pycnometer. Viscosity measurements are usually made by measuring flow times through a capillary viscometer. In addition, a well-regulated constant temperature bath ($\pm 0.05°C$), clean solutions (preferably filtered through a Millipore filter), and a method to accurately measure the flow times are needed. The concentration should be the same as used in the ultracentrifuge. The volume needed is 3-8 ml, depending on the type of viscometer used.

### 6.5b  Sedimentation Equilibrium

Before a method for sedimentation equilibrium is chosen, you must have some idea of the purity and the size of the material to be studied. Generally a sedimentation velocity experiment will provide this information.

The choice between schlieren and interference optics and between high and low speed methods can be made on the basis of what information is wanted, how much material is available, and the properties of the protein.

Should the velocity experiment indicate a molecular weight of less than 15,000, the Yphantis method is not recommended since the high rotor speeds necessary cause window distortions and the interference patterns are not legible.

The Yphantis method may be advantageous when a small amount of high molecular weight contaminant is present. The rotor speed is set for the lighter component, and the lowest molecular weight can be determined with little interference from the contaminant(s) which is preferentially sedimented to the bottom of the cell. If low molecular weight contaminants are present, a low speed method should be used.

## 6.6 ANALYSIS AND INTERPRETATION OF RESULTS

### 6.6a Sedimentation Velocity

The pattern obtained with a protein preparation during a sedimentation velocity experiment is frequently used as a criterion of the homogeneity of size of the preparation if you obtain only one symmetrical boundary throughout the experiment. Schlieren optics are almost always used in these experiments, since the differential curve obtained is very sensitive to the presence of irregularities in the boundary.

When you are examining a protein preparation for homogeneity, you should actually be trying to detect contaminating proteins. Consequently, you should use a double sector cell, since the baseline it provides blanks out variations in the schlieren pattern due to redistribution of solvent constituents and helps reveal small amounts of more slowly or faster sedimenting materials.

The more concentrated the material you are examining, the more chance you have of seeing contaminants with sedimentation constants different from the major component (21). The optics limit the concentration you can use to 15-20 mg/ml and the lower limit of detection of a boundary is usually about 0.5-1 mg/ml of protein.

Note that smaller proteins diffuse more rapidly and that the self-sharpening of the boundary due to the concentration dependence of the sedimentation constant also will affect the usefulness of this method as a criterion of purity. A more rapidly spreading boundary can better conceal smaller boundaries due to contaminants.

A single boundary will also be given by two or more macromolecules which have similar sedimentation constants. This can sometimes be detected by a more rapid broadening of the boundary than one would expect for a single macromolecule of that size. Quantitating this is difficult, however (2).

Broadening of the boundary can also result from some polymerization reactions if these are rapid. Often this phenomenon is associated with tailing of a boundary: the lower macromolecule concentrations are more dissociated. Note that tailing also may be the result of an impurity of lower sedimentation constant.

A protein which is chemically homogeneous can give one or more boundaries if it undergoes complex association and dissociation reactions (25).

Sometimes "spikes" are seen in these experiments. This is usually from convection and can occur because the solution wasn't thermally equilibrated before the experiment, because of the Johnston-Ogston effect, or sometimes because of a high concentration dependence of the sedimentation constant of the macromolecule.

### 6.6b  Sedimentation Equilibrium

The slope of a plot of ln $c$ versus $r^2$ is used to calculated the molecular weight (equation 6-23, section 6-2c (ii), above). If the macromolecule is homogeneous, the line may be straight. However, an upward curvature in this plot is an indication of heterogeneity (or association). A downward curvature indicates non-ideality of the solution. Molecular weights from sedimentation equilibrium experiments should be extrapolated to zero protein concentration to correct for non-ideality and also to check for any association or dissociation occurring. This is seldom necessary when the Yphantis method is used.

Heterogeneity, non-ideality, or charge effects are discussed in detail elsewhere (2, 3). Charge effects and non-ideality will, in most cases, contribute negligible error: less than 2% in $M_w$ with the experimental conditions usually employed in sedimentation equilibrium experiments. In some cases, with material such as nucleic acids, polysaccharides, or high molecular weight synthetic polymers, experimental conditions and interpretation of data will have to be altered to account for deviations from ideal behavior and effects of the charge of the macromolecule. Low concentrations (less than 2 mg/ml) used in experiments with proteins usually render the non-ideality corrections negligible, and the use of low molecular weight electrolyte, at an ionic strength of about 0.1 and a pH near neutrality, suffices to mask the charge effects.

### 6.6c  Density Gradient Sedimentation Equilibrium

Possible complications which would change the apparent density of the DNA should always be considered in the evaluation. These include the effect of protein binding, divalent cation binding, and the presence of some constituent other than the four common bases, i.e., 5-methyl cytosine or carbohydrate.

Mixtures of materials of identical buoyant densities but having different molecular weights form symmetrical but non-Gaussian bands. Buoyant density heterogeneity is sometimes indicated by skewed bands.

## 6.7 SUMMARY OF ADVANTAGES AND DISADVANTAGES OF METHODS

### 6.7a Sedimentation Velocity

The sucrose gradient method depends on several assumptions: that the partial specific volumes and frictional ratios of the unknown and marker proteins are the same or nearly so, and that the continuous variation in solvent density and viscosity does not distort the relative positions of marker and unknown proteins too much.

Martin and Ames (4) showed these factors do not markedly affect the reliability of sedimentation constants calculated by this method, at least for "globular" proteins. While the sucrose gradient method is not as precise as the boundary method ($\pm$5-10$\%$ versus $\pm$2-3$\%$), reproducibility is reasonably good (21). The resolution of the sucrose gradient method is not as good as the boundary method, since the swinging bucket rotors cannot be run at as high a speed as the analytical rotor.

Furthermore, the sucrose gradient method can be used even with crude extracts if enzyme activity is measured. If confusion exists about which of several molecular weight forms of an enzyme are active, the sucrose gradient method can resolve the uncertainty. Also, very small amounts of protein can be used.

### 6.7b Sedimentation Equilibrium

The Yphantis method is the best available method of molecular weight determination if the protein under study is homogeneous. It requires very small amounts of protein, is relatively fast, the data are easy to calculate, and since a wide range of protein concentration is covered no extrapolation to zero protein concentration is necessary. The shortness of the region where the protein concentration varies greatly limits the usefulness of the method if other proteins are present, however.

The equilibrium density gradient technique has not been used extensively with proteins. For densities of nucleic acids the method is very sensitive, so the chief experimental problem is making the right concentration of cesium chloride. Note that salts other than cesium chloride can be used.

This is not a good method for measuring molecular weights. Density heterogeneity, complex formation in the high concentration of salts used, and preferential interactions of the solution components greatly complicates such estimates.

### 6.7c Determinations of Diffusion Constants

The method described in 6.4f works reasonably well; its major drawback is in the calculations of areas. If a computer is available, these will not be as much trouble (Appendix 15). Proteins with high sedimentation constants are difficult to examine by this method, since they will sediment appreciably at the rotor speeds needed to form the boundary.

### 6.7d  Determination of Extinction Coefficients

The method of Babul and Stellwagen (14) works well, if temperature equilibrium is achieved before boundary formation. Proteins which have an absorption band near the wavelength (548 nm) at which the refractive index is measured, or which have a significant carbohydrate, lipid, or nucleic acid content can be examined by this method only if you have a way of calculating their refractive index (14). This of course requires knowledge of the exact composition of the protein, as well as the molar refractive indices of the chromophore, the carbohydrate, etc.

## REFERENCES

1. Svedberg, T. and K. O. Pederson, *The Ultracentrifuge*, Oxford University Press, New York (1940).
2. Schachman, H. K., *Ultracentrifugation in Biochemistry*, Academic Press, New York (1959).
3. Tanford, C., *Physical Chemistry of Macromolecules*, Wiley, New York (1961).
4. Martin, R. G. and B. N. Ames, *J. Biol. Chem.*, **236**, 1372 (1961).
5. Cohen, R., B. Giraud, and A. Messiah, *Biopolymers*, **5**, 203 (1967).
6. Kuntz, I. D., Jr., T. S. Brassfield, G. D. Law and G. V. Purcell, *Science*, **163**, 1329 (1969).
7. Edelstein, S. J. and H. K. Schachman, *J. Biol. Chem.*, **242**, 306 (1967).
8. Gordon, J. A. and J. R. Warren, *J. Biol. Chem.*, **243**, 5663 (1968).
9. Yphantis, D. A., *Biochem.*, **3**, 297 (1964).
10. La Bar, F. E., *Biochem.*, **5**, 2368, (1966).
11. Archibald, W. J., *J. Phys. and Colloid Chem.*, **51**, 1204 (1947).
12. Ehrenberg, A., *Acta Chem. Scand.*, **11**, 1257 (1957).
13. Schumaker, V. N. and H. K. Schachman, *Biochim. Biophys. Acta*, **23**, 628 (1957).
14. Babul, J. and E. Stellwagen, *Anal. Biochem.*, **28**, 216 (1969).
15. Meselson, M., F. W. Stahl and J. Vinograd, *Proc. Nat. Acad. Sci. U. S.*, **43**, 581 (1957).
16. Vinograd, J. in *Methods in Enzymology*, Vol. VI, S. P. Colowick and N. O. Kaplan, Eds., Academic Press, New York (1963), p. 854.
17. Schildkraut, C. L., J. Marmur and P. Doty, *J. Mol. Biol.*, **4**, 430 (1962).
18. Model E Instruction Manual E-I-M-3 May, Beckman-Spinco, Palo Alto, Calif. (1964).
19. DeSa, R. J., *Anal. Biochem.*, **35**, 293 (1970).
20. Lamers, K., F. Putney, I. Z. Steinberg and H. K. Schachman, *Arch. Biochem. Biophys.*, **103**, 379 (1964).
21. Brewer, J. M., L. Ljungdahl, T. E. Spencer and S. H. Neece, *J. Biol. Chem.*, **245**, 4798 (1970). (*See also* Ljungdahl et al., *J. Biol. Chem.*, **245**, 4791 (1970).)

22. Schachman, H. K. in *Methods in Enzymology*, Vol. IV, S. P. Colowick and N. O. Kaplan, Eds., Academic Press, New York (1957), p. 32.
23. Cohn, E. J. and J. T. Edsall, *Proteins, Amino Acids and Peptides*, Rheinhold, New York (1943), p. 373.
24. Linderstrom-Lang, K. and H. Lanz, Jr., *Compt. Rend. Trav. Lab. Carlsberg Ser. Chim.*, **21**, 315 (1938).
25. Frieden, C., *Ann. Rev. Biochem.*, **40**, 653 (1971).

# 7 | ABSORPTION AND FLUORESCENCE

## 7.1 INTRODUCTION

Measurement of the absorbence of substances for analysis of their properties or determination of their concentration is the most widely used analytical procedure in biochemistry. Absorbence of light by a substance is characterized by two parameters: the wavelength of maximum absorption, and the extent of absorption, i.e., the extinction coefficient. Both parameters are determined by the structure and composition of a substance, though the environment of the substance can also have some characteristic effects on both. The fluorescence process is the reverse of absorption and tends to be more affected by the environment of the substance. Consequently, considerable information about both the substance and its environment can be obtained by examination of its absorption and fluorescence.

In this chapter, we consider what information is available in the absorption and fluorescence of compounds and how to extract it. Then we describe instruments for measurement of absorption and fluorescence in terms of their components and operating parameters. Finally we discuss the use of absorption and fluorescence in measurement of an important environmental effect, the binding of a ligand by proteins.

## 7.2 THEORY

### 7.2a Properties of Light

Observation of a natural phenomenon inevitably adds to or subtracts energy from the system, and the resulting limits on accuracy of observation have been defined by the Heisenberg relations ("uncertainty principle"). Applied to light, these have shown that light cannot consist of a single frequency in any wave train (beam) of finite length (1). Thus completely monochromatic light is not possible—or at least not measurable. And similarly, a wave train encountering any restriction on its size, such as a slit which is small (or of the same order of magnitude) relative to the average wavelength of the light, will be seen to consist of a collection of waves traveling in slightly different directions. This is the basis of the phenomenon of diffraction, and illustrates the composition of actual beams of light.

The phenomenon of interference, that is of the interaction of light with itself, shows that light exhibits a periodicity of structure and consequently can be described mathematically by wave equations. Studies of the interaction of light with matter have shown that light also behaves as though it has particulate properties, that these "particles" (called photons) have an energy associated with them, and that the energy varies with an experimentally determinable quantity called the wavelength.

Light in the visible and ultraviolet regions of the spectrum is most important for fluorescence spectroscopy, and the properties and effects of visible and ultraviolet light will be dealt with nearly exclusively. In these regions, the wavelengths are 200 to 800 millimicrons ($m\mu$) or nanometers (nm) long (2). The velocity of light in a vacuum ($c$) is $3 \times 10^{10}$ cm/second so if the wavelength is known the frequency of the wave train may be calculated from the relationship

$$c = \lambda v \tag{7-1}$$

where $\lambda$ is the wavelength and $v$ the frequency. For blue-green light (500 nm wavelength) the frequency would be:

$$v = \frac{c}{\lambda} = \frac{3 \times 10^{10} \text{ cm/sec}}{500 \text{ nm}} = \frac{3 \times 10^{10} \text{ cm/sec}}{500 \times 10^{-7} \text{ cm}} = 6 \times 10^{14} \text{ sec}^{-1}$$

The amount of energy in one unit or photon of 500 nm wavelength is given by Planck's relationship:

$$E = hv = 6.6 \times 10^{-27} \text{ erg-sec} \times 6 \times 10^{14} \text{ sec}^{-1} = 3.95 \times 10^{-12} \text{ ergs} \tag{7-2}$$

where $E$ is the energy in ergs and $h$ is Planck's constant: $6.6 \times 10^{-27}$ erg-sec.

A mole of photons, $6.023 \times 10^{23}$ of them, is defined as an Einstein ($\bar{E}$). The energy of one mole of photons with a wavelength of 500 nm is $N \times E$:

$$\bar{E} = 6.02 \times 10^{23} \times 3.96 \times 10^{-12} \text{ erg} = 24 \times 10^{11} \text{ ergs}$$

and since 1 kilocalorie (kcal) is $4.18 \times 10^{10}$ ergs, there are 60 kcal of energy in one mole of blue-green light.

Experimentally, it has been found that light "waves" can be described by the same equations (Maxwell's equations) as radio waves, X-rays, etc. When they interact with matter, these forms of energy all behave as though they consist of an electric field and a magnetic field which alternate with the frequency of the light. The magnetic field is at right angles to the electric field and is 180° out of phase with it. Both fields are at right angles to the direction of propagation of the light; that is, light acts as a transverse wave, like all "electromagnetic" radiation.

### 7.2b  Interaction of Light with Matter

Since matter is apparently comprised of charged particles, and since an alternating electric field represents an electric dipole, it is reasonable that the light should induce similar dipoles in matter, changing the normal distribution of electrons and protons. These induced dipoles will be oriented roughly parallel to the electric field of the light. Plane polarized light, whose electric vector oscillates in the plane of polarization, would induce dipoles oscillating in roughly the same plane.

The energy of the light is transferred into these dipoles, which in turn become sources of light themselves. This "restored" light will travel symmetrically in all directions perpendicular to the plane of the electric field of the original light. This "restoration" process is called scattering and occurs in a very short time—about $10^{-16}$ seconds, or about the time required for radiation of the frequency of the light to "pass by" a molecule. Because of the dipolar nature of the process, the scattered light is polarized in roughly the same plane as the original, incident light.[1]

The "roughly" in the preceding sentences is used because the magnetic fields in the matter and the light tend to produce a rotation in the plane of

[1] The light scattered in the same direction as the incident light combines with it, and since the polarity of the induced dipoles must be opposite that of the electric field, the "restored" light is out of phase by 180° to the original. This produces a light beam with a different "group velocity."

The velocity of light measured by physical instruments is the velocity of the wave group, the group velocity. In groups consisting of a fairly narrow range of frequencies, this corresponds to the velocity of the "beat wave" or outer envelope of the sum of the amplitudes of the constituents of the group. The group velocity may be greater or less than the velocity of any single constituent of the group (which is $c$, a constant). In other words, the refractive index—the experimentally measured ratio of the velocity of a wave group in material to that in a vacuum—may be greater or less than 1.0 (1, 3).

polarization of the light. However, the net rotation of light passing through a solution of some substance is zero, if the substance contains a center of symmetry, since the rotation induced by a molecule oriented in one way would be cancelled by rotations induced by an oppositely oriented molecule. The absence of a center of symmetry makes it impossible for all induced rotations to be cancelled out (3).

The quantum-mechanical description of scattering and absorption are formally similar—the incident light deforms the electron clouds of a molecule. With scattering, this deformation is restored by the inherent electrical forces in the molecule, and the energy is released as scattered light of the same wavelength. This process can be described mathematically as an elastic collision between the light and the molecule and is called Rayleigh scattering. The extent of scattering is proportional to $1/\lambda^4$ for scattering centers which are small relative to the wavelength of the light.[2]

If the collision of light is "nonelastic," the emerging radiation will be of a different wavelength because of mixing the electromagnetic energy with the rotational and vibrational energy of the colliding molecule. This nonelastic collision is less probable in occurrence and is called Raman scattering. It is important to the fluorescence spectroscopist because it introduces spectral artifacts, particularly in dilute solutions (4).

In the case of absorption of ultraviolet or visible light, the energies associated with light of this spectral region are sufficient to produce a new distribution of an electron—promoting it to a more energetic orbital. This new distribution is called an electronic "excited state." The excited state is relatively long-lived since most electronic excited states persist for $10^{-9}$ seconds or more, some six or seven orders of magnitude longer than the time needed to produce the initial distortion.

### 7.2c Absorption

#### (i) Effects of absorption on matter.

The difference between absorption and scattering is thus due to the existence of these "excited states" (also called "stationary states"), which allows the energy put into the molecule by the light to be trapped, producing a "permanently" different electronic state of the molecule.

So scattering and absorption are competitive processes. The extent of scattering increases going towards shorter wavelengths as $1/\lambda^4$ until an absorption band is reached, then it drops as absorption increases, to rise again as absorption diminishes.[3]

---

[2]Dispersion is defined as a dependence of the group velocity of light on wavelength. The dispersion increases as an absorption band is approached going toward shorter wavelengths, since the scattering and its contribution increases as $1/\lambda^4$.

[3]This loss of scattering to the absorption process keeps the refractive index finite at an absorption band (see footnote 1).

The trapping or absorption process is thought to take about the same time as scattering; it is so fast that most chemical processes and even nuclear motion cannot occur in this time period. This is called the "Franck-Condon principle." We can diagram the energy transition of the electron versus the internuclear distance of a simple diatomic molecule (Figure 7.1).

The ordinate describes the potential energy of the electron as it goes from

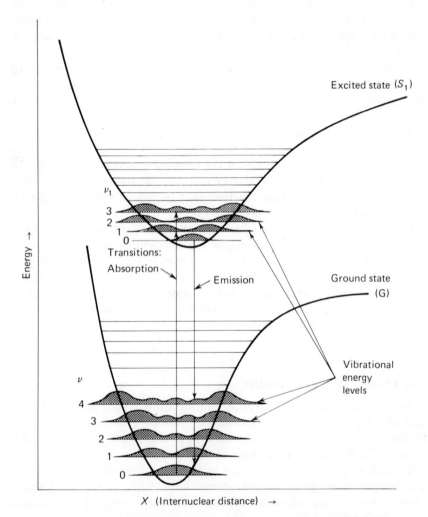

**Figure 7.1** Representation of probable electronic and vibrational transitions for a molecule from the ground state to the lowest excited state. The shaded areas represent the probabilities of electron location. (Modified from G. M. Barrow, *Introduction to Molecular Spectroscopy*, McGraw-Hill, New York (1962) (5). Reprinted by permission of the publishers.)

one state (ground) to another (excited). In such an electronic transition the quantum term $n$ changes: the electron changes its orbital.

For each electronic state, there also exist various vibrational and rotational modes which may be occupied. For our purposes only the vibrational modes need be considered. The transition for a molecule from the ground state to the lowest excited state, corresponding to the lowest energy absorption band, is called $S_1$ (Figure 7.1).

The horizontal lines represent the vibrational levels of the molecule. The shaded areas give the probability of the electrons' location on a particular vibrational level as a function of the distance between nucleii. The probability of particular transitions from G to $S_1$ depends on the probabilities of where the electron is in both the ground and excited states. Since the Franck-Condon principle states that there is no change in the nuclear coordinates during the transition, a vertical line is drawn describing the transition. Those transitions which have the highest probability will describe the excitation energy most likely to be absorbed.

While it is thought that nuclear motion cannot occur during absorption, it does occur after absorption has occurred, as the molecule relaxes into the $S_1$ excited state (see 7.2e). Studies of the excited states of compounds have shown that their chemical and physical properties, such as bond angles and distances, $pK$'s and absorption spectra, often differ significantly from the properties of the parent ground state compounds.

This is also illustrated in Figure 7.1; the two energy wells are displaced from one another, reflecting the differences in the electronic structure of these states. The profile of the excited state is broader, shallower, and displaced towards a larger internuclear distance than the ground state, representing a "looser" molecule (Figure 7.1).

### 7.2d  Factors Affecting Absorption

The speed of the absorption process limits the factors which can affect it. Most chemical processes require considerably longer than $10^{-15}$ seconds to occur, so the chemical environment has much less effect on the absorption than the structure and composition of the molecule, as these determine the existence and nature of the "stationary states."

#### (i) Internal effects

*1. Effects on the wavelength of absorption maximum.* The area or distance an electron may cover in a molecule affects the energy required to move it from one state to the next. The results of the "particle in a box" calculation from quantum mechanics show that the energy of an electron is inversely proportional to the square of its distance of allowed travel (6):

$$E_n = \frac{h^2}{8\,ml^2}n^2 \qquad (7\text{-}3)$$

$m$ is the mass of the electron, $l$ is the length of its "box," and $n$ is an integer $(1, 2, 3, \ldots)$.

$$E_{n+1} - E_n = \Delta E = h\nu = \frac{h\lambda}{c} = \frac{h^2}{8\,ml^2}(2n + 1) \qquad (7\text{-}4)$$

$$\lambda_{\max} = \frac{8\,ml^2}{hc(2n + 1)} \qquad (7\text{-}5)$$

The wavelength of maximum absorption ($\lambda_{\max}$) for a given transition from ground state to some excited state is proportional to the square of the length of the electronic system. The shift of the $\lambda_{\max}$ of the transition to the first excited state in the benzene (263 nm) napthalene (312 nm), anthracene (379 nm) series is an example of this.

The composition of molecules tends to affect energy levels, since atoms of greater mass tend to require smaller energies for promotion of electrons. This is seen in the absorption maxima of imidazole (207 nm in ethanol) and thiazole (240 nm) (7).

*2. Effects on the extinction coefficient.* Since absorption is a dipolar process, it is reasonable that the orientation of a molecule relative to the direction of the exciting electric field should affect the probability of absorption or the "extinction coefficient." Studies of the absorption of polarized light by crystals have shown this is so. For example, the two lowest energy tran-

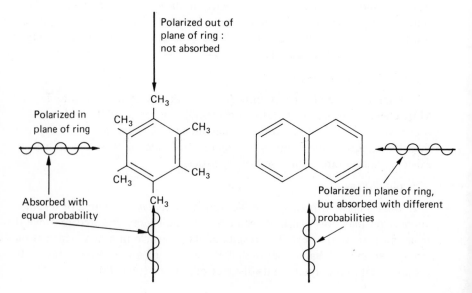

**Figure 7.2**   Polarization of absorbence of hexamethylbenzene and napthalene.

sitions in hexamethylbenzene and napthalene apparently are excited only by light whose electric vector is in the plane of the ring, as shown in Figure 7.2. Because the benzene ring is otherwise symmetrical, no further change in extinction coefficient with rotation of the ring is seen. Napthalene, however, shows some change in absorption on rotation of the ring ("dichroic" effect), because of the decreased symmetry of its ring.

The polarization also varies with the particular transition. The transitions in benzene and napthalene just mentioned are called $\pi$-$\pi$* transitions because the electron is promoted from a $\pi$ orbital to a $\pi$* ($\pi$ antibonding) orbital. In cases where the electron promoted is from a nonbonding orbital of a ring nitrogen, oxygen, sulfur, etc., to a $\pi$* antibonding orbital (called an n-$\pi$* transition) the electric vector producing maximum absorption is perpendicular to the plane of the ring (7, 8).

The symmetry of the molecule determines the symmetry (defined as a change in sign on reflection, rotation, etc.) of the mathematical functions describing the orbitals of the electrons on the molecule. Mathematically, the probability of a transition ($\epsilon$) is expressed as an integral:

$$\epsilon \propto \left( \int \phi_G T \phi_S \, dV \right)^2 \qquad (7\text{-}6)$$

where $T$ is the direction of the electric vector (polarization) of the incident light, $\phi_G$ is the function ("wave function") describing the behavior of the electron in its ground state, and $\phi_S$ the excited state. $V$ stands for the volume. The integral and thus the probability of transition ("extinction coefficient") is largest when the product of these three functions mathematically has the same sign everywhere.

For example, the $\pi$ and $\pi$* wavefunctions change sign on reflection in the plane of the molecule—the two functions are called "odd." Light polarized in the plane of the molecule is mathematically "even" since its direction doesn't change on reflection in that plane. Consequently, the absorption of light polarized in the plane ($T$ "even") should be strong, since the product of the "odd" $\pi$ and $\pi$* functions is also "even." But the absorption of light polarized perpendicular to the plane of the molecule ($T$ "odd") should be weak or "forbidden" in producing $\pi$-$\pi$* transitions, since the integral of the resulting "odd" function over all space is zero (7, 8). See Figure 7.3.

The extinction coefficients of the longest wavelength absorption bands of molecules generally increase with decreasing symmetry. The rationalization of this tendency (7) is that the higher the symmetry, the more planes of reflection, axes of rotation, etc., the molecule has. The wavefunctions can be "even" (symmetric) or "odd" (antisymmetric) with respect to each plane or axis of symmetry, so more wavefunctions of different symmetry classification can be formed. $\epsilon$ is greatest when the product of the ground and excited state

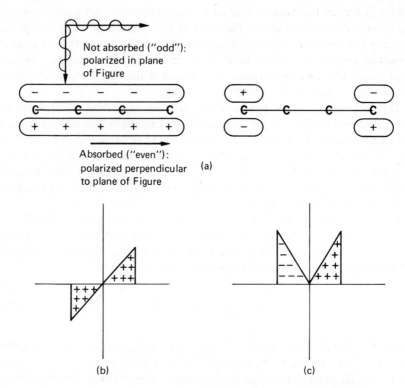

**Figure 7.3**  Polarization of absorbence of benzene for $\pi$-$\pi^*$ transitions. (a) Representation of $\pi$ and $\pi^*$ orbitals of benzene seen from end on; (b) an "even" function—its integral over all space is positive; (c) an "odd" function—its integral over all space is zero (7).

wavefunctions (and $T$) over all the molecule is positive. This is increasingly unlikely in molecules of greater symmetry, so more transitions turn out to be "forbidden."

The longest wavelength absorption band of benzene is in fact "forbidden" by symmetry considerations. However, it is measurable ($\epsilon = 100$), though small. This is rationalized by saying that vibrational distortions (stretching of bonds) make the molecule less symmetrical. A study of benzene derivatives whose vibration is hindered has shown that derivatives with greater vibrational freedom have higher extinction coefficients, in fact (7).

**(ii) Effects of the environment on absorption**

These are weaker than structure and composition effects, but are important nonetheless.

*1. Effects on absorption band width.* A solution is a population of molecules interacting with one another, and these molecules can have an average energy difference of $kT$ between each other. To a limited extent, absorption spectra will reflect the distribution of the thermal collision energy among the molecules. At room temperature for one mole, this would be $NkT$, and since

$$Nk = R = 1.987 \text{ cal/mole-degree} \qquad (7\text{-}7)$$

the energy difference would be about 600 cal/mole. At 500 nm this alone would produce variations in wavelength of

$$\frac{600 \text{ cal/mole}}{60,000 \text{ cal/mole}} = 1\%, \text{ or 5 nm}$$

However, vibrational and rotational motions, caused largely by collisions with solvent molecules, produce the broad absorption bands seen with all but the simplest molecules. The absorption spectra of diatomic molecules in the gas phase at low pressures are line spectra. Increasing the pressure, which increases the collision frequency, causes these lines to broaden. Increasing the complexity of the molecule will broaden the lines still more, and dissolving it in a solvent will broaden the lines to bands. Note that this effect does not affect the average energy of the transition, only the distribution of energies.

*2. Effects on the absorption maximum.* The absorption maximum reflects an energy difference between two states. To alter it, some interaction must occur preferentially with one of the states. Only interactions which are very rapid, such as short-range electronic effects like London-van der Waal's forces, would be expected to affect absorption spectra.

The $\lambda_{max}$ of compounds in solution generally occurs at longer wavelengths than in the gas phase. This is called the "universal red shift." It is thought that the London-van der Waal's interactions between the solvent and compounds are greater when the compounds are in an excited state, since their electrons are then presumably farther, on average, from their nucleii (9).

n-$\pi$* transitions in a compound are defined and detected partly because the $\lambda_{max}$ of the longest wavelength absorption band occurs at shorter wavelengths ("blue shift") in hydroxylic solvents (water, ethanol, etc.) than in nonpolar solvents. The unpaired electrons of O, N, S, etc., in the compound can hydrogen bond with the hydroxylic solvents, but not with nonhydroxylic ones. This lowers the energy of the ground state more than the excited state, where the unpaired electrons are not as available to interact with any solvent, since one has been promoted to a different state. The energy of the transition consequently increases in hydroxylic solvents (8). Another factor which increases this effect is that the ground state has a higher dipole moment than the $\pi$* excited state, and solvents with higher dipole moments (such as water or ethanol) interact more strongly with the ground state than the excited state.

### (iii) Effects on the extinction coefficient

The well-known hypochromicity of polynucleic acids, particularly DNA, is an example of this. The extinction coefficient of native DNA at 260 nm is about 25% lower than the sum of the extinction coefficients of its constituent nucleic acids, while "melted" or disorganized DNA has an extinction much closer to that of its constituents. In other words, the presence and orientation of a second nucleotide base affects the transition probability of the first. This effect is rationalized in Figure 7.4a and b.

The ovals in Figure 7.4 represent nucleic acid bases oriented as in native DNA, the arrows the directions of the dipoles produced in the molecule by light. When one molecule absorbs light, this induces a dipole and a small change in the ground state symmetry of the adjacent ones. The change in symmetry produces a greater difference between the ground and excited state symmetries. Since the probability of absorption is greatest when the symmetry of the two states is most alike, the extinction coefficient of native DNA is reduced.

This effect has a very short range, since it falls off rapidly with increasing polynucleic acid chain length.

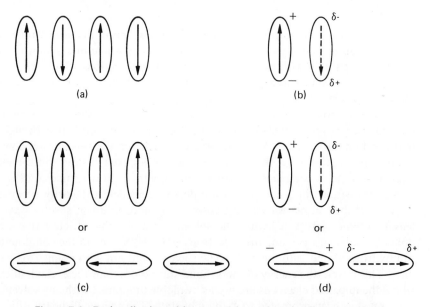

**Figure 7.4** Rationalization of hypo- and hyperchromic effects in ordered polymers. (a) "Head to tail" arrangement of absorbing units in native DNA; (b) showing how absorption by one induces a dipole in a neighboring unit; (c) "head to head" arrangement in helical polypeptides; (d) absorption by one unit induces a dipole in the next. For a different explanation, see reference 10.

A smaller effect can be detected in FAD and DPNH, showing that the interaction of unlike compounds can affect extinction coefficients.

Hyperchomicity can also result from the relative orientations and proximity of chemical compounds. The n-$\pi^*$ transition of amide groups of poly-L-glutamic acid is more probable when the polymer is in a helical conformation than when it is in a random configuration, and a similar rationalization can be employed (Figure 7.4c and d).

### 7.2e  Loss of Excitation Energy

A molecule in an electronic excited state has far more energy than it is ever likely to obtain by collisions with thermally activated solvent molecules. Thus it tends to lose its energy rather quickly—in about $10^{-9}$ seconds.

The loss of energy is believed, partly from studies of fluorescence (see Sec. 7.2b), to occur in at least two steps. There are also believed to be several competitive routes of energy loss, with certain routes more favored in some molecules than others.

The first loss of energy yields a molecule in the first (lowest energy) excited singlet[4] state, $S_1$. This process is evidently very fast, occurring in perhaps $10^{-12}$ seconds. Although a considerable loss of energy is usually involved, no visible or ultraviolet radiation is emitted, so these transitions are called "radiationless."

This energy is lost largely by dissipation in relatively small amounts. This can occur by what is called external conversion, which may mean: excitation of a solvent molecules to higher translational, vibrational or rotational states; or internal conversion, which is thought to involve excitation of internal vibrational and rotational transitions.

The amount of energy dissipated in this way depends on the relatively small differences between rotational, vibrational and translational energy levels. For example, the amount of energy that can be dissipated in one vibrational mode is roughly one-fifteenth of the energy involved in an electronic transition, so this method of dissipation of electronic excitation is most feasible in flexible molecules with many possible vibrational modes. And the energies involved in rotational and translational transitions are even smaller.

The second stage in energy loss by electronically excited molecules occurs from state $S_1$. For unknown reasons, this process is much slower than the initial loss of energy and involves different mechanisms. Both external and internal conversion can occur, but are apparently relatively minor factors. Many chemical changes, such as loss of a proton or electron by an excited molecule, are fast enough to dissipate considerable energy from state $S_1$.

In addition, deactivation can occur by energy transfer, which produces another electronically excited molecule. This can occur over 10–50 Å if con-

[4]So called because the electron's spin doesn't change.

ditions are right. There are also substances such as iodide anion or oxygen which are very efficient absorbers ("quenchers") of this energy. There are also two types of radiative transitions observed in a minority of molecules. In both, energy is lost by emission of visible or ultraviolet light.

In the case of phosphorescence, the excited electron reverses its spin, which puts the excited molecule in a "triplet" state. Relaxation to the ground state can occur by internal conversion, etc., or by emission of light after a second spin reversal. Since the probability of spin reversal is generally low, the triplet state, once attained, tends to persist for long times—up to minutes. Or the excited $S_1$ molecule may emit a photon and return directly to the ground state. This is called "fluorescence." These deactivation pathways of a molecule are illustrated in Figure 7.5.

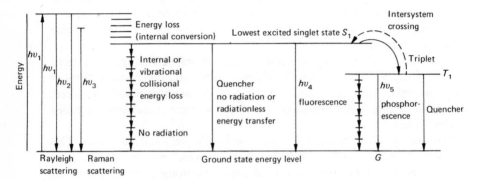

**Figure 7.5**   Possible deactivation pathways of a molecule in an electronic excited state. (From A. J. Pesce, C. G. Rosen and T. L. Pasby, *Fluorescence Spectroscopy*, Marcel Dekker, New York (1971). Reprinted by permission of the publisher.)

### 7.2f  Characteristics of Fluorescence

#### (i) Relation to other processes

Fluorescence is a spontaneous deactivation process, a relaxation. Spontaneous energy loss by a molecule is a first-order process, so is characterized by a first-order rate constant $k$. The reciprocal of any first-order rate constant is the "lifetime" of the process, called in the case of fluorescence the "lifetime of the excited state," $\tau$.

Since one is generally dealing with populations of molecules, it is of interest to see if lifetimes of these populations are unique or if they are populations themselves.

*1. Determination of lifetimes.* $\tau$ can be estimated from the absorption spectrum, based on the equation of Perrin-Forster (11).

$$\frac{1}{\tau_n} = 2.88 \times 10^{-9} n^2 \bar{r}_a \int E_{\bar{r}} \, d\bar{r} \tag{7-8}$$

where $\tau_n$ is the natural lifetime of the dipole oscillator, $n$ the refractive index of the medium surrounding the oscillator, $\bar{r}_a$ the wave number of the absorption maximum, and $E_{\bar{r}}$ the molar absorption coefficient of the last absorption band. The most difficult quantity to estimate is $E_{\bar{r}}$ because of overlapping of other transitions on the short wavelength side. In some cases, such as dimethylaminonapthalene or anilinonapthalene sulfonic acids (DNS or ANS) and other naphthalene derivatives, two electronic transitions actually make up one band. The calculation is simplified if one assumes the band to be Gaussian. In this instance only the absorption maximum and half-band width are needed in the calculation

$$\int E_{\bar{r}} \, d\bar{r} = \sigma E_m \sqrt{\pi} \tag{7-9}$$

where $\sigma$ is the half-band width and $E_m$ is the absorption coefficient at the maximum.

This assumes that the probability of emission and absorption are equal. If the quantum yield in a situation were less than the maximum value for that compound, the lifetime may be obtained by multiplying the theoretical lifetime by the quantum yield:

$$\tau_e = \tau_n Q \tag{7-10}$$

The lifetime may also be calculated from quenching or depolarization experiments (see 7.2f(ii)).

Direct methods of lifetime measurement are of two types: pulse and modulation (12). The former uses very short pulses of light for excitation and measures the decay of intensity of fluorescence directly. Assuming that the process is monomolecular,

$$-\frac{d[A^*]}{dt} = k_f [A^*]; \; [A^*] = [A_0^*] \, e^{-t/\tau_0} \tag{7-11a and b}$$

where $[A^*]$ is the number of excited molecules, $k_f$ is the rate of fluorescence, $\tau_0$ is the lifetime or $1/k_f$, $A_0^*$ is the number of molecules in the excited state at time zero and $t$ is time. This method is still technically difficult for lifetimes shorter than 1 nanosecond ($10^{-9}$ sec).

The modulation technique uses light modulated at a high frequency with a Kerr cell or sound waves and measures the phase shift of the emitted light relative to the incident light (13). In theory the latter method gives only an average lifetime and less information on heterogeneity of lifetimes than the former method, but this is disputed (13). What is apparent is that calculated lifetimes are often in good agreement with measured ones.

The question of whether unique lifetimes exist for populations is still unsettled, but available evidence indicates this is so, at least to a first approximation.

*2. Relation of lifetime to quantum yield.* The observed quantum yield of fluorescence is the result of competitive processes involving nonradiative transitions to the ground state and the radiative transition of emission. The quantum yield $q$ is thus a function of the rate of these competitive processes, and may be expressed as

$$q = \frac{\lambda}{\lambda + k} = \frac{\text{number of quanta emitted}}{\text{number of quanta absorbed}} \qquad (7\text{-}12)$$

where $\lambda$ is the rate of emission, and is simply the reciprocal of $\tau_0$, and $k$ is a rate term including all nonradiative relaxation processes. Note the quantum yield is independent of the number of excited molecules.

If there were no other factors involved, the probabilities of fluorescence emission and absorption should be roughly proportional to one another. Molecules with high extinction coefficients should have a high quantum yield and a short lifetime of excited state.

*3. Significance of the magnitude of excited state lifetimes to the quantum yield.* In general, however, fluorescence phenomena must be represented by a four-level scheme rather than a simple two-level scheme (9). (See Figure 7.6.) Vibrational relaxation and excited state interactions tend to make the fluorescence transition $(k_e)$ independent from absorption.

The molecules in states G and $S_1$ are a population. Those in $v_0^1$ of $S_1$ are not in thermodynamic equilibrium with the others, since they can disappear by radiationless transitions $k_r$, which molecules in the ground state (G) cannot do. The fraction of molecules reaching $v_0^1$ of $S_1$ that do not emit is $k_r/(k_e + k_r)$. Now $k_r$ will in general depend on an activation energy that must be acquired during the excited state:

$$k_r = Ae^{-E/RT} \qquad (7\text{-}13)$$

$k_e$ is $1/\tau_e$, that is, about $10^8$/sec. $A$ will be about $10^{12}$—a value which is typical of monomolecular processes—so $k_e$ will equal $k_r$ when the activation energy is about 5,000 cal/mole. Any interaction with the surroundings which yields or requires this energy will have a large effect on the fluorescence while having relatively little effect on absorption. This means that fluorescence can provide information not yielded by absorption measurements.

This can also be seen from a consideration of the relatively long lifetimes of excited state of molecules. In $10^{-9}$ seconds, some chemical processes (in-

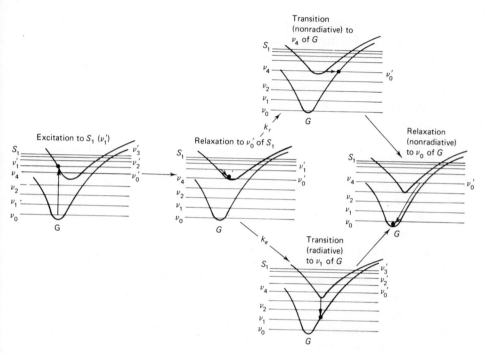

**Figure 7.6** Representation of deactivation kinetics of a molecule in an electronic excited state. [Modified from G. Weber and J. F. W. Teale, in *The Proteins*, Vol. III, H. Neurath, ed., Academic Press, New York (1965), Chapter 17.]

cluding diffusion-controlled ones), nuclear motion (which occurs in about $10^{-11}$ seconds), etc., can occur and hence affect the quantum yield of fluorescence and lifetime of excited state. This is why only a minority of molecules are fluorescent.

*4. Factors affecting quantum yields.* Factors affecting quantum yields can be separated into internal and external (environmental). Internal factors are largely questions of how many modes of vibrational and rotational transitions are available for dissipation of electronic energy. Flexible molecules such as carotenoids are nonfluorescent; most fluorescent substances are atoms or fairly rigid molecules. External factors are of more interest to biochemists.

While $k_e$ may not be strongly influenced itself by the environment, $k_r$—the rate of the radiationless competing processes—often is. For example, the fluorescence of many compounds decreases as the temperature increases; with tryptophan, the fluorescence decreases 2% per degree (14). If $k_r$ is strongly temperature-dependent, the quantum yield will be also.

"Quenching" of fluorescence means decreasing the natural quantum yield of a substance. External factors can produce quenching by two basic mechanisms: transfer of matter (chemical reaction) and transfer of energy.

a. Transfer of matter. The matter is generally a proton or electron. Quenching by proton or electron transfer requires an acceptor. Weak anions such as acetate can quench tyrosine, but strong acids, which are poor proton acceptors, can not (15).

b. Transfer of energy. This can occur by two main mechanisms, one dependent on diffusion and requiring contact (orbital overlap) between donor and acceptor, while the other is a long-range resonance process–contact between fluorophore and quencher does not occur. This quenching process is called "resonance energy transfer," though energy is always transferred in quenching.

"Energy transfer" requires an appropriate acceptor and donor, though the acceptor can be the same substance as the donor (11). A very efficient long-range energy transfer can occur through a dipole–dipole resonance mechanism, and consequently shows a dependence on relative orientation of donor and acceptor. The efficiency of long-range energy transfer increases with increasing overlap between the absorption spectrum of the acceptor and the emission spectrum of the donor, and decreases with the sixth power of the donor-acceptor distance. This makes energy transfer measurements extremely sensitive to donor-acceptor distance. When donor and acceptor are the same substance, energy transfer can be detected by a decrease in polarization of fluorescence with increasing concentration of the substance. Note, however, a substance can be a very efficient donor and have a very low quantum yield of fluorescence in the absence of any acceptor. An example of this is the adenine in NADH or NADPH. It can transfer its excitation energy, but not lose it by emitting.

If contact is necessary for quenching, the complex between fluorophore and quencher may exist for a relatively long or short time.

In "collisional" quenching, the complex between fluorophore and quencher exists for a considerably shorter time than the lifetime of excited state, while in "complex" quenching, the association of the two molecules persists for a much longer time than the excited state does. Intermediate types of quenching, in which the lifetimes of excited state and complex are similar, are possible, and the properties of this type of quenching are also intermediate between complex and collisional quenching. In addition, complex quenching may occur with the fluorophore in the ground state ("static" quenching) or the excited state.

Iodide anion is the classical example of a collisional quencher. It deactivates excited molecules on contact (collision) with high efficiency. This effectively shortens the lifetime of excited state, since a new means of radiationless deactivation is added to those already present.

Without quencher:

$$q = \frac{k_e}{k_e + k_r} \qquad (7\text{-}12)$$

With quencher, the quantum yield is lowered proportionally to the concentration of quencher,

$$q = \frac{k_e}{k_e + k_r + k[Q]} \qquad (7\text{-}14)$$

$$\frac{q}{q_1} = 1 + \frac{k[Q]}{k_e + k_r} = 1 + k[Q]\tau^1 = \frac{\text{unquenched fluorescence } (F_0)}{\text{quenched fluorescence } (F)}$$

$$(7\text{-}15)$$

since we define $\tau^1$ as the observed fluorescence lifetime, equal to $1/(k_e + k_r)$. This is the Stern-Volmer equation. A plot of $F_0/F$ versus concentration will give a straight line of slope $k\tau$.

Complex or "static" quenching is found in the quenching of the flavin in FAD by the adenine ring. When the adenine is protonated at low pH, the complex breaks up, and the quantum yield of the flavin increases (16). The complex forms with the ground state of the fluorophore, with a dissociation constant $K_Q$:

$$F + Q \underset{\overset{K_Q}{\rightleftarrows}}{} FQ \qquad (7\text{-}16)$$

This effectively reduces the concentration of fluorophore.

The lifetime of excited state is not affected since only uncomplexed molecules fluoresce. It is easily shown (4) that:

$$\frac{F_0}{F} = 1 + [Q] \times K_Q \qquad (7\text{-}17)$$

Again, a plot of $F_0/F$ versus quencher concentration yields a straight line of slope $K_Q$.

Experimentally, complex and collisional quenching can be distinguished by measuring the effect of the quencher on the polarization of the fluorescence of the fluorophore, since the polarization is related to the lifetime of the excited state (see equation 7-18, below). Collisional quenching increases the polarization as it shortens the lifetime; complex quenching has no effect.

At high concentrations of fluorophore a nonfluorescent ground state dimer can form between two molecules of fluorophore. This process is similar to complex quenching, and can be detected by a change in the absorption spectrum, due to complex formation, by difference spectral techniques [7.4a(ii)].

Another possible mechanism of quenching at high fluorophore concentrations is the formation of dimers comprised of a molecule in the ground state and one in the excited state. These are called excimers. This decreases the population of monomers and consequently the fluorescence intensity if the excimers are non- or weakly fluorescent. Sometimes the excimers emit, but at longer wavelengths than the monomers. Note that in the case of excimer formation, there can be no change in the absorption spectrum.

Apparent quenching is often observed when the fluorescence intensity of high concentrations of fluorophore are examined. If there is significant overlap between the absorption and emission spectra, the light emitted at shorter wavelengths may be reabsorbed by ground state molecules, thus exciting them. This is called "trivial reabsorption of emitted light" and results in an apparent preferential quenching of the fluorescence emitted at wavelengths closer to the absorption band. This is actually a rather important phenomenon; equations for correction for reabsorption are given in reference 17.

### (ii) Relation to absorption

Fluorescence is in a sense the reverse of absorption. It occurs as a result of a transition between states, which leads to emission of light centered around a particular wavelength, the emission maximum. This is also characteristic of the molecule and its environment.

Some energy is usually lost, since emission seems to occur from the lowest vibrational level of the lowest excited state. This is seen as longer wavelengths of emission than excitation. However, the amount of energy loss again depends on both internal and external factors, such as solvent effects. Internal factors affecting the emission maximum are not well defined. Simple molecules or atoms can emit with no loss of energy ("resonance radiation"). Some large molecules, such as fluorescein, show considerable overlap between their absorption and emission spectra, while others, such as "dansyl," exhibit a very large "Stokes shift."

Again, external factors are more interesting to biochemists. These generally follow rules similar to those for absorption, only the effects are more pronounced: compounds usually have emission maxima at shorter wavelengths in hydrophobic solvents than hydrophilic. The latter would interact more strongly with excited states, since they generally have greater dipole moments than the parent ground states.

The hypothesis that relaxation from higher excited singlets to $S_1$ is much faster than all other processes is partly based on the generally observed characteristics of fluorescent molecules.

The shape and position of emission bands are usually independent of the wavelength of excitation, and the efficiency of fluorescence should be the same no matter how the molecule was excited—the quantum yield should be independent of the exciting wavelength. This is true for most compounds. A major exception to this observation is azulene, in which the absorption band of longest wavelength appears not to have a strong interaction with those of shorter wavelength.

The experimental observation for many aromatic compounds is that the fluorescence spectra are mirror images of the absorption spectra. This is thought to occur because excitation of molecules at room temperature is predominantly from the lowest vibrational level of the ground state. Consequently, the vibrational structure seen in absorption spectra reflects the excited state geometry (2). In general, there are different probabilities (extinction coefficients) of obtaining vibrational substates of the first excited singlet state (see Figure 7.1). The lowest vibrational level of the excited state is also favored at room temperature. That is, the vibrational structure observed in emission spectra reflects ground state vibrational levels. (See Figure 7.7.) The probability distribution of emission to the various vibrational levels of the ground state will be similar to the probability distribution of absorption, if the geometry of the excited molecule is sufficiently similar to the unexcited one (which is generally the case).

If a fluorescent substance is excited with polarized light and the polarization

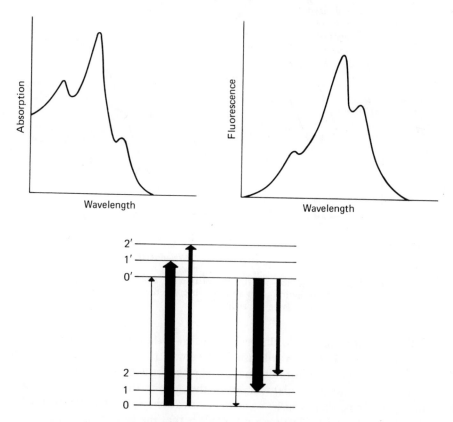

**Figure 7.7**   Rationalization of the mirror image relation between absorption and emission spectra. The thicknesses of the vertical arrows represent transition probabilities.

of the fluorescent light is measured, there will be some relation between the orientation of the light that the molecule prefers to absorb and the orientation of the light that the molecule will emit. In other words, the fluorescence is polarized. In the case of phenol in a rigid glass (Figure 7.8), the polarized exciting light picks out those molecules oriented properly for absorption. "$p$" refers to the polarization of the emitted light, and $\lambda$ to the wavelength of the exciting light. The spectrum is what would be expected from two transitions at right angles to each other.

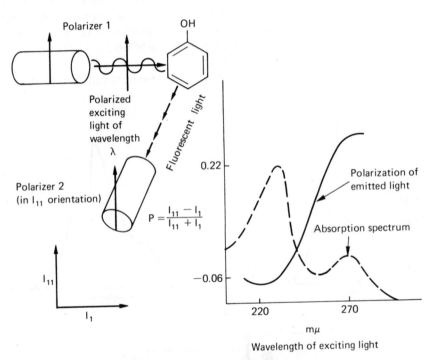

**Figure 7.8**   Fluorescence–polarization of phenol. The molecule is oriented for excitation (left) and the observed fluorescence–polarization spectrum is on the right.

If the polarization of the fluorescence of a substance is analyzed, however, the observed polarization will usually be lower than the theoretical. There may be several reasons for this. In the first place the angle of emission (relative to the axes of the molecule) will not be the same as the angle of absorption. In other words, there will be some depolarization even if no other cause for depolarization exists. The reason for this is that the geometry of the excited molecule may be very different from that of the ground state, and the rate of relaxation to the excited state geometry is usually fast relative to $1/\tau$.

The theoretical maximum polarization if the directions of absorption and emission coincide is 0.50, so evidently the direction of emission in phenol is at some angle from the absorption (Figure 7.8), and therefore phenol in its excited state is thought to have a different shape from phenol in its ground state.

Some possible sources of depolarization have rates which depend on experimental conditions. Two of the most important are a Brownian rotational diffusion of the molecule between excitation and emission, and energy transfer between molecules. The former will depend on the size of the molecule, on the viscosity and temperature of the solution, and on the lifetime of the excited state:

$$\left(\frac{1}{p} - \frac{1}{3}\right) = \left(\frac{1}{p_0} - \frac{1}{3}\right)\left(1 + \frac{RT\tau}{\eta V}\right) \qquad (7\text{-}18)$$

$V$ is the volume of the molecule and $p_0$ is the polarization in the absence of rotational diffusion—the angle between absorption and emission oscillators. Reducing the lifetime $\tau$ of the excited state will raise the polarization, since the molecule has, on the average, less time to rotate away from the orientation in which it was excited before emitting the light. Energy transfer is discussed above. The dependence on so many parameters has allowed polarization of fluorescence to be used in measurements of the microscopic viscosity about a fluorophore, electronic structure of fluorophores, size and rigidity of macromolecules conjugated to fluorophores, and lifetimes of excited states (9, 12).

## 7.3 INSTRUMENTATION

### 7.3a Photometers or Spectrophotometers

Simple spectrophotometers consist of a light source, a monochromator, an adjustable slit and a detector and readout device, as shown in Figure 7.9.

### (i) The light source

The best light source provides maximum light energy in the range of interest. This enables the operator to work with narrower slits, which provides

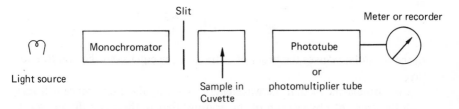

**Figure 7.9** Schematic representation of a spectrophotometer.

more monochromatic light, and reduces the amplification needed for signals from the detector. This can provide more information about the absorption spectrum, a closer following of Beer's law (see Sec. 7.4a) and greater stability of readings.

For general use, a lamp that puts out as continuous and featureless a spectrum as possible is preferred. Currently, hydrogen or deuterium lamps are used for ultraviolet work and tungsten lamps for the visible region. A lamp that can be used in both regions is sometimes advantageous and Xenon arcs are favored at present. However, Xenon lamps are somewhat expensive and must be replaced every few hundred hours of operation. Also, because of their high operating temperature and internal pressure, they are somewhat dangerous.

For special wavelengths, low pressure mercury arcs are used. With these lamps, the emission energy is concentrated into a few lines (e.g., 366 and 548 nm) instead of a continuum (see Figure 7.12, Sec. 7.4b).

### (ii) The monochromator

The monochromator should feature high spectral purity of the dispersed light, high dispersion (angular separation of the spectrum), a high light-gathering capacity and high efficiency (little loss of light).

The primary choice is between a prism and a grating instrument. The operation of a prism depends on the variation of refractive index of the glass with wavelength, which increases with nearness to an absorption band (see 7.2b). Consequently, prisms have a lower wavelength limit of usefulness and also a nonlinear relation between dispersion and wavelength.

A diffraction grating consists of a regular array of parallel lines or grooves on a flat plate of glass or metal (1). Figure 7.10 illustrates the operation of a diffraction grating with diffraction occurring forward (transmission) or backwards (reflection). The path difference $\Delta S$ for diffraction at an angle $\theta_2$ from two or more adjacent lines is

$$2d(\sin \theta_2 - \sin \theta_1) \qquad (7\text{-}19)$$

For constructive interference,

$$\Delta S = m\lambda \quad (m \text{ an integer}) \qquad (7\text{-}20)$$

So

$$\sin \theta_2 - \sin \theta_1 = \frac{m\lambda}{2d} \qquad (7\text{-}21)$$

In the formula, $m$ gives the order of diffraction; $m$ equals 0 is the undiffracted ray.

The formula shows that the wavelength surviving the interference will vary with the angle of observation of the grating; that is, there is a dispersion. It

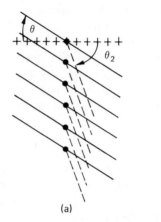 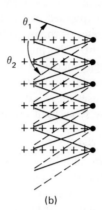

(a)                                    (b)

**Figure 7.10**  Operation of a diffraction grating, either by (a) transmission or (b) reflection. The dashed line is the direction of the diffracted light. The line of crosses marks a perpendicular to the array of diffraction elements (points). [From R. W. Ditchburn, *Light*, Wiley Interscience, New York (1953).]

also shows that different orders of spectra, corresponding to $\lambda$, $\lambda/2$, $\lambda/3$, will overlap: diffracted light of say 500 nm wavelength will be contaminated with light of 250 nm, 167 nm, etc. The intensity of succeeding orders fall off roughly as $m^2$, but often these contaminating orders of light must be removed with filters (nonfluorescent).

The ability of a grating to disperse light increases with the number of diffraction elements per inch (decreasing $d$)—the number of lines or grooves ruled on the plate per inch. The shape of the grooves on a plate can determine how much of the incident light of a given wavelength will appear in the first order spectrum, that is, the efficiency of a grating. Gratings are manufactured to have the greatest efficiency at a given wavelength, at which the grating is said to be "blazed."

Gratings generally feature a greater light-gathering capacity and a higher dispersion than prisms and linear dispersion. On the other hand, they are less efficient than prisms, higher order spectra can be troublesome, there may be "ghosts" in the spectrum due to imperfections in the grating, and their higher light-gathering capacity is accompanied by a greater amount of stray (white) light coming through the monochromator.

Stray light can be a serious problem in spectrophotometers, especially when measuring solutions with high absorbencies (18). An optical density of 1.0 means that only 10% of the light coming through the "blank" (solvent) is transmitted through the sample (see 7.4a). So if the monochromator also transmits 1% stray light which is not absorbed, the instrument will read 89% transmission. The apparent optical density will be 0.96, an error of 4%.

The light-gathering capacity of a monochromator, like most optical instruments, is expressed in $f$ numbers, which are the ratio of focal length to aperture diameter of the instrument. A monochromator with a low $f$ number has a high light-gathering capacity, but it also transmits more stray light. Since stray light increases much faster than light-gathering capacity, good spectrophotometers use monochromators with high $f$ numbers. The Cary 15, for example, uses an $f/8$ grating monochromator, sacrificing light-gathering power for low stray light levels.

### (iii) The detector

The detector should be as sensitive as possible, cover as wide a spectral range as uniformly as possible, provide a linear output with incident light intensity, and provide as low a "dark current" as possible. Sensitivity is greatly increased by using a photomultiplier tube, which is a phototube with 5-20 amplification stages built in. Photomultiplier tubes provide a total signal amplification of as much as $10^8$. The efficiency of detection of light depends on the coating used in the phototube, since different coatings are most sensitive in different spectral regions (see Figure 7.11, Sec. 7.3b).

Photomultipliers consist of a photoemissive cathode, a system for collection of the photoelectrons, dynodes for amplification and an anode (19). The photocathode consists of a photosensitive film on a supporting surface. The collection system focuses the photoelectrons on the first dynode, and the efficiency with which it can do this and keep sharp pulses of photoelectrons arriving as pulses are important factors. The dynodes are also called "secondary emission electrodes"; they are often a cesium-coated alloy of silver and magnesium. The anode is usually a wire mesh, placed directly in front of the last dynode.

### (iv) Meter or recorder

Photomultipliers are current sources, and since it is easier for meters or recorders to measure voltage changes the circuitry must be designed for this. The meter or recorder should of course provide a linear response, but since all light-detecting instruments measure transmitted light, conversion to the logarithm (7.4a) must be made graphically or electronically.

### 7.3b Fluorometers

Fluorometers consist of the same components, but since fluorescence usually occurs at longer wavelengths than absorption, provision must be made for elimination of the exciting light. This implies a different geometry—the detector is usually at 90° or less to the incident beam—and the use of filters or a second monochromator to help eliminate scattered exciting light (Figure 7.11).

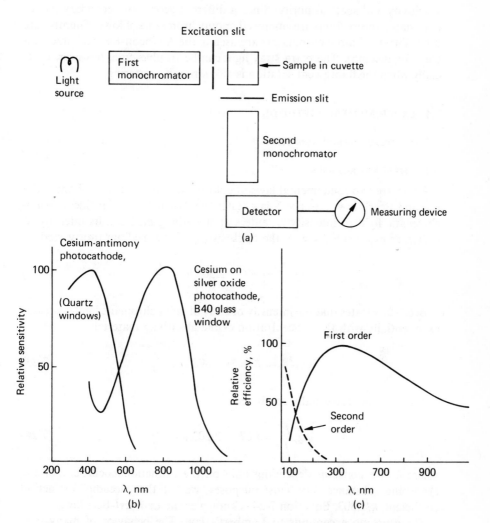

**Figure 7.11** (a) Schematic representation of a spectrofluorometer; (b) spectral response curves of two different types of photomultiplier tube [from "Electron Tubes," Amperex Data Handbook, Amperex Electronic Corp., Hicksville, New York (1969)]; (c) spectral efficiency of a diffraction grating blazed at 300 nm [from J. F. James and R. S. Sternberg, *The Design of Optical Spectrometers*, Chapman and Hall, London (1969)].

For fluorometers, the choice of monochromators is also a choice between sensitivity and spectral purity. Since a different geometry, secondary monochromator and/or filters are employed, monochromators of lower $f$ number are used. Two $f/3.5$ monochromators are used in the Amincon-Bowman fluorometer, for instance. The extra stray light can be troublesome, however, especially when the fluorescent solution is also strongly scattering.

## 7.4  EXPERIMENTAL METHODS

### 7.4a  Measurement of Absorbence

#### (i) Definitions and units

There are two fundamental laws of colorimetry: Bouguer's (or Lambert's) law and Beer's law. Lambert's law states that when a ray of incident monochromatic light of intensity $I_0$, enters an absorbing medium, its intensity, $I$, decreases exponentially with the thickness, $l$, of the medium transversed.

$$\log_e \frac{I_0}{I} = Kl = 2.303 \log_{10} \frac{I_0}{I} \tag{7-22}$$

Beer's law states that the intensity of a ray of monochromatic light decreases exponentially with the concentration of the absorbing material.

$$\log_e \frac{I_0}{I} = K'C = 2.303 \log_{10} \frac{I_0}{I} \tag{7-23}$$

The two laws are usually combined:

$$\log_e \frac{I_0}{I} = kCl = 2.303 \log_{10} \frac{I_0}{I} \tag{7-24}$$

where $k$ is a constant depending only on the extinction coefficient of the absorbing substance. For most purposes, we use the "decadic" extinction coefficient, $k/2.303$. Equation 7-24 is known as the Lambert-Beer law.

There are no exceptions to Lambert's law. The behavior of many substances, however, is not adequately described by Beer's law. Discrepancies can occur when a solute ionizes, dissociates, or associates in solution. Also, if the light is not monochromatic, the law may not hold.

In absorbing solutions, light of different wavelengths is transmitted in different proportions. Beer's law holds only when the band width (range of frequencies) of the incident light is small, relative to the width of the absorption band, so that transmittance is approximately constant over all the frequencies of the incident light. This is tested by plotting $\log I_0/I$ or $\log T$ (transmittance) versus concentration. A straight line passing through the origin indicates conformity. If a substance does not conform, a standard calibration curve must be constructed, as in the case of protein determination by the Lowry reaction (see Appendix 6).

The value $\log_{10} I_0/I$ or $-\log T$ is called optical density (OD) or absorbence (A). The transmittance $(I/I_0)$ is usually less than 1.0, since it represents that fraction of light passed by a solution.

In practice, it is impossible to measure $I$ and $I_0$ as absolute values because of solvent absorption, light scattering, etc., so most measurements actually show:

$$\text{Transmittance} = \frac{T \text{ sample}}{T \text{ blank}} \qquad (7\text{-}25)$$

where $T$ blank is that of the solvent alone and $T$ sample is the transmittance of the solvent plus the absorbing material. In this way, all effects of solvent are essentially cancelled out, since the concentration of the solvent in both solutions is nearly the same (about $56\ M$ in the case of water). However, where the absorption of the solvent is large or where the concentration (mole fraction) of solute is large, this may not be so. It is best to remember that you are always measuring a difference spectrum, between the solute and the displaced solvent.

Each absorbing substance at a given wavelength has its own $k$ value or extinction coefficient. The molar extinction of a substance is numerically equivalent to the optical density of a $1\ M$ solution of a substance, with a light path of 1 cm. Since optical density is dimensionless, the extinction, $E$, is in terms of liters/mole/cm or $M^{-1}\ cm^{-1}$.

The molar extinction ($E_{1cm}^{1M}$) is equivalent to that of 1 millimole of substance /ml where the light path is 1 cm. The millimolar extinction coefficient is in terms of liters/millimole/cm and is one-thousandth of the molar extinction coefficient.

### (ii) Methods

For most purposes a "single beam" spectrophotometer can be used. The limits of 100% and 0% transmittance (zero and infinite absorbance) are set with a blank or reference solution and an opaque block or closed shutter.

Sometimes one wants to detect interactions between substances which produce very small absorbance changes. For this, a "double-beam" instrument is best. Spectrophotometers such as the Cary 15 are double-beam instruments —the absorbence of the sample is plotted continuously, relative to the absorbence of a blank. In other words, double-beam instruments measure directly the difference spectrum between the sample and blank (reference).

Two solutions containing equal concentrations of the absorbing substance are measured against each other over the desired wavelength range to obtain a baseline. If the interacting compound or "perturbant" doesn't absorb over the wavelength region, it is mixed with the sample while an equal volume of solvent is added to the reference. If there is a change in absorbence, a "difference spectrum" is obtained.

If the perturbant absorbs, two pairs of cells can be used. Perturbant is added to the reference cell without the material whose spectrum is being measured, and solvent is added to the other reference curvette, with the material. The perturbant is added to the sample cell with the material, and solvent is added to the other.

The difference spectrum itself may or may not be subjected to analysis: most often it is used to demonstrate an interaction between an absorbing substance and the perturbant, generally to measure the amount of perturbant bound (21) or of the absorbing substance (22).

### 7.4b   Measurement of Fluorescence

#### (i) Theory

The fluorescence intensity of a molecular species is dependent upon the amount of the substance present and, with the light sources commonly in use, upon the intensity of the exciting light (that is, on the light which is absorbed by the substance).

The fraction of incident light absorbed is constant for a given concentration of absorbing substance, no matter what the intensity of the incident light is (Beer's law). Since the amount of light emitted by a fluorescent substance depends on the amount of light absorbed, the fluorescence will be proportional to the incident light intensity while the optical density will not. All light sources vary greatly in the number of photons emitted per unit time according to the wavelength. An illustration of this is the profiles of the Xenon and mercury arcs as shown in Figure 7.12. Thus, the spectral energy distribution of the light source is not nearly as important in absorption measurements as it is in fluorescence work.

For the same reason, variations in lamp output at a given wavelength with time are not important in absorption (at least in double beam spectrophotometers), but are directly related to fluorescence intensity. Such variations may occur because of fluctuations in line voltage, but in addition, some lamps show a systematic variation in output with time. The output of the Xenon arc decreases steadily for several hours after ignition, before finally achieving stability. For this reason, more recent fluorometers monitor the incident light intensity with a second photomultiplier and display the results as the ratio of emitted to incident light intensity.

The efficiencies of transmission by the monochromator and detection by the photomultiplier vary with wavelength, and these factors are also important in measurements of fluorescence intensities as a function of wavelength (Figure 7.11).

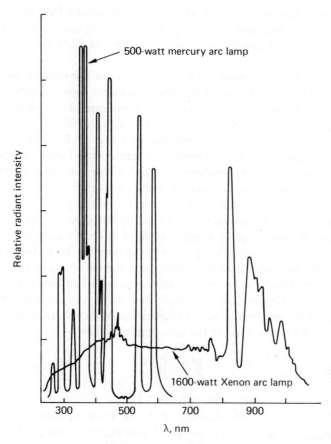

**Figure 7.12**   Emission profiles of a line source (a 500-watt mercury arc lamp) and a continuous source (a 1600-watt Xenon arc). Note the greater energy in the mercury emission lines.

### (ii) Methods

*1. Measuring relative fluorescence intensities.* This is essentially a matter of replacing one solution with another, but care should be taken that shifts in emission wavelengths are not occurring. The emission spectra of selected samples should be taken if this is suspected. Care should also be taken that all solutions are measured at the same temperature and that the absorbencies of the samples are not too high (see 7.5b). A standard sample should be re-checked occasionally to guard against variations in incident light intensity.

*2. Measurement of the quantum yield.* The central problem in measuring quantum yields is that it requires comparison of the total number of quanta of polychromatic light emitted with the number of quanta of monochromatic light absorbed. The number of quanta of monochromatic light incident on the sample can be measured with a thermopile or chemical actinometer and the number absorbed can be calculated from this and the optical density of the substance.

Three factors must be considered with respect to the fluorescence quantum yield measurement. In the first place, the fluorescent light is emitted in all directions, though not equally in all directions because of the shape of the cuvette and the refractive index of the solvent. This introduces problems of geometry in the measurement of the emitted light. Second, the different wavelengths of the emitted light are transmitted and detected with varying efficiencies by the emission monochromator and photomultiplier. Third, the different wavelengths are of different intensities to begin with, so the number of quanta emitted must be obtained by an integration process.

Experimentally, problems with the incident light and the geometry of the apparatus can be eliminated by using a standard of known quantum yield in the same solvent. The integration process is the same for both standard and sample, but the transmission efficiency and detection sensitivity must be determined first. Note that if the emission spectra of standard and sample have the same emission maximum, half-bandwidth and shape, the quantum yield of the sample can be obtained by direct comparison of the relative fluorescence intensities (usually measured at the emission maximum):

$$\frac{Q_1}{Q_2} = \frac{F_1}{F_2} \qquad (7\text{-}26)$$

However, this is seldom the case (Figure 7.13).

To determine the overall efficiency of the emission measuring system of the fluorometer, a lamp of known color temperature (measured with an optical pyrometer) can be used if the color temperature of the source normally used is not known. The spectral energy distribution of the lamp is given by Wein's law (2).

A simpler, though approximate, method can be used instead. Aluminum foil, which has an approximately constant reflectance over the visible and ultraviolet, is placed in the cuvette holder to reflect the incident light into the emission monochromator slits. The meter reading gives the incident light intensity ($I$) times the efficiencies of the excitation ($E_x$) and emission ($E_m$) monochromators[5] and the detector ($P$):

$$F(\lambda) = \text{meter reading with both monochromators set}$$
$$\text{for the same wavelength} = I(\lambda)E_x(\lambda)E_m(\lambda)P \qquad (7\text{-}27)$$

A series of such readings are taken over the spectral range of interest.

[5]This is also true if a filter is used instead of the second monochromator; the filter is removed when the aluminum foil is set in place.

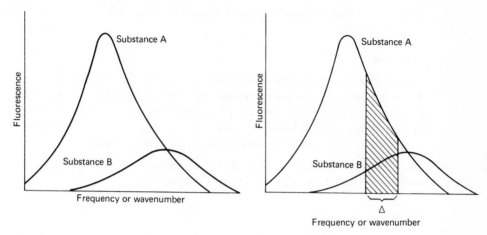

**Figure 7.13** Comparison of relative quantum ($Q$) and fluorescence ($F$) yields of two substances when a spectral shift in emission occurs. Note that the comparison is valid only when the abscissa is *wavenumber* or *frequency* (17).

Then the aluminum foil is replaced by a "quantum counter": a cuvette containing a fluorescent dye so concentrated that all the incident light of all wavelengths is absorbed.[6] "Front surface" (see 7.5b) detection may have to be used. As a result, the fluorescence at a given wavelength is proportional only to the intensity of the exciting light. Readings are taken varying the excitation wavelength, at a constant emission wavelength.

$$F'(\lambda) = \text{meter reading with the emission monochromator}$$
$$\text{set for the fluorescence emission maximum of the}$$
$$\text{dye} = I(\lambda)E_x(\lambda) \cdot E_{m\lambda'}P_{\lambda'} \qquad (7\text{-}28)$$

$E_{m\lambda'}$ is the efficiency of the emission monochromator and $P_{\lambda'}$ that of the photomultiplier, both at $\lambda'$. These factors are constant. So, dividing $F(\lambda)$ by $F'(\lambda)$ for each value of $\lambda$ should give the relative efficiency of the emission measuring system as a function of $\lambda$:

$$\frac{F(\lambda)}{F'(\lambda)} = \frac{E_m(\lambda)P}{E_{m\lambda'}P_{\lambda'}} = \frac{E_m(\lambda)P}{\text{constant}} = P(\lambda) \qquad (7\text{-}29)$$

Correction of the emission spectra of sample and standard using the values of $P(\lambda)$, converting the spectra to functions of frequency or wavenumber, and integrating will give a reasonably accurate value for the quantum yield of the sample.

---

[6]Rhodamine B will work reasonably well, but since it aggregates at concentrations above $10^{-5}M$, other dyes should be used for accurate work.

### 7.4c  Use of Fluorometry in Studies of Binding of Small Molecules by Proteins

**(i) Theory**

Binding problems may be divided into two categories: the determination of stoichiometry, the maximum number of moles of ligand bound per mole of protein; and the determination of the equilibrium constants describing the interaction. The information required in each case and the overall design of meaningful binding experiments are described in this section.

*1. Description of ligand binding.* The mathematical description of a simple equilibrium (Equation 7-30) is given in Equation 7-31.

$$PX \rightleftharpoons P + X \tag{7-30}$$

$$K = \frac{[P][X]}{[PX]} = \frac{(1 - \phi)}{\phi}[X] \tag{7-31}$$

where $[P]$ is the concentration of free protein, $[P_0]$ the total protein concentration, $[X]$ the concentration of free ligand, $[X_0]$ the total ligand concentration, and $[PX]$ is the concentration of complex. Whenever a protein possesses more than one binding site per molecule, the concentration $[P]$, $[P_0]$, and $[PX]$ represent equivalents per liter. In addition, $\bar{n}$ is the average number of moles of ligand bound per mole of protein, $N$ the maximum number of moles of ligand bound per mole of protein over a certain range of ligand concentration, and $\phi$ is $[PX]/[P_0]$, the fractional saturation of protein with ligand, and equals $\bar{n}/N$.

The simplest case of ligand binding is described by a single dissociation constant $K$. Analysis of the behavior of such a system will enable us to predict conditions appropriate for determinations of stoichiometry and of equilibrium in both simple and complex systems.

*2. Determination of stoichiometry.* Stoichiometry is often established by measurements of fractional saturation $\phi$ as a function of the total ligand concentration $[X_0]$ added to a solution of fixed protein concentration $[P_0]$. Correct interpretation of such data depends upon understanding of the role of the protein concentration, relative to the equilibrium constant (see Figure 7.14).

The plots of $\phi$ *versus* $[X_0]/[P_0]$ shown in Figure 7.14 were calculated from the following form of the Equation 7-31:

$$\frac{[X_0]}{[P_0]} = \frac{K}{[P_0](1 - \phi)}\phi + \phi \tag{7-32}$$

Whenever $[P_0] \gg K$, the curves are composed of two linear segments de-

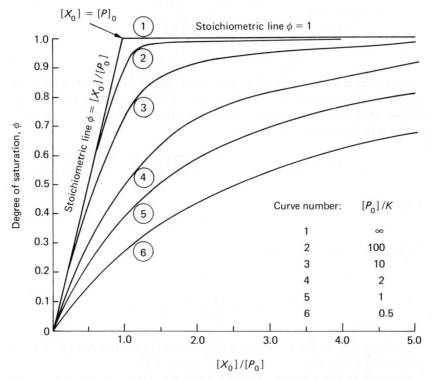

**Figure 7.14**  Effect of protein concentration on results obtained by titration of a protein with a ligand, $K$ being constant. Calculated from Equation 7-31.

scribed by the equations $\phi = [X_0]/[P_0]$ and $\phi = 1$. The intersection of these lines, or end point of the titration, is used in the calculation of combining weight and hence the average number of binding sites possessed by each protein molecule.

It must be realized that this method is valid only when the stoichiometric plot contains *well-defined* linear portions such as those in Figure 7.14.

*3. Determination of the protein-ligand equilibrium constant(s).*

a. The role of protein concentration in the multiplication of errors: Evaluation of $K$ requires knowledge of the concentration of three components—free ligand, free protein, and protein-ligand complex. Many experimental techniques, however, do not provide direct values of concentration. The differential methods (see 7.4c(ii)) involve correlation of some measurable change accompanying binding with the degree of saturation $\phi$. The concentration of the individual components must be calculated from $\phi$ and the known values of $[X_0]$ and of $[P_0]$.

$$[X] = [X_0] - \phi[P_0] \tag{7-33}$$

$$[P] = (1 - \phi)[P_0] \tag{7-34}$$

$$[PX] = \phi[P_0] \tag{7-35}$$

Assurance that these differences give meaningful values of $K$ demands consideration of potential experimental errors. Weber has presented a mathematical analysis of this problem (23). The analysis is based on the concept of probability of binding, which is defined as

$$p = \frac{\text{actual concentration of protein-ligand complexes}}{\text{maximum possible concentration of protein-ligand complexes}} \tag{7-36}$$

When $[X_0] \geq P_0$,

$$p = [PX]/[P_0] - \phi \tag{7-37}$$

On the other hand, when $[X_0] \leq [P_0]$,

$$p = \frac{[PX]}{[X_0]} = \frac{\frac{[P_0]}{K}(1 - \phi)}{\frac{[P_0]}{K}(1 - \phi) + 1} \tag{7-38}$$

The results of Weber's analysis, which relates the relative error in $K$ to the probability of binding, are presented in Figure 7.15.

The error in $K$ is a minimum when $p$ is 0.5, whereas it increases very sharply in the range above 0.8 and below 0.2. A few calculations using the above equations indicate that *a protein–ligand equilibrium can be accurately determined from measurements of p only when the protein concentration is comparable in magnitude to, or smaller than, K*. This is in contrast to the conditions for stoichiometric titration, where the protein concentration is much greater than $K$ and essentially all ligand is bound; that is, $p = 1$.

Preliminary experiments to establish favorable protein concentrations are consequently necessary to evaluate $K$. Since specific interactions in biological systems are frequently characterized by values of $K$ of less than $10^{-6}\ M$, selection of an appropriate protein concentration and analytical method requires care. The probability of binding and experimental error must both be considered whenever evaluating the reliability of a binding study.

b. Presentation of the results: Early studies of the binding of oxygen by hemoglobin indicated that results could be described by an empirical expression known as the Hill equation:

$$K' = \frac{(1 - \phi)}{\phi}[X]^j \tag{7-39}$$

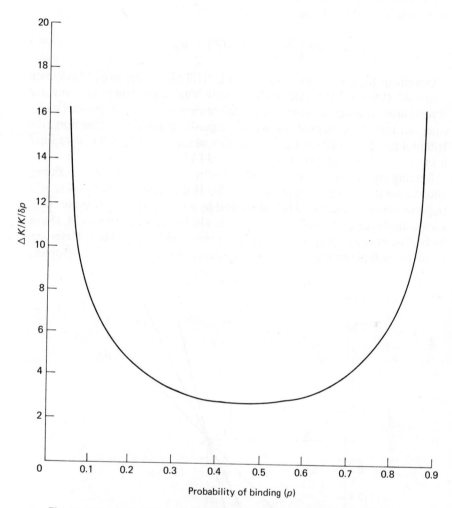

**Figure 7.15** Relative error in $K$ ($\Delta K/K$) as a function of probability of binding. $\delta p$ is the uncertainty in $p$. [Modified from G. Weber in *Molecular Biophysics*, B. Pullman and M. Weissbluth, Eds., Academic Press, New York (1965). Reprinted by permission of the publisher.]

The exponent $j$, which has no simple physical significance, is called the "Hill constant" or "order" of the binding reaction. The Hill reaction is often written in logarithmic form.

$$\log \frac{\phi}{1 - \phi} = j \log [X] - \log K' \qquad (7\text{-}40)$$

Graphs of $\log \phi/(1 - \phi)$ versus $\log [X]$, "Hill plots," are used to calculate $j$ and $K'$ (Figure 7.16). Although an individual equilibrium may involve several intermediates, or even intrinsically different species of protein molecule, our analytical methods do not distinguish among them. Therefore, the Hill plot has been used for the presentation of all types of binding data since it involves only the known quantities $\phi$ and $[X]$.

Binding equilibria may be divided into three categories: simple, multiple, and cooperative. The average value of the Hill constant permits us to distinguish among these. Equilibria described by a unique dissociation constant, such as the binding of NADH by horse liver alcohol dehydrogenase (24, 25), or by beef heart lactic dehydrogenase (25), correspond to $j$ equal to 1. Heterogeneous or multiple equilibria, such as the interaction of haptens with $\gamma$-globulin

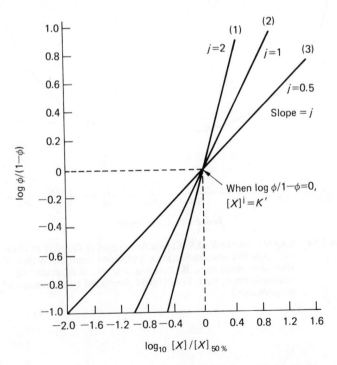

**Figure 7.16**   Hill plots. For normalization $[X]$ was divided by $[X]_{50\%}$, which represents the free ligand concentration at 50% saturation.

(26) are approximated by values of $j$ less than 1. In cases of cooperative binding, interaction of the protein with each successive mole of ligand appears to enhance its affinity for the next and thus $j$ is greater than 1. The equilibrium of oxygen with hemoglobin is a classical example of cooperative binding (27).

Other graphical methods have been used in binding studies. However, since the values of $[X]$ must cover a large concentration range, the free ligand concentration should be presented in an intelligently selected logarithmic scale. Plots of $-\log [X]$ versus $\phi$ or $\bar{n}$ are recommended since they reflect both the stoichiometry and the nature of the equilibrium (Figure 7.17). This graphical method is used to describe the dissociation of complex ions, acids, and bases.

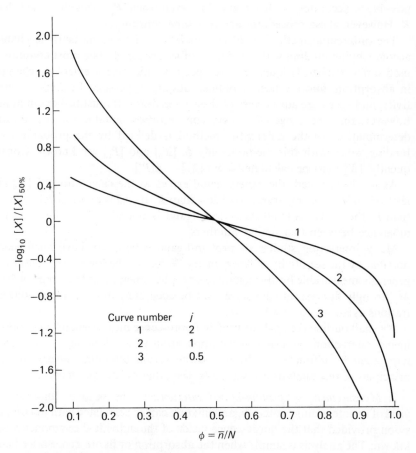

**Figure 7.17** Plots of $-\log [X]$ versus $\phi$, showing simple, multiple, and cooperative binding curves.

### (ii) Experimental procedures

*1. Types of procedure.* Two basic types of experimental procedure have been used in measurements of ligand binding by proteins. On the basis of the information provided by each, we shall call these the partition method and the differential method.

In the partition method, the protein-ligand system is physically separated from a phase or solution which contains free ligand (or protein) alone at the equilibrium concentration. Examples of this method include: the gasometric techniques used in studies of the equilibria of respiratory proteins with gases; ultracentrifugation, in which the centrifugal field ideally produces sedimentation of the protein-ligand complex in a solvent containing free ligand; equilibrium dialysis; and gel filtration (Chapter 3). Since the free ligand concentration $[X]$ and fractional saturation $\phi$ are measured independently, it is possible to accurately evaluate equilibria even when $[P_0]$ is much greater than $K$. However, these procedures are very time-consuming.

The differential method involves correlation of some measurable change accompanying binding with the degree of saturation $\phi$. The most commonly used differential techniques are the spectroscopic measurement of changes in absorption, fluorescence, or optical rotatory dispersion. Flexibility, sensitivity, and speed are advantages of these procedures. It must be kept in mind, however, that the range of protein concentration suitable for equilibrium determination by the differential method is defined by the probability of binding, since with this method, only $\phi$, $[X_0]$ and $[P_0]$ are known. Consequently, $[X]$ must be calculated from $[X_0] - \phi[P_0]$.

As has been noted, the experimental errors in the calculation of $[X]$ are least when $[P_0]$ is comparable to or less than $K$. If $[P_0]$ is considerably greater than $K$, then essentially all the added ligand is bound and $[X]$ becomes the difference between two similar numbers.

Many interesting studies of protein-ligand interaction have been made possible by the sensitivity of fluorometry. Whereas the lowest concentration generally measurable by absorption spectrophotometry is in the range of $10^{-6}$ $M$, strongly fluorescent substances can be accurately determined at concentrations as low as $10^{-8}$ to $10^{-9}$ $M$.

We shall outline the analysis used in fluorescence measurements of protein-ligand interaction, and its specific application to the binding of 1-anilino-napthalene-8 sulfonate (ANS) by bovine serum albumin. Several of the principles of this analysis are shared by the other differential techniques.

*2. Measurement of protein-ligand interaction.* The relative fluorescence yield of an interacting protein-ligand mixture can be correlated with its composition provided that the fluorescence yields of the individual components are known. The analysis is simple when the absorption or fluorescence arise from only one moiety of the complex. The derivation presented assumes that only $X$ and $PX$ contribute, and $P$ neither absorbs nor fluoresces. The three principles of this analysis are that:

a. $X$ and $PX$ have different relative fluorescence yields ($RFY$).

RFY of free ligand at known concentration $[X_0] = F_0 = k_0[X_0]$ (7-41)

RFY of bound ligand at known concentration
$$[X_0] = F_{00} = k_{00}[X_0] \qquad (7\text{-}42)$$

$k_0$ and $k_{00}$ are proportionality constants: the fluorescence quantum yields of free and bound ligand, respectively.

b. $[X_0] = [X] + [PX]$ (7-43)

That is, the total ligand concentration is the sum of the bound and unbound ligand concentrations.

c. The relative fluorescence yield of a mixture is composed of the additive contributions of its components.

Now the $RFY$ of a mixture of free and bound ligand at a total concentration $[X_0]$ is

$$F_{obs} = k_0[X] + k_{00}[PX] \qquad (7\text{-}44)$$

Combination with Equations 7-41 to 7-43 gives

$$F_{obs} = k_0[X_0] + (k_{00} - k_0)[PX] \qquad (7\text{-}45)$$

Division by $F_0$ yields

$$\frac{F_{obs}}{F_0} = 1 + \frac{(k_{00} - k_0)}{k_0}\frac{[PX]}{[X_0]} \qquad (7\text{-}46)$$

$$= 1 + \left(\frac{F_{00}}{F_0} - 1\right)\frac{[PX]}{[X_0]} \qquad (7\text{-}47)$$

Therefore the expression relating to the relative fluorescence yield to the fraction of ligand bound ($f$) is:

$$f = \frac{F_{obs} - F_0}{F_{00} - F_0} = \frac{\dfrac{F_{obs}}{F_0} - 1}{\dfrac{F_{00}}{F_0} - 1} \qquad (7\text{-}48)$$

The calculation of concentrations is then straightforward

$$[PX] = f[X_0] \tag{7-49}$$

$$\phi = f\frac{[X_0]}{[P_0]} = \frac{\bar{n}}{N} \tag{7-50}$$

$$[X] = (1 - f)[X_0] \tag{7-51}$$

$$[P] = [P_0] - f[X_0] \tag{7-52}$$

When ANS is bound by serum albumin, the ligand fluorescence is enhanced more than two hundred-fold. In such a favorable case the expression for $f$ becomes simply:

$$f = \frac{F_{obs}}{F_{00}} \tag{7-53}$$

**(iii) Some practical considerations**

*1. Determination of $F_{00}$.* Quantitative evaluation of any property distinguishing the protein-ligand complex is possible only when the contributing moiety is present exclusively in the complexed form. Thus the relative fluorescence yield of bound ligand ($F_{00}$) must be measured under conditions producing complete binding of the ligand within the limits of experimental error. When $X_0$ is less than $P_0$, the fraction of ligand which is bound ($f$) is given by:

$$f = \frac{\dfrac{[P_0]}{K}(1 - \phi)}{\dfrac{[P_0]}{K}(1 - \phi) + 1} \tag{7-38a}$$

Since $f$ approaches unity whenever $[P_0]$ is much greater than $K$, $F_{00}$ is determined by measurement of the fluorescence of a fixed concentration of ligand as a function of the total protein concentration. The relative fluorescence yield which is independent of $[P_0]$ corresponds to $F_{00}$.

*2. Selection of wavelengths for excitation and fluorescence measurements.* Overlooked variations in optical density may cause discrepancies in the observed fluorescence yields. The absorption spectra of free and bound ligand must be determined. If appreciable spectral changes accompany binding, monochromatic light of a wavelength corresponding to the isosbestic point should be chosen for excitation (Figure 7.18).

The fluorescent standard should have the same absorbence as that of the ligand at $\lambda_{isosbestic}$. Fluorescence measurements should be undertaken in a spectral region where the difference between the relative fluorescence yields

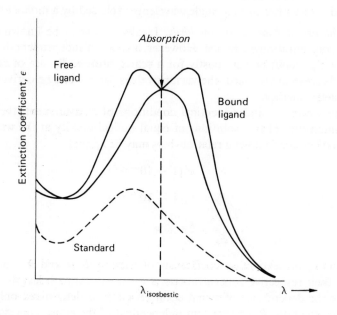

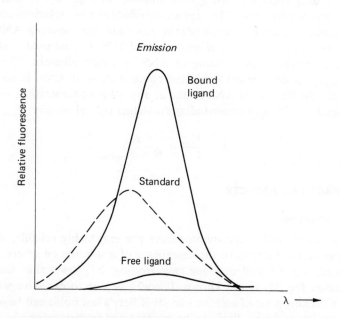

**Figure 7.18** Wavelength selection for excitation and emission measurements.

of free and bound ligand is maximal. This may be either within a broad region isolated with a filter or at a single wavelength selected by a monochromator.

3. *The use of fluorescent standards.* Ideally, $F_{00}$ must be known for each ligand concentration examined. However, a series of independent determinations of $F_{00}$ could be very costly for a scarce protein. By use of an appropriate fluorescent standard, the various values for $F_{00}$ can be derived from a single determination.

If one observes, under identical conditions of measurement, the fluorescence intensities of two solutions of equal optical density at the wavelength of excitation, the following relationships may be applied.

$$F_A = k'_A(1 - 10^{-\epsilon_A c_A d}) \tag{7-54}$$

$$F_B = k'_B(1 - 10^{-\epsilon_B c_B d}) \tag{7-55}$$

$$\frac{F_A}{F_B} = \frac{k'_A}{k'_B} = R \tag{7-56}$$

$\epsilon_A$ and $\epsilon_B$ are extinction coefficients of substances $A$ and $B$, and $C_A$ and $C_B$ are their respective concentrations. $d$ is the optical path length.

Since the dependence of signal upon geometry is determined only by absorption, the ratio $R$ is a constant independent of the actual concentrations.

The fluorescence and absorption spectra of a good standard overlap the corresponding spectra of the system studied. See Figures 7.13 and 7.18. In cases where the free ligand has appreciable fluorescence, solutions containing free ligand alone are used as standards. Since unbound aqueous ANS is only slightly fluorescent, solutions of quinine in $0.2\ M\ H_2SO_4$ are used as standards in measurements of the binding of ANS by serum albumin. The ratio $R$ ($F_{00}/F_{\text{quinine}}$) is determined at a fixed concentration of ANS. If we wish to determine the fraction of ANS bound at any other concentration, we simply compare $F'_{\text{obs}}$ with its corresponding (matched optical density) $F_{\text{quinine}}$:

$$f = \frac{F_{\text{obs}}}{F_{00}} = \frac{F'_{\text{obs}}}{R \times F'_{\text{quinine}}} \tag{7-57}$$

## 7.5 PRACTICAL ASPECTS

### 7.5a Absorption

Most commercial spectrophotometers are reasonably reliable, though a colleague recently measured the absorbence of a solution on several spectrophotometers and found that one gave readings 5% lower than the others. The reason for this is not known. It could be due to excessive stray light levels. Checking a set of solutions to see if Beer's law holds can be useful.

The major problems likely to be encountered in measuring absorbencies are from sample preparation—dilution errors, excessive scattering from turbidity or lint, etc. Care in sample preparation is the only answer to these problems.

### 7.5b Fluorescence

Mainly because of its greater sensitivity, fluorescence measurements are far more likely to be affected by artifacts.

#### (i) Incident intensity fluctuations

Lamp intensity variations due to age of the lamp or to line voltage fluctuations can best be overcome with a stable power supply for the lamp, a stable ac power supply to the instrument and/or the use of the ratio of emitted to incident light intensity. Otherwise, line voltage fluctuations can be a big problem, especially in the summertime.

#### (ii) Analysis of concentrated solutions

Several pitfalls caused by high optical density complicate fluorescence measurements of concentrated solutions. We assume that the wavelength calibration is correct for the fluorometer.

1. One must demonstrate the absence of appreciable absorption in the region of fluorescence measurement since overlap of the absorption and emission spectra can result in reabsorption and an apparent decrease in fluorescence yield.

2. The dependence of signal upon geometry must always be kept in mind since only a fraction of the emitted light is within the field of the detector. Total absorption of the incident light occurs within a thin layer of solution of high optical density. The resulting fluorescence, which is restricted to a layer of solution adjacent to the irradiated cuvette surface, is detectable by surface measurement but not by right-angle measurement (Figure 7.19). The highest optical density used in a right-angle instrument should be less than 0.2.

#### (iii) Fluorescent impurities

The lowest measurable concentration is usually determined by the magnitude of background fluorescence. Pure reagents must be used in the preparation of solutions and inadvertent contamination must be scrupulously avoided in order to minimize the level of impurities (Chapter 1). Rubber stoppers, body surfaces, stopcock grease, filter paper, and untreated dialysis tubing are some sources of contamination. All solutions should be prepared and stored in clean glass-stoppered containers. Any turbidity, which causes errors by scattering, should be removed by centrifugation or filtration through a Millipore filter.

The contribution of impurities is estimated by measurement of the fluorescence of the solvent. If the background is only a small fraction of the signal, the readings are corrected by subtraction of the "blank."

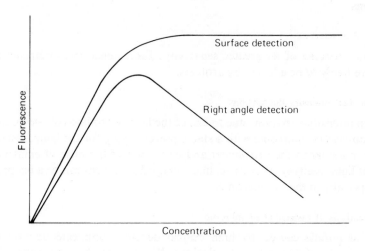

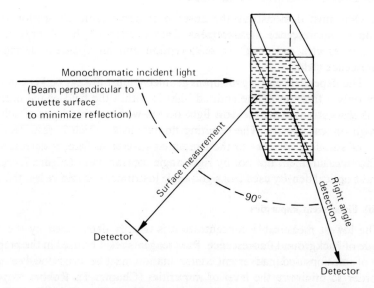

**Figure 7.19** Dependence of fluorescence on geometry. With surface measurement, the emission is proportional to $(1 - 10^{-\epsilon cl})$, where $\epsilon$ is the extinction coefficient of the substance, $c$ the concentration and $l$ the pathlength.

### (iv) Deterioration of solutions

Several sources of instability may cause deterioration of the sample either before or during measurement. Photochemical decomposition and loss of both protein and ligand by adsorption to glass are common problems. The solutions should be stored in the dark and protected from prolonged exposure to the exciting light. Losses by adsorption to glass in dilute solutions of either protein or ligand alone are revealed by lack of proportionality between fluorescence and concentration. Application of a silicone coating to the cuvettes and to all glass surfaces helps to prevent adsorption (Chapter 1). Very dilute solutions should be prepared immediately before measurement.

### (v) Temperature effects

The fluorescence enhancement factor is very temperature dependent, as is fluorescence generally. Be sure all solutions are at the same temperature before measurement; a thermostatic cuvette holder is provided with some recent fluorometers.

## 7.6 SUMMARY

Spectroscopic techniques use electromagnetic radiation as a "probe" to characterize substances or to follow changes in concentration. Like all electromagnetic radiation, light is characterized by its wavelength (frequency), intensity, and polarization (28). This constitutes the information potentially available with spectroscopic techniques, since interaction of light with matter will change one or more of these characteristics. In absorption measurements, we commonly observe changes in intensity of our "probe" as the light is sent through a solution. In fluorescence measurements, the material changes some fraction of the absorbed light to light of another wavelength. Because of the time delay involved in fluorescence it is sensitive to processes occurring much more slowly than those which affect absorption. These factors make fluorescence a more specific and informative property than absorption. Measurements involving polarization of absorption and fluorescence can make use of additional information potentially present in light and, for this reason, are becoming more widely used.

## REFERENCES

1. Ditchburn, R. W., *Light*, Interscience, New York (1953).
2. Seliger, H. H. and W. D. McElroy, *Light: Physical and Biological Action*, Academic Press, New York (1965).
3. Van Holde, K. E., *Physical Biochemistry*, Prentice-Hall, Englewood Cliffs, New Jersey (1971).

4. Pesce, A. J., C. G. Rosen and T. L. Pasby, *Fluorescence Spectroscopy—An Introduction for Biology and Medicine*, Marcel Dekker, New York (1971).
5. Barrow, G. M., *Introduction to Molecular Spectroscopy*, McGraw-Hill, New York (1962).
6. Moore, W. J., *Physical Chemistry*, 4th ed., Prentice-Hall, New Jersey (1972).
7. Jaffe, H. H. and M. Orchin, *Theory and Applications of Ultraviolet Spectroscopy*, Wiley, New York (1962).
8. Kasha, M. in *Light and Life*, W. McElroy and B. Glass, Eds., Johns Hopkins Press, Baltimore (1961) p. 31.
9. Weber, G. and J. F. W. Teale in *The Proteins*, Vol. III, H. Neurath, Ed., Academic Press, New York (1965), Ch. 17.
10. Kasha, M., *Radiation Research*, **20**, 55 (1963).
11. Forster, A., *Fluoreszenz Organisches Verbindungen*, Vendenboeck und Ruprecht, Gottingen (1954).
12. Brand, L. and B. Witholt in *Methods in Enzymology*, Vol. XI, C. H. W. Hirs, Ed., Academic Press, New York (1967), p. 776.
13. Spencer, R. D. and G. Weber, *Ann. N. Y. Acad. Sci.*, **158**, 361 (1969); also *J. Chem. Phys.*, **52**, 1654 (1970).
14. Galley, J. A. and G. M. Edelman in *Quantum Aspects of Polypeptides and Polynucleotides*, M. Weissbluth, Ed., Biopolymers Symposium #1, Interscience, New York (1964), p. 367.
15. Weber, G. and K. Rosenheck in *Quantum Aspects of Polypeptides and Polynucleotides*, M. Weissbluth, Ed., Biopolymers Symposium #1, Interscience, New York (1964), p. 333.
16. Weber, G., *Biochem. J.*, **47**, 114 (1950).
17. Parker, C. A. *Photoluminescence of Solutions*, Elsevier, New York (1968).
18. Cook, R. B. and R. Jankow, *J. Chem. Educ.*, **49**, 405 (1972).
19. "Electron Tubes," Amperex Data Handbook, Amperex Electronic Corp., Hicksville, New York (1969).
20. James, J. F. and R. S. Sternberg, *The Design of Optical Spectrometers*, Chapman and Hall, London (1969).
21. Kayne, R. and C. H. Suelter, *J. Amer. Chem. Soc.*, **87**, 897 (1965).
22. Herskovits, T. T. in *Methods in Enzymology*, Vol. XI, C. H. W. Hirs, Ed., Academic Press, New York (1967), p. 748.
23. Weber, G. in *Molecular Biophysics*, B. Pullman and M. Weissbluth, Eds., Academic Press, New York (1965), p. 369.
24. Theorell, H., *Adv. in Enzymol.*, (F. F. Nord, Ed.), **20**, 31 (1958).
25. Anderson, S. R. and G. Weber, *Biochemistry*, **4**, 1948 (1965).
26. Nisonoff, A., F. C. Wissler and D. L. Woernley, *Arch. Biochem. Biophys.*, **88**, 241 (1960).
27. Rossi-Fanelli, A., E. Antonini and A. Caputo, *Adv. in Protein Chemistry*, **19**, 73 (1964).
28. Feofilov, P. P., "The Physical Basis of Polarized Emission," Consultants Bureau, New York (1961).

# 8 | RADIOACTIVITY AND COUNTING

## 8.1 INTRODUCTION

In the last few decades it has been of great assistance to biochemists to be able to discriminate by physical methods between substances which are chemically indistinguishable. In this chapter we discuss the uses of discrimination between stable isotopes and unstable ones.

We shall describe why some atomic nucleii are unstable, according to current theories of nuclear structure, and what the results of this instability are. Then we shall discuss some of the instrumentation used to measure amounts and types of the products of nuclear instability.

The experimental applications of radioactivity to biochemists are discussed in general terms. Since measurement of radioactivity involves counting, we describe the somewhat specialized statistics involved.

263

## 8.2 ORIGIN AND NATURE OF RADIOACTIVITY (1–7)

### 8.2a Nuclear Structure

To a first approximation, atomic nuclei are composed of protons and neutrons, such that:

$$A = Z + N \qquad (8\text{-}1)$$

with: $Z$ the atomic number (number of protons), $N$ the number of neutrons, and $A$ the mass number. Neutrons and protons together are called "nucleons." The convention for showing the gross nuclear composition of a particular isotope is, for example for phosphorous-32 ($^{32}P$), written $^{32}_{15}P_{17}$, where 32 is $A$, 15 is $Z$, and 17 is $N$.

It is now believed that neutrons and protons in a nucleus are interchangeable, by release and capture of $\pi^+$ mesons, particles which are associated with or perhaps constitute nucleons (1). This constant exchange of these particles seems analogous to the sharing of electrons by atoms in a chemical bond, in which delocalization provides the energy for chemical bonding.

While the $\pi^+$ mesons may function as a kind of nuclear "glue," there are about thirty other "fundamental" particles now known, and it is quite likely that nucleons have a structure and composition themselves. For our purposes, it is of more relevance that nucleons in a nucleus appear to have some kind of structural arrangement which bears some similarities to the arrangement of the electrons around a nucleus. For example, it has long been known that certain numbers ("magic numbers") and combinations of nucleons are more stable than others, which suggests an energy level, orbital-type arrangement. Furthermore, if unstable nuclei are oriented (as in some chemical compounds or with a magnetic field), there are certain preferred directions for emission of radiation. This seems very like the directional characteristics of some electronic orbitals like the d orbitals.

A variety of experimental evidence, much of it from studies of natural and induced radioactivity of various elements, is consequently interpreted in terms of a "shell-like" nucleus, in which the neutrons and protons are arranged in orbitals much like the s, p, d, etc., orbitals of electrons. While this model is under constant refinement, the language and many of the concepts derived from atomic and molecular spectroscopy (see Chapter 7) have been extensively used to correlate much of the data on nuclear structure (2). A major difference between nuclear reactions and electronic reactions (or roughly the difference between physics and chemistry) lies in the many orders of magnitude greater energies (and hence greater velocities) of the former. See Table 8.1.

Empirically, the stability of a nuclide is determined by its neutron to proton ratio. For the light nuclides this is one, but for the heavier nuclides the ratio

**TABLE 8.1** ELEMENTARY PARTICLES ASSOCIATED WITH NUCLEAR STRUCTURE

| Name | Rest Mass, electron units | Occurrence | Electronic Charge | "Function" |
|---|---|---|---|---|
| Proton | 1836 | Nucleus | +1 | Neutralizes electron charge |
| Neutron | 1839 | Nucleus | 0 | Dilutes + charge |
| Electron | 1 | Extra-nuclear | −1 | Chemical interactions |
| $\mu^+$ meson | 207 | from $\pi^+$ meson decay | +1 | Possibly an excited state of a positron |
| $\mu^-$ meson | 207 | from $\pi^-$ meson decay | −1 | Possibly an excited state of an electron |
| $\pi^+$ meson | 273 | Nucleus | +1 | Nuclear "glue" between protons and neutrons |
| $\pi^0$ meson | 264 | Nucleus | 0 | Nuclear "glue" between like particles |
| $\pi^-$ meson | 273 | Nucleus | −1 | Nuclear "glue" between unlike particles |
| Neutrino | 0 | From $\beta^-$ decay | 0 | Conserve energy in $\beta^-$ decay |
| Antineutrino | 0 | Antiparticle | 0 | |
| Positron | 1 | Antiparticle | +1 | Carries away +1 charge in $\beta^+$ decay |

increases, apparently because the nuclear charge density becomes too high and must be diluted by additional neutrons (see Figure 8.1). Thus for $^2$H, $^4$He, $^{12}$C, and $^{16}$O, the ratio of $N/Z$ is 1; but for $^{200}$Hg the ratio is 1.5, and for $^{208}$Pb it is 1.53.

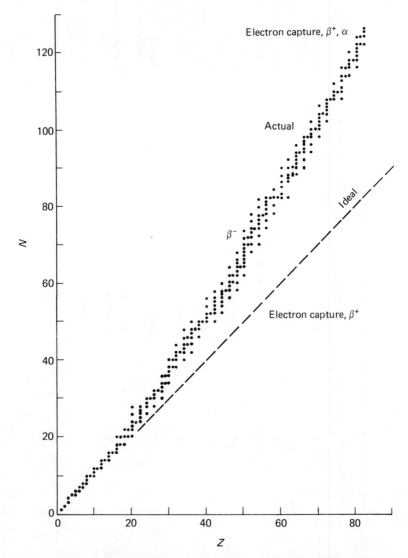

**Figure 8.1** Actual versus ideal nuclear compositions. Note how deviations in certain regions are associated with particular types of emission. [Modified from R. T. Weidner and R. L. Sells, *Elementary Modern Physics*, Allyn and Bacon, Boston (1960). Reprinted by permission of the publisher.]

For smaller nuclides, deviation from this ratio in either direction is associated with instability, a tendency to spontaneously readjust the ratio with loss of energy to give a more favorable ratio.

The mass of any nucleus is always less than the sum of the masses of its constituent protons and neutrons. For example, in the case of deuterium:

| | |
|---|---|
| Mass of neutron | 1.008665 amu[1] |
| Mass of proton | 1.007825 |
| | 2.016490 |
| Mass of deuterium nucleus | 2.014102 |
| Mass deficit, or "binding energy" | 0.002388 amu |

This missing mass is the origin of the nuclear binding energy. From Einstein's relation for the interconversion of mass and energy, we may calculate the energy, in Mev, equivalent to 1 amu.

$$E = mc^2$$
$$= \frac{1 \cdot 10^{20} \cdot 9}{6.025 \cdot 10^{23} \cdot 1.6 \cdot 10^{-6}}$$
$$= 931 \text{ Mev}$$

$1 \text{ Mev} = 1.6 \cdot 10^{-6} \text{ erg}$

$E$ = Energy    (8-2)

$m$ = mass in gm

$c$ = speed of light, $3 \cdot 10^{10}$ cm/sec

$N$ = Avogadro's number, $6.025 \cdot 10^{23}$

Accordingly, the binding energy for deuterium is $0.002388 \times 931 = 2.23$ Mev.

Carswell (1) points out that this is essentially the amount of energy necessary to produce free nucleons, to break up the $\pi$ meson sharing. It seems quite analogous to the fact that breaking up a molecule of hydrogen gas, for example, into two molecules of atomic hydrogen also requires energy to disrupt the electron sharing that occurs in a molecule.

### 8.2b  Origin and Characteristics of Radioactivity

The origin of radioactivity is nuclear disintegration. When this happens, three main types of radiation occur: alpha emission, beta emission (and neutrino emission), and gamma emission.

### (i) Alpha Emission

The alpha particle is a helium nucleus. It is composed of two neutrons and two protons and has two positive charges.

Alpha emission generally occurs only in nuclides of mass greater than 140. It may occur even though the neutron/proton ratio is favorable for stability,

---

[1]One atomic mass unit (amu) is one-twelfth the mass of a $^{12}$C carbon atom, which in 1961 was assigned the mass value of 12.000000. This is presently used as the standard in nuclear physics.

since loss of an alpha particle lowers the Coulombic repulsion between protons, while not affecting the nuclear binding forces as much (the alpha particle has about as much binding energy per nucleon as a nucleus of mass greater than 140). Alpha emission does not have neutrino emission associated with it and is thus monoenergetic.

This type of emission is not often encountered in biochemical tracer experiments.

### (ii) Beta Emission

Beta particles are electrons and occur in two forms: positrons ($\beta+$) and negatrons ($\beta-$). We will defer positron emission for consideration with the electron capture process. $^3$H, $^{14}$C, $^{32}$P, $^{36}$Cl all decay by $\beta^-$ emission. $\beta^-$ emission leads to a decrease in the neutron to proton (n/p) ratio, and is favored for elements with lower atomic numbers and n/p ratios above 1.0.

Some calculations of typical $\beta^-$ energies are given: These mass values are based on $^{16}$O $= 16.000000$ (8).

Consider the decay of tritium:

$$^3_1H_2 \longrightarrow {}^3_2He_1 + \beta^- + \nu \tag{8-3}$$

The neutron to proton ratio drops from 2.0 to 0.5. The calculated energy loss is:

$$\begin{aligned}
\text{Mass:} \quad &{}^3_1H_2 \quad 3.017005 \text{ amu} \\
\text{Mass:} \quad &{}^3_1He_1 \quad 3.016086 \\
\hline
\text{Difference:} \quad &0.000019 \times 931 \text{ Mev} = 0.0177 \text{ Mev}
\end{aligned}$$

This is close to the actual maximum energy observed, but the actual energies of the emitted $\beta$ particles form a continuous distribution, with average energies which are much lower, as shown in Figure 8.2.

The energy mean for $\beta^-$ emission is often about one-third of the maximum (theoretical) energy.

Theory predicts bell-shaped emission spectra, at least for elements of low atomic numbers, and this is generally realized. Elements of higher atomic numbers tend to have rather flat emission profiles.

Very few disintegrations are accompanied by $\beta^-$ emission with the energy expected from the mass loss. This discrepancy led Bohr to propose that the principle of the conservation of energy did not hold in $\beta$ decay. In order to retain the principle, Pauli in 1930 proposed that a particle of no charge and zero rest mass shared the disintegration energy with the $\beta$ particle. This particle is now called the neutrino ($\nu$). These particles pass through matter with almost no effect on it. Thus they proved extremely difficult to detect.

In 1942, however, Allen obtained evidence of nuclear recoil from neutrino emission during the decay of $^7$Be by electron capture (see 8-2b, vi, below),

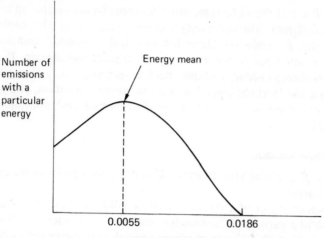

**Figure 8.2** $\beta$ emission spectrum of tritium. [From C. H. Wang and D. L. Willis, *Radiotracer Methodology in Biological Science,* Prentice-Hall, New Jersey (1965). Reprinted by permission of the publisher.]

and in 1955 Reines and Cowan effected the conversion of a proton to a neutron and a positron using the energy of the neutrino:

$$p + v \longrightarrow n + \beta^+ \tag{8-4}$$

For $\alpha$ emitters there is an inverse relation between emission energy and decay time $(t_{1/2})$ (see Chapter 7). The situation in $\beta$ emission is more complex: nuclear spin and parity changes affect $\beta^-$ disintegration lifetimes as well as the energies of the particular transitions. In 1934, Fermi proposed a theory in which more than a dozen nuclear transformation types were recognized. In a simplified version of the theory, there is within each reaction series a constant $(k)$:

$$\lambda = k(E_{max})^5 = \frac{0.693}{t_{1/2}} \tag{8-5}$$

where $\lambda$ is the decay constant and $t_{1/2}$ is the time in days for half of the molecules to decay.

$$k_{^{14}C} = \frac{\lambda}{t_{1/2}(E_{max})^5} = \frac{0.693}{5568 \cdot 365(0.155)^5} = 8.57 \cdot 10^{-2}$$

$$k_{^{32}P} = \frac{\lambda}{t_{1/2}(E_{max})^5} = \frac{0.693}{11.3(1.71)^5} = 8.33 \cdot 10^{-2}$$

$$k_{^{35}S} = \frac{0.693}{87.1(0.167)^5} = 15$$

$^{14}C$ and $^{32}P$ are of the same type, but $^{35}S$ is not; its disintegration is accompanied by different spin and parity changes. This theory also explains the shape of the $\beta^-$ emission curve for so-called "allowed" transitions, in which the emitted particles carry away no angular momentum. If angular momentum changes occur, asociated nuclear spin and parity changes generally increase the "forbiddenness" of a given atomic transition, and the half-life lengthens much as in the cases of fluorescence and phosphorescence (Chapter 7).

### (iii) Positron emission

Positron ($\beta^+$) emission increases the n/p ratio since a positron is a positively charged electron.

The positron is one of a class of particles collectively called "antimatter." Interaction of a particle of "antimatter" with one of "matter" results in annihilation of both, with emission of energy equivalent to their combined masses. When the kinetic energy of the positron is spent it may combine with a negatron (electron) to form a new atom-like substance, "positronium." Both subsequently are annihilated with two $\gamma$ photons of 0.51 Mev being formed.

Positron emission can occur only when the decay energy is greater than 1.02 Mev—the energy equivalent of the rest mass of two electrons. One may view this process as the intranuclear formation of the equivalent of a negatron (electron), which converts a proton to a neutron (for example, by electron capture, which is often an alternate process). Positron emission is not commonly encountered in biochemical tracer work.

### (iv) Gamma Emission

$\gamma$-rays are photons like X-rays, visible light, etc., and travel with the speed of light. The term is often also used to designate rays of extra-nuclear origin and of lower energy. X-rays have wavelengths of 1 to 100 A° and energies of 0.12 to 12.4 kev. However, the $\gamma$-rays we are concerned with are of intranuclear origin, from nucleii in excited states, and have higher energies: from 0.0124 to 12.4 Mev and wavelengths of 1 to $10^{-3}$ A°.

Emission of a $\beta^-$ particle can lead to an excited daughter nucleus which decays by $\gamma$ emission. An isotope commonly encountered in biochemistry which decays this way is $^{131}I$ (see Figure 8.3).

$$^{131}_{53}I_{78} \longrightarrow {}^{131}_{54}Xe_{77} + \beta^+ + \nu + \gamma \qquad (8\text{-}6)$$

Its theoretical decay energy is 0.97 Mev. $^{131}I$ can be considered a $\beta^-$ emitter, but its additional $\gamma$ emission poses special handling problems.

$\gamma$ emission, like $\alpha$ emission, gives radiation of discrete energies. Lifetimes of $\gamma$ transitions are dependent on the transition energy, the nuclear mass, and the nuclear spin and parity change. Loss of energy by $\gamma$ emission is usually

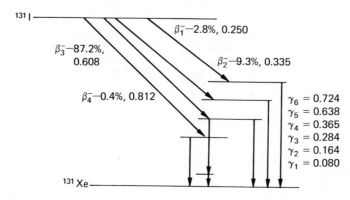

**Figure 8.3**  Conventional representation of the decay of $^{131}$I. By convention, the arrows go to the left for electron capture, $\beta^+$ or $\alpha$ emission and to the right for $\beta^-$ emission.

in competition with loss by "internal conversion." "Internal conversion" shortens $\gamma$ lifetimes.

### (v) Internal Conversion

In lieu of emitting a $\gamma$-ray the energy can be transferred directly to an orbital electron which is then emitted—$^{133}$Ba decays in part this way, as does $^{125}$I.

### (vi) Electron Capture

$\beta^-$ emission leads to a decrease in n/p ratio, and $\beta^+$ emission an increase. The n/p ratio may also be increased by capturing an orbital electron. Since the electron captured is usually in the K orbital, this process is sometimes called K capture. The empty K orbital is filled by one of the higher electrons of the daughter nuclide with X-ray emission characteristic of that (the daughter) nuclide.

$\gamma$ emission may or may not also occur from the nucleus following electron capture. Consider the decay of $^7$Be which was used in the search for neutrino recoil:

$$^7_4Be_3 + e \longrightarrow {}^7_3Li_4 + \nu + \gamma \tag{8-7}$$

Mass $^7_4Be_3$        7.019158 amu
Mass $^7_3Li_4$        7.018232
Binding energy:   0.000926·931 = 0.862 Mev

In this process, n/p increases from 0.75 to 1.33.

However, the capture follows two paths, as shown in Figure 8.4.

The excited lithium produced will emit very "soft" X-rays—54.7 ev. Thus despite the rather high decay energy, 90% of the beryllium molecules convert

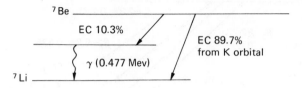

Figure 8.4    Representation of the decay of $^7$Be.

to lithium with only soft X-ray radiation—and a neutrino which was detected with difficulty.

An isotope decaying by electron capture which is starting to be used in biological work is $^{125}$I. It has the advantage that it decays with a half-life of 60 days (see Figure 8.5), which is much longer than $^{131}$I.

$$^{125}_{53}\text{I}_{72} + \text{e} \longrightarrow {}^{125}_{52}\text{Te}_{73} + v + \gamma_1 \tag{8-8}$$

The decay energy is 0.15 Mev, but the electron orbital rearrangement will lead to X-ray radiation of 0.031817 Mev, characteristic of tellurium. Thus the total radiation—neglecting the neutrino—is again of rather low energy. See Table 8.2.

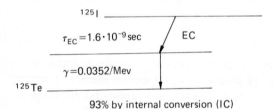

Figure 8.5
Representation of the decay of $^{125}$I.

### 8.2c    Interaction of Radiation with Matter

#### (i)  Alpha radiation and its interactions with matter

Alpha emission in itself is not accompanied by neutrino emission. Thus it is possible to show (4) that for elements of $A$ greater than 82 and $Z$ greater than 200:

$$K_\alpha = \frac{A-4}{A}Q = \frac{1}{2}m_\alpha v_\alpha^2 \tag{8-9}$$

where $A$ is the atomic number, $Q$ is the disintegration energy, $K_\alpha$ is the kinetic energy of the $\alpha$ particle, $m_\alpha$ is the mass of the $\alpha$ particle, and $v_\alpha$ is the velocity of the $\alpha$ particle.

Since $Q$ is a precise figure, $K_\alpha$ and $v_\alpha$ are precise figures—i.e., all alpha particles from a given decay reaction have the same (or a limited number of

**TABLE 8.2** ELEMENTARY PARTICLES AND RADIATIONS EMITTED DURING NUCLEAR REACTIONS (2)

| Name Symbol | Alpha particle $\alpha$ | positron $\beta^+$ | electron $\beta$ | photon $\gamma$ | neutrino $\mu$ | antineutrino $\bar{\mu}$ |
|---|---|---|---|---|---|---|
| Rest Mass* | 4.003842 | 0.0005486 | 0.0005486 | 0 | $<2 \times 10^{-7}$ | $<2 \times 10^{-7}$ |
| Nature | $_2\text{He}^4$ | Positron | Electron | Electromagnetic (EM) radiation | EM radiation emitted in $\beta^-$ decay | EM radiation emitted in $\beta^+$ decay |
| Charge | $2^+$ | $1^+$ | $1^-$ | 0 | 0 | 0 |
| Speed | depends on energy; $1\text{--}2 \times 10^9$ cm/sec | depends on energy | depends on energy | speed of light ($c$) | $c$ | $c$ |
| Energy | 4–9 Mev (single value) | 0.5–2 Mev (spectrum of energies) | 0.5–2 Mev (spectrum of energies) | 0.05–2 Mev (discrete energies) | zero to same as $\beta^-$ | zero to same as $\beta^+$ |

*Assuming the mass of $^{12}\text{C}$ to be 12.000000.

specific) energies and are emitted with the same velocity (or velocities) relative to the decaying particle (at rest).

If $\alpha$ particles of several energies are emitted from a substance, the resulting nucleii can be left in excited states which decay to the ground state by $\gamma$ emission.

Although $\alpha$ particles are emitted with high energies (4 to 8 Mev), their velocities are lower than those of $\beta$ particles because of their greater mass. The velocities of $\alpha$ particles upon emission range from 1.0 to $2.2 \cdot 10^9$ cm/sec depending on the source.

For $^{212}_{83}$Bi, $K_\alpha$ is 6.09 Mev:

$$K_\alpha = \tfrac{1}{2} m_\alpha v_\alpha^2 \tag{8-10}$$

$$v_\alpha = \sqrt{\frac{2K_\alpha}{m_\alpha}} = \sqrt{\frac{2 \times 6.09 \times 1.6 \times 10^{-6}}{\dfrac{4}{6.025 \times 10^{23}}}} = 1.7 \times 10^9 \text{ cm/sec}$$

Since the energy equivalent of an $\alpha$ particle is 3720 Mev ($4.004 \times 931$ Mev), the mass will not be effectively changed by an additional 6.09 Mev. By comparison, the $\beta^-_{\max}$ energy of $^{14}$C, 0.155 Mev, represents a significant increase in the energy equivalent of the electron, which is 0.511 Mev. Therefore, the mass must be adjusted:

$$K_\beta = mc^2 - m_0 c^2 \tag{8-11}$$

where $m_0 = 9.1 \times 10^{-28}$ g

$$m = \frac{K_\beta}{c^2} + m_0 = \frac{0.155 \times 1.6 \times 10^{-6}}{(3 \times 10^{10})^2} + 9.1 \times 10^{-28} \text{ g} = 11.86 \times 10^{-28} \text{ g}$$

which is the relativistic mass of an electron with an emission energy of 0.155 Mev.

The initial velocity of such a particle is also readily calculated:

$$K_\beta = \tfrac{1}{2} mv^2 \tag{8-10}$$

$$v = \sqrt{\frac{2K_\beta}{m}} = \sqrt{\frac{2 \times 0.155 \times 1.6 \times 10^6}{11.86 \times 10^{-28}}} = 2.04 \times 10^{10} \text{ cm/sec}$$

which is higher than the velocities of most $\alpha$ particles.

By comparison, for a molecule such as $H_2$, the most probable kinetic velocity at 0°C is:

$$v = \sqrt{\frac{2RT}{m}} = \sqrt{\frac{2 \times 8.314 \times 10^7 \times 273}{2.015}} = 1.5 \times 10^5 \text{ cm/sec}$$

α particles react primarily with the electrons in matter, losing energy by producing excitation and ionization of molecules. The ranges of α particles in air vary from 2.5 to 8.6 cm, depending on their energies, that is, on the source. As a rough approximation, the energy of the α particle in Mev equals the range in air in centimeters. Since the energies of all α particles from a given source are the same or nearly so, all the particles have nearly the same range; that is, they all run out of energy after covering the same distance (see Figure 8.6). This is because it takes about 32 ev of energy to ionize a molecule

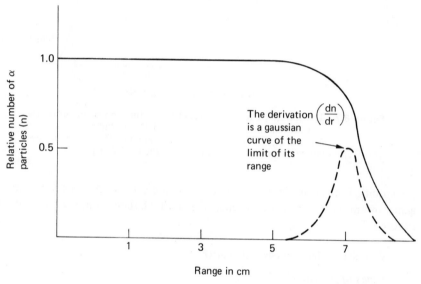

**Figure 8.6**   The range of α particles. [From C. H. Wang and D. L. Willis, *Radiotracer Methodology in Biological Science*. Prentice-Hall, New Jersey (1965). Reprinted by permission of the publisher.]

of any gas by any charged particle (4). So the total number of ion pairs produced by a charged particle is roughly proportional to its energy. A 4 Mev α particle with a range of about 5.0 cm will produce on the average:

$$\frac{4 \times 10^6}{32 \times 5} = 2.5 \times 10^4 \text{ ion pairs/cm}$$

Because the α particle travels more slowly than the β particle, it spends a proportionately greater time in the vicinity of a molecule and thus is more apt to ionize it. And in general, the slower the α particle is moving, the more ionization it produces per centimeter of path length. During the movement of the α particles in Figure 8.6, their energies are decreasing. Consequently,

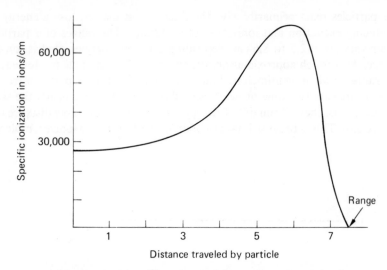

**Figure 8.7**  The ionizing effect of α particles as a function of the distance
they travel. [From C. H. Wang and D. L. Willis, *Radiotracer
Methodology in Biological Science*, Prentice-Hall, New Jersey
(1965). Reprinted by permission of the publisher.]

the specific ionization (ions/cm) increases as the velocity decreases, reaching
a maximum as the α particles reach the limit of their range, as shown in
Figure 8.7.

### (ii) β⁻ radiation interactions with matter

*1. Types of interactions.*

a. Rutherford scattering. The β particle is deflected by an atomic nucleus
in an "elastic collision"—the electron loses no energy. This is what causes
"back scattering" (see 8-4d, (i)1(b) below).

b. Bremsstrahlung ("braking radiation"). This is the way X-rays are usually
prepared. The β particle "collides inelastically" with the nucleus (is decel-
erated in its electric field), and the energy lost by the β particle appears as a
photon (X-ray). The radiation is continuous with superposed peaks due to
interaction with orbital electrons. The energy of the β particle determines the
maximum energy and part of the shape of the spectrum, while the target
material contributes only to the shape of the spectrum. Since the probability
of Bremsstrahlung production increases with atomic number, it is advisable
to shield β radiation with materials of low atomic number (e.g., plastics).
Generally little of the energy of the β particles appears as Bremsstrahlung.
See Figure 8.8.

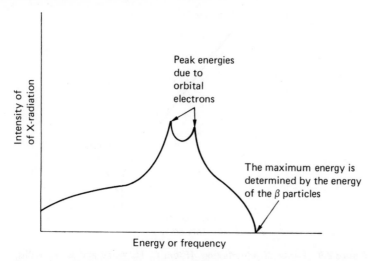

**Figure 8.8**  Emission spectrum of X-radiation produced by Bremsstrahlung. [Modified from R. T. Weidner and R. L. Sells, *Elementary Modern Physics*, Allyn and Bacon, Boston (1960). Reprinted by permission of the publisher.]

c. Ionization. Since like charges repel each other, a $\beta^-$ particle passing by an atom or molecule may cause an electron to be ejected. The $\beta$ particle loses energy:

$$\Delta E = d + \tfrac{1}{2}mv^2 \qquad (8\text{-}12)$$

where $d$ is the binding energy of the electron. This energy can reappear as a photon when the positive ion reabsorbs an electron.

Since the electron often is not ejected but merely excited and can lose its energy as heat, much energy is dissipated in this way.

2. *Absorption of $\beta$ particles in matter.* As it moves through matter, the $\beta$ particle gradually loses energy by the above processes until it comes to "rest." Since in radioactive decay the initial $\beta$ particles span a continuous energy spectrum, the detailed treatment is complex, but the overall result is simple. The intensity (i.e., the number of $\beta$ particles with sufficiently high kinetic energy to be detected) falls off nearly exponentially, like light absorption, but unlike light absorption there is a distance beyond which the intensity is zero (see Figure 8.9). The ranges of $\beta$ particles of various energies can be calculated from the following empirical relations:

1. For $E_{max}$ of less than 0.6 Mev:

$$\text{range} = R = 407(E_{max})^{1.38} \text{ mg/cm}^2 \qquad (8\text{-}13)$$

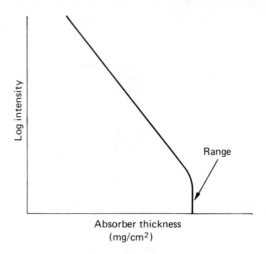

**Figure 8.9**  Range of $\beta^-$ radiation. [From C. H. Wang and D. L. Willis, *Radiotracer Methodology in Biological Science*, Prentice-Hall, New Jersey (1965). Reprinted by permission of the publisher.]

For $^{35}$S, $E_{max} = 0.167$ Mev

$$R_{calc} = 34.5 \text{ mg/cm}^2 \qquad R_{obs} = 34 \text{ mg/cm}^2$$

This corresponds to 0.14 mm of glass (density 2.5 gm/cm²)

2. For $E_{max}$ greater than 0.6 Mev:

$$\text{range} = R = 542E_{max} - 133 \text{ mg/cm}^2 \qquad (8\text{-}14)$$

For $^{32}$P, $E_{max} = 1.71$ Mev

$$R_{calc} = 795 \text{ mg/cm}^2 \qquad R_{obs} = 800 \text{ mg/cm}^2$$

This corresponds to 3 mm of glass.

Because of their lower energies but greater velocities, $\beta$ particles produce much less ionization in a given volume of material than $\alpha$ particles; their "specific ionizations" are about 1/200-1/700 of $\alpha$ particles.

### (iii) Interactions of γ-radiation with matter

1. In order of increasing importance, the interactions are:

a. Rayleigh scattering involves an "elastic collision" of a photon with the nucleus. There is no energy change, but simply a change in direction, usually of only a few degrees.

b. Photodisintegration is important only at energies of more than 10 Mev. The photon is completely absorbed by the nucleus. The excited nucleus ejects a neutron, proton, or an $\alpha$-particle.

c. In nuclear resonance scattering (Mossbauer effect), a photon can be absorbed by a nucleus and then reemitted, unchanged, in a new direction if the energy spacing in the nucleus is exactly equal to that of the photon.

d. Bragg scattering is an interference effect and consequently is important only with crystals and at certain angles of incidence with $\gamma$ radiation of very low energies (X-rays). X-radiation is used in determinations of crystal structure.

e. The "photoelectric effect" is an "inelastic collision" of a photon with complete energy transfer to an electron, resulting in its ejection (ionization). This is the most important process for photons of low energies. It takes up to about 0.1 Mev to eject an electron. Usually a $K$ electron is involved, and the probability of this interaction increases with increasing atomic number.

f. Compton scattering is most important for $\gamma$ rays of 0.5 to 1.0 Mev, since it falls off more slowly with photon energy than the photoelectric effect. A photon and an electron "collide inelastically," the photon giving up only a part of its energy, and the photon emerges in a new direction at a reduced energy (frequency). The energy appears as a recoil energy in the electron.

g. If its energy is greater than 1.02 Mev, the $\gamma$-ray can be converted to a $\beta^+$ and $\beta^-$ in a strong electromagnetic field such as that which surrounds the nucleus. This process is called "pair production." The $\beta^+$ annihilates with an electron from the medium to produce two colinear 0.51 Mev photons (8.2b iii, above).

2. The number of incident photons is cut exponentially by the photoelectric effect, and there is no clear-cut range. Absorption of $\gamma$ radiation in matter is consequently like light absorption, at least to a first approximation:

$$\frac{-dI}{dx} = upI \tag{8-15}$$

where $p$ is the density of the absorber, in mg/cm$^2$.

The absorption efficiency $u$ is called the "mass absorption coefficient," and is the linear absorption coefficient divided by the density of the absorber.

$$\log \frac{I_0}{I} = upx \tag{8-15a}$$

However, degradation of the initial $\gamma$ radiation by Compton scattering and pair production causes deviations from the exponential decrease in intensity. Pair production is also responsible for the fact that the efficiency of absorption of $\gamma$-radiation by a substance tends to increase with increasing energy of the radiation (Figure 8.10).

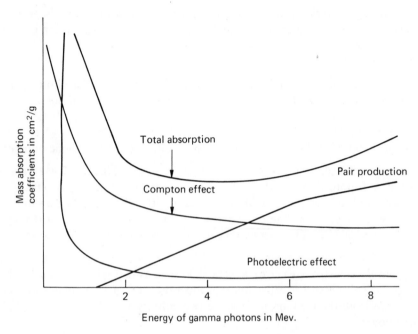

**Figure 8.10**   The efficiency of various mechanisms of absorption of $\gamma$-radiation as a function of energy of the radiation. [From C. H. Wang and D. L. Willis, *Radiotracer Methodology in Biological Science*, Prentice-Hall, New Jersey (1965). Reprinted by permission of the publisher.]

The absorbing power of a substance is not strictly proportional to its density. In general $u$ increases as the energy of the $\gamma$-ray decreases, except for absorbers of high atomic number, where pair production becomes important. For some purposes, a function related to the reciprocal of the mass absorption coefficient, called the "half-thickness," is used (Figure 8.11).

The half-thickness value of lead for photons of energy equal to 1 Mev is about 10 gm/cm². The energy of $^{131}$I $\gamma$ emission is 0.72 Mev. The density of lead is 11.35 gm/cm², so a lead plate 1 cm thick will absorb about half the radiation from $^{131}$I.

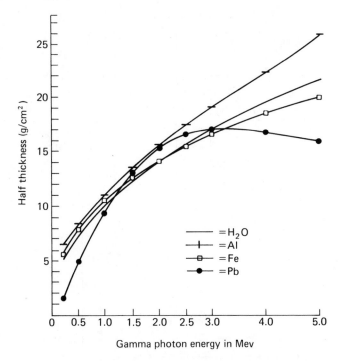

**Figure 8.11** "Half-thickness" values of various materials as a function of
$\gamma$-radiation energy. [From C. H. Wang and D. L. Willis, *Radiotracer Methodology in Biological Science*, Prentice-Hall, New Jersey (1965). Reprinted by permission of the publisher.]

## 8.3 INSTRUMENTATION

The instruments covered are limited to those used with isotopes commonly
employed in biochemical experiments.

### 8.3a Geiger and Proportional Counters

These measure increases in conductivity due to ionization of gas produced
by radiation (Figure 8.12). After exposure to radiation, the tube is filled with
ion pairs: electrons and positive ions.

The fate of these ions depends on the voltage applied to the detection tube
(Figure 8.13). In the figure, the amount of conductivity change produced by
a single "event" (incidence of an $\alpha$, $\beta$, or $\gamma$) is plotted as a function of the voltage between the center wire and the walls of the detection tube.

At low voltages (region 1), many ion pairs formed by interaction with a
single $\alpha$ or $\beta$ particle recombine before reaching the electrodes. The ions

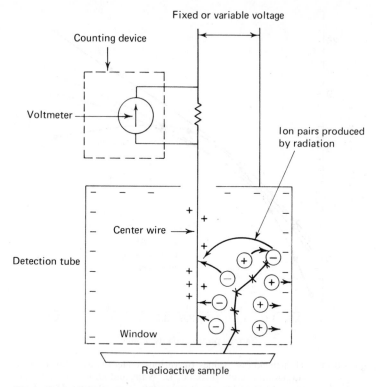

**Figure 8.12**    Schematic profile of a simple ionization detector. [Modified
from R. T. Overman and H. M. Clark, *Radioisotope Techniques*,
McGraw-Hill, New York (1960).]

cannot move rapidly enough in the low electric field to reach the center wire
or wall before recombination. At higher voltages (region 2), all ions produced
are collected. Note the much smaller pulse heights produced by β particles:
because of their greater velocities, these produce less ionization in passing
through the detection tube. In region 3, electrons produced by ionizing radiation are accelerated sufficiently to produce further ionization. The total ions
are proportional to the initial ions produced. This is called the region of proportionality and is the region in which proportional counters are operated.

   At still higher voltages (region 4) positive ions, which move more slowly
than electrons, accumulate around the anode (the center wire) and are neutralized by electrons from other ion pairs before collection. The total ions are
consequently no longer proportional to the initial ions. This is called the region of limited proportionality. Region 5 is called the Geiger-Muller region:
the pulses of ions collected are independent of the ions initially formed
because of a "cascade effect"—secondary, tertiary, etc. ionization occurs.

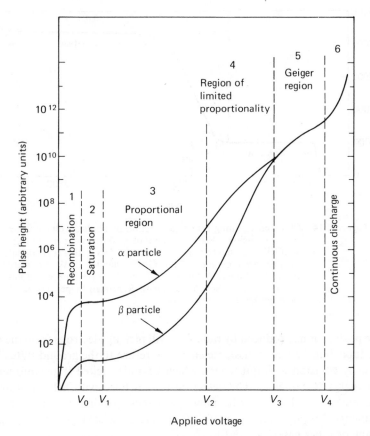

**Figure 8.13**   Pulse heights for $\alpha$ or $\beta^-$ particles as functions of voltage applied
to the detection tube of a Geiger or proportional counter. [From
G. Friedlander, J. W. Kennedy and J. M. Miller, *Nuclear and
Radiochemistry*, 2nd ed., John Wiley and Sons, New York
(1964). Reprinted by permission of the publisher.]

In region 6, continuous discharge due to ionization of the gas in the tube is
encountered, which makes these voltages useless for counting purposes.

So increasing the pulse height, within limits, makes an "event" easier to
detect. But to count "events" you want proportionality between "events" and
pulses. The result of increasing the applied voltage in terms of counts per
minute for a source of mixed $\alpha$ and $\beta$ radiation is shown in Figure 8.14. The
voltage should naturally not be increased above the plateau voltage for the
type of radiation you are counting. Note that this is not the same curve as
Figure 8.13. Figure 8.13 shows the ions produced per discharge, Figure 8.14
the discharge detected.

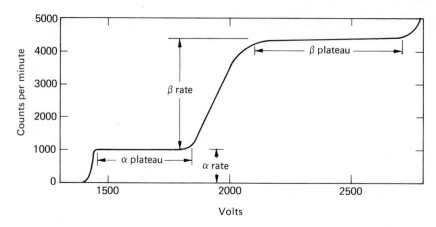

**Figure 8.14**  Observed counting rate as a function of applied voltage for $\alpha$ and $\beta^-$ emission. "Plateau" slopes of 3–10 per cent per 100 volts for Geiger counters and 1 per cent or less with proportional counters are common. [From G. Friedlander, J. W. Kennedy and J. M. Miller, *Nuclear and Radiochemistry*, 2nd ed., John Wiley and Sons, New York (1964). Reprinted by permission of the publisher.]

The positive ions produced by radiation will pick up electrons from the wall of the tube, but they must dissipate energy to return to the ground state. This is done by collision with quenchers, which degrade excited states into vibrational energy (Chapter 7). The detection tube is filled with helium or argon plus a quencher—chlorine or bromine (permanent) or organic materials (not permanent). Proportional counters often employ a steady flow of argon-methane or other gases.

The resolving time with proportional counters is generally about 1 microsecond, which is considerably better than that of Geiger counters (300 microseconds). The resolving time limits the accuracy of either counter at high counting rates, since two disintegrations occurring within the resolving time would be counted as one. This "simple coincidence correction" for Geiger counters is 0.5% per 1000 cpm. This is from:

$$\text{correction} = \frac{\text{cpm}}{60} \times \text{resolving time} = \frac{10^3}{60} \times 3 \times 10^2 \times 10^{-6} = \frac{10^{-2}}{2}$$
$$= 0.5\% \qquad (8\text{-}16)$$

Geiger and proportional counters are generally 100% efficient for $\alpha$, nearly 100% efficient for "hard" (high energy) $\beta$ and 1-2% efficient for $\gamma$ radiation. While proportional counters are superior in most respects to Geiger counters, the latter do produce pulses large enough to be detected without the need for elaborate and costly electronics. Thus Geiger counters are simpler, cheaper, and more rugged.

### 8.3b Scintillation Counters

In Geiger and proportional counting, it is the primary and secondary ions resulting from the ionizing radiation that are detected. In scintillation counting, light emitted by a fluorescent material which has been excited by the radiation is detected. There are basically two types of scintillation counters:

1. Crystal scintillation counters employ a detector which is a crystal of sodium iodide sensitized usually by thallium iodide. The crystal is sometimes formed with a hole into which a tube containing the sample is placed. This allows high geometric efficiency in counting. This is of value primarily for $\gamma$ emitters, since $\beta$ and $\alpha$ particles will not normally penetrate the container to reach the crystal. Even so, the radiation must be sufficiently energetic. $^{55}$Fe, which decays by electron capture to manganese, with evolution of 6 kev $\gamma$ rays, cannot be counted by crystal scintillation counters; liquid scintillation counters must be used. The primary advantage of crystal scintillation counting is that essentially no preparation of the sample is necessary.

2. Liquid scintillation counting is used for $\alpha$, $\beta$, or sufficiently weak $\gamma$ radiation such that it can not efficiently escape from the container. The energy from the decay radiation excites the solvent (usually toluene), and for $\beta$ particles the range in this solvent is usually less than 1 cm. Within about $10^{-9}$ seconds, the excited solvent molecules transfer their energy to the fluorescent compound (fluor). The fluor may emit the energy as a photon or may transfer it to a secondary fluor which will emit the photon in a wavelength more suitable to the spectral sensitivity maximum of the photomultiplier tube. Only about 5% of the initial energy of the radiation is emitted as light. Most of the balance is lost by quenching, i.e., degraded to heat.

#### (i) The fluor

"PPO" (2,5 diphenyloxazole) is the most popular primary fluor, but its emission maximum is at 380 nm which is below the region of maximum sensitivity of the photocathode used in the instrument. Therefore it is commonly used with a secondary fluor, "POPOP" (2,2-phenylene bis-(5-phenyloxazole)) (Figure 8.15). POPOP, with an emission maximum at 420 nm, is the most widely used secondary fluor. More recently, "dimethyl POPOP," with an emission maximum at 430 nm, has gained favor as a secondary fluor. This is also more soluble in the usual scintillation solvents, and since its emission maximum is at longer wavelengths absorption by the solvent is less.

The counting efficiency initially increases with fluor concentration until a plateau is reached. At this plateau, the efficiency is independent of the fluor

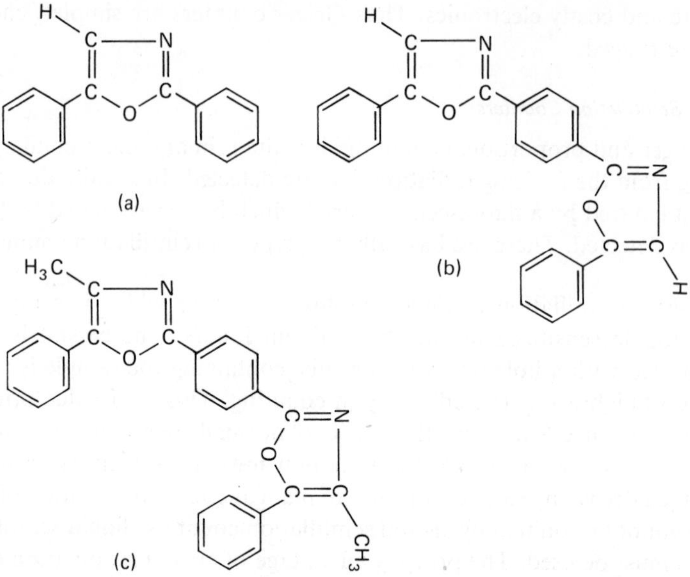

**Figure 8.15**  Chemical structure of widely used primary and secondary fluors:
(a) 2,5 diphenyloxazole (PPO); (b) 2,2-phenylene bis-(5-pheny-
loxazole) (POPOP); (c) "Dimethyl POPOP."

concentration. Some fluors, however, are not sufficiently soluble to reach
this concentration. With sufficiently soluble fluors, the efficiency may fall
at higher concentrations due to self quenching, i.e., the fluor reabsorbs the
light.

### (ii) The photomultiplier tube

The photocathode of the photomultiplier is constructed of material with a
low "work function" (commonly cesium-antimony), and thus electrons are
also readily emitted thermally (Chapter 7). Although thermal emission can
be reduced by cooling the photomultiplier tube, the effect is still troublesome
because the rate of emission is time-variant. For this reason, in most commercial scintillation counters, the vial containing the sample is viewed by
two photomultiplier tubes, and only processes occurring simultaneously in
both are counted.

### (iii) The counter[2]

The output from the photomultiplier tube is proportional to the incident
light, which in turn is proportional to the energy of the initiating $\beta$-particle.

[2]This description is based on the characteristics of both Nuclear-Chicago and Packard
scintillation counters, but should be generally applicable.

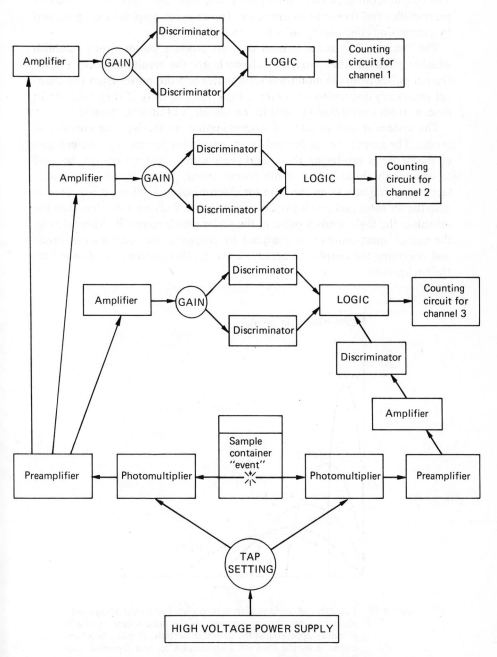

**Figure 8.16** Schematic representation of a three channel scintillation counter.
(Adapted from a brochure of the Nuclear-Chicago Co.)

The output from each photomultiplier ("analyzer" and "monitor") is fed to a preamplifier and thence to an amplifier. The analyzer amplifiers are governed by attenuation controls ("gain"). See Figure 8.16.

The "monitor" channel is used only to produce pulses which establish whether pulses in the "analyzer" channels are the results of decay events. Output pulses from the monitor discriminator and the pulses from the channel (analyzer) discriminators enter a logic circuit. Only if they coincide in time, as from a scintillation event in the sample, will they be counted.

The system is also capable of discrimination on the basis of energies of events. The amplitudes of the pulses are nearly proportional to the energies of the particles producing them, and the discriminator system can be used to count only those pulses of the proper energy range. If the logic system receives an impulse from the monitor discriminator $A$ saying it was greater than the set value and one from the discriminator $B$ saying it was less than the set value, the logic sends a pulse to the scaler which counts it. Alternatively, the energy spectrum can be analyzed by changing the discriminator levels and observing the output. Generally, however, this system is used to reduce the background.

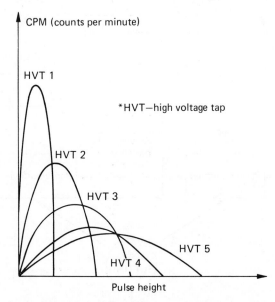

Figure 8.17   The effect of various high voltages applied to the photomultipliers of a scintillation counter on the pulse spectrum due to $\beta^-$ decay appearing at the amplifier outputs. (From a brochure prepared by the Packard Instrument Co., and reprinted with their permission.)

The output pulse height is a sensitive function of the voltage applied to the photomultiplier tubes (about $1\%$ per volt). Higher voltages increase the effective pulse heights which appear at the discriminators, as shown in Figure 8.17.

## 8.4 EXPERIMENTAL PROCEDURES

Consideration will be restricted to the information needed for use of the isotopes listed in Table 8.3.

**TABLE 8.3**

| Isotope | Energy of Radiation, Mev | Half-life | Type of Radiation |
|---------|--------------------------|-----------|-------------------|
| $^3H$ | 0.0186 | 12.26 yr | $\beta^-$ |
| $^{14}C$ | 0.155 | 5720 yr | $\beta^-$ |
| $^{32}P$ | 1.71 | 14.3 days | $\beta^-$ |
| $^{35}S$ | 0.167 | 87.1 days | $\beta^-$ |
| $^{36}Cl$ | 0.714 | $3 \times 10^5$ days | $\beta^-$; EC($\beta^+$): X-ray |
| $^{125}I$ | 0.0352 | 60 days | $\gamma$; EC($\beta^-$): X-ray |
| $^{131}I$ | 0.60; 0.37; 0.08 | 8.06 days | $\beta^-$; $\gamma$ |

Biochemical experiments involving radiochemistry are of two major types, kinetic and equilibrium. Since an enormous variety of experiments have been performed using radioactive isotopes, we will discuss each type only in general, giving references to a few especially instructive cases.

### 8.4a  Kinetic Experiments

In these experiments, an isotopic form of some compound is introduced into a system at time zero. Later, the distribution of the isotope is determined, either to measure the kinetics of the uptake, loss, or conversion of the original compound or to see what the original compound has been converted into. It is usually assumed that the radioactive compound is chemically indistinguishable from the "cold" compound, and this is generally the case—though not always, because of the isotope effect.

This can be a major effect, especially with the lighter elements; even with heavier ones, a sufficiently long series of reactions can markedly change the ratio of isotope to stable element, since each reaction will tend to enrich or diminish the fraction of the isotope present to some extent. Of course, if only bonds far from the "tagged" atom(s) are involved, little or no isotope effect will be seen.

Some examples of these types of experiment are given in references 9 to 11.

### 8.4b   Equilibrium Experiments

This type of experiment is perhaps less widely used, but is still important—the carbon-14 "dating" technique is an equilibrium technique, for example. Equilibrium experiments can involve "labeling" a specific group in a molecule to make its recovery easier, or it can involve measurement of the amount of a certain element or compound, either directly or by an "isotope dilution" experiment.

There are several types of isotope-dilution methods (12).

In order to assay a sample containing $x$ g of a compound, $y$ g of its isotopic form with an initial specific activity $SA_{initial}$ is added. After mixing, some pure compound is separated, and its final specific activity $SA_{final}$ is determined. Since total activity is constant,

$$(x + y)SA_{final} = ySA_{initial} \qquad (8\text{-}17)$$

Or

$$x = y\left(\frac{SA_{initial}}{SA_{final}} - 1\right) \qquad (8\text{-}17a)$$

Only two specific activities and the weight of the added labeled compound must be known.

If it is necessary to measure the total radioactive material in a system, an inverse dilution technique is used. In order to determine $x$ g of compound with a specific activity $SA_i$, $y$ g of pure unlabeled compound (carrier) is added, a quantity of the isotopic mixture is isolated, and its specific activity, $SA_f$, determined. The amount of the original radioactive compound is found from:

$$x = y\frac{1}{\dfrac{SA_i}{SA_f} - 1} \qquad (8\text{-}18)$$

Or

$$x = y\frac{SA_f}{SA_i} \qquad (8\text{-}18a)$$

Double-dilution is often used when an isolated labeled compound must be carried through a complex procedure after initial extraction. For example, a $^{32}P$-labeled compound is added to a mixture to be assayed, then a mixture containing this compound is extracted, and it is chemically reacted, say with a $^{3}H$ compound. The doubly labeled mixture is then purified. Carrier may be added here to facilitate purification. The pure compound is collected, dissolved in a suitable scintillation mixture, and the contents of the two isotopes are measured simultaneously (see 8.4e).

The percentage of $^{32}$P recovery is taken as an index of recovery; from the tritium activity the quantity of compound present is calculated from the specific activity of the chemical used for the reaction. From these two figures the initial content of compound is calculated:

$$\text{compound recovered} = {}^{32}\text{P cpm recovered}/{}^{32}\text{P cpm added} \qquad (8\text{-}19)$$

$$\mu\text{mol compound} = ({}^{3}\text{H cpm}/{}^{3}\text{H eff.})(1 \ \mu c/3.7 \times 10^{4} \ \text{dpm})$$
$$\times \ (1/\text{sp. act.}(\mu c/\mu\text{mol})) \qquad (8\text{-}20)$$

$$\mu\text{mol compound} = \mu\text{mol } {}^{3}\text{H labeled compound/recovery of compound} \qquad (8\text{-}21)$$

### 8.4c  Setting up Experiments

The first thing to do in setting up an isotope experiment is to calculate how much isotope you will need. You will have to consider not only volumes and aliquot sizes, but counting efficiencies and counting statistics (see 8.6a).

There are several general pitfalls to avoid. The first is that your isotope may be contaminated with several chemical forms—$^{32}$PO$_4$ with $^{32}$P-pyro-phosphate, for example. Most isotope companies are fairly reliable, but "zero time" controls can be very helpful in detecting this kind of problem. Note the distinction between chemical purity of an isotope (only one chemical form is present) and radiochemical purity (only one isotopic chemical species is present). For some experiments, you can get by with radiochemically pure material, but for many you will need chemically pure material. Remember also that these compounds break down with time, so they should be used within a reasonable period.

The second problem can occur, especially in tracer experiments, from failure to realize that "pools" of varying sizes of most metabolites exist. So the isotope will be diluted to varying extents as it is converted from one compound to another. The presence of several pathways, exchange reactions, etc. can also confuse the results. The best way around this problem is to study only small sections of metabolic pathways at a time, and if possible to obtain estimates of "pool" sizes from isotope dilution experiments. In addition, a common procedure is to starve the organism before the experiment, to cut down pool sizes (9).

Related to this is an occasionally ignored distinction between *total* activity of a sample and its *specific* activity. The former is a function of many processes, including exchange rates and pool sizes, both of which can change

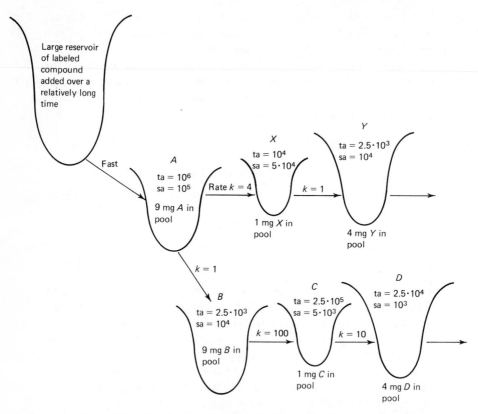

**Figure 8.18**  Showing the effect of (a) relative rates of a series of reactions in a metabolic pathway, and (b) of relative "pool" sizes of intermediates in a metabolic pathway on the total ("ta") and specific ("sa") activities of the intermediates.

drastically during an experiment. The specific activity of a sample is a somewhat truer measure of metabolic conversions (see Figure 8.18).

### 8.4d   Preparation of Samples for Counting

#### (i) Proportional counters

*1. Problems with proportional counting.* By "problems," we mean situations which alter the counting rate from the true one, changing the efficiency of counting. Efficiency ($E$) is defined as:

$$E = \frac{\text{counts per minute (cpm)}}{\text{disintegrations per minute (dpm)}} \qquad (8\text{-}22)$$

a. Geometry. If the sample were a point source, the geometry of the

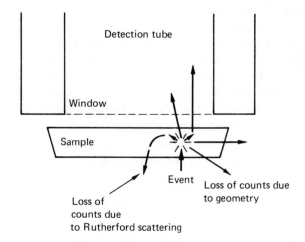

**Figure 8.19** Illustration of the effect of the geometry of the counting arrangement with a proportional counter and scattering of $\beta^-$ radiation by the sample on counting efficiency. [From C. H. Wang and D. L. Willis, *Radiotracer Methodology in Biological Science*, Prentice-Hall, New Jersey (1965). Reprinted by permission of the publisher.]

counting arrangement must be considered. See Figure 8.19. The counting rate is reduced by a factor due to the geometry:

$$\text{Geometry factor} = G = \frac{\text{portion of area of sphere enclosed by window}}{\text{area of sphere}}$$

$$= \frac{S'}{4\pi d^2} = \frac{2\pi d^2(1 - \cos\theta)}{4\pi d^2} = \frac{1 - \cos\theta}{2}$$

$$= \frac{1}{2}\left(\frac{h}{\sqrt{h^2 + d^2}}\right) \tag{8-23}$$

since $\cos\theta = h/d$.

If the sample is not too close to the window:

$$S' = A = \pi u^2 = \text{area of window} \tag{8-24}$$

Thus

$$\frac{\pi u^2}{4\pi d^2} = \frac{u^2}{4d^2} = \frac{u^2}{4h^2} \tag{8-25}$$

This is the inverse square law for the distance of the point source from the tube.

Most sources actually counted are not point sources, which further complicates the situation and makes geometry corrections more difficult.

b. Scattering. The geometry of the counting arrangement is important in another way when Rutherford scattering of the radiation by the sample is considered. This tends to lower the absolute apparent radioactivity of a sample (see Figure 8.19 above). It is dependent on several factors: the atomic number of the scattering material, the thickness of the scattering material, the energy of the radiation, and the physical geometry (see Figure 8.20). The scattering is the same for $^{90}$Sr and $^{14}$C, but the weaker $^{14}$C $\beta^-$ particles are absorbed more readily during the process.

c. Absorption. Absorption of radiation by the sample is a problem which can affect the relative as well as the absolute apparent radioactivity. At a

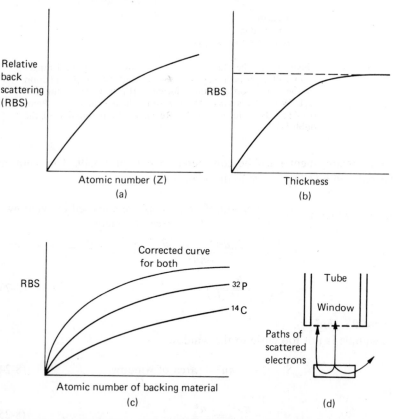

**Figure 8.20**   Factors involved in scattering of $\beta^-$ radiation by a sample: (a) atomic number of scattering material; (b) thickness of scattering material; (c) energy of the radiation; (d) the physical geometry. [(a) and (b) are from C. H. Wang and D. L. Willis, *Radiotracer Methodology in Biological Science*, Prentice-Hall, New Jersey (1965). Reprinted by permission of the publisher.]

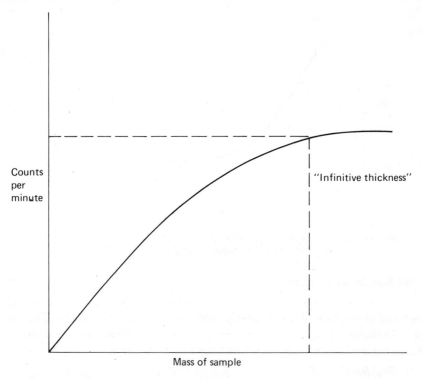

**Figure 8.21**  Effect of amount of the sample on the observed counting rate.

given activity of the sample, the counts will vary with the amount of sample (see Figure 8.21).

*2. Solutions.*

a. Geometrical factors are seldom a problem if one is comparing several samples at a constant geometry.

b. Scattering cannot be avoided, but is often minimized by either using an infinitely thin sample or an infinitely thick sample. Losses of counts from scattering the radiation away from the counter window can be reduced by counting with the sample as close to the counter tube as possible.

c. Absorption problems can be minimized by counting samples of constant masses. At constant sample volume or mass, the counting rate should be a linear function of the specific activity of the sample. If this is not possible, a calibration curve of mass versus activity can be constructed and used to correct the observed counting rates (see Figure 8.22).

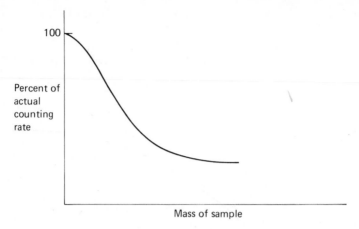

**Figure 8.22**   Calibration curve of sample mass versus observed counting rate, for correcting self-absorption effects.

### (ii) Scintillation counters

These are now favored in work with biological systems because of the high self-absorption of the relatively "soft" $\beta^-$ emitting $^3$H and $^{35}$S. The use of scintillation counters reduces or eliminates problems with geometry, scattering and absorption but introduces some new ones.

### 1. Problems.

a. For greatest efficiency, the radioactive sample must be dissolved in the scintillation fluid. In order to solubilize the fluors involved in light emission, scintillation fluids are generally nonpolar in character. One therefore can count nonpolar samples efficiently, for such materials readily dissolve in the scintillation fluid. Polar (e.g., water soluble) and insoluble solid samples present counting problems.

b. High "background" counts or serious losses of counts can result from the sample vials. Glass contains $^{40}$K ($\beta^-$, 1.33 Mev), or may be rendered phosphorescent by strong sunlight. This may require hours or even days to decay.

c. Since fluorescence is actually being measured, quenching can become a severe problem (Chapter 7). There are two types of quenching. Chemical quenching is proportional to the concentration of quenchers. TCA, $HClO_4$, $H_2O$, pyridine, and various other compounds are quenchers. Color quenching, by the sample or sample support, can also occur. The fluors PPO and POPOP emit a bluish light. Accordingly samples containing complementary yellowish colors, either dissolved in the fluid or adhering to a paper or Millipore support, effectively reduce counting efficiency by shifting the spectrum to a lower intensity. (see Figure 8.23a).

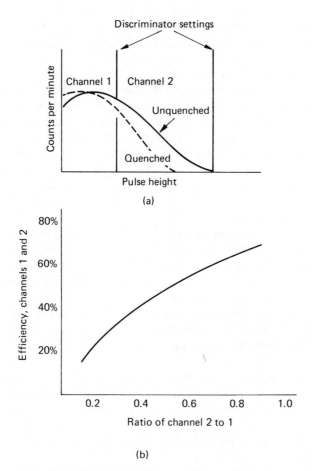

**Figure 8.23** Illustrating the "channels ratio" method of correction for quenching: (a) discriminator settings and effect of quenching on $^{14}C$ spectrum; (b) efficiency curve for $^{14}C$. (Modified from a brochure prepared by the Packard Instrument Co., and reprinted with their permission.)

d. Chemiluminescence will give "counts" which are spurious. Chemiluminescence can occur when strong acids or bases are neutralized in the scintillation fluid. This effect can give very large "counts" and last for hours.

*2. Solutions.*

a. Water soluble, nonvolatile samples are either dried down in scintillation bottles, dried down on Millipore filters in scintillation bottles, or counted at a lower efficiency in a scintillation system that will tolerate (i.e., dissolve) some water. Insoluble solids are either collected on Millipore filters and counted on these or are suspended uniformly by means of some emulsifier or inert supporting medium. Precipitates and insoluble materials can often be retained in suspension for counting by addition of 3-5% silica, e.g., "Carb-O-Sol." Acidic materials like proteins or peptides can be solubilized with a strong base such as Hyamine 10X hydroxide. Usually 1-2 ml of a 1 $M$ solution is used with up to 6 ml of ethanol and diluted to 15 ml with a toluene solution of the fluor. A more recent quaternary ammonium base, "NCS," reportedly gives as good solubilizations as Hyamine with less quenching (13).

Several scintillation "cocktails" are available (14, 15). For nonpolar organic compounds, mix 4 g PPO plus 0.1 g of POPOP per liter of reagent grade toluene (i.e., 15.2 g ppO and 0.379 g POPOP per 8 pints of toluene). This system will not tolerate water or water dissolved materials. Millipore filters are translucent in this system if they are free of water; if they appear white, they need further drying.

A good system for routine counting of $^{14}C$-containing aqueous solutions is prepared from 4 g PPO, 0.1 g POPOP, 125 ml dioxane, 125 ml anisole, 750 ml diglyme (diethylene glycol dimethyl ether, "Ansul 141 ether"). Sink disposal is easy because of the lack of naphthalene (it plugs up the drain).

A system was developed which has good counting efficiency for $^3H$ samples in water (it is poor for nonpolar organic compounds) and is easy to dispose of down the sink (16,17). A sample in 0.5 ml of water is mixed with 6 ml absolute ethanol and then added to 8 ml of 0.8% PPO, 0.01% POPOP in toluene.

Another system, which can accommodate up to 0.5 ml of water per 15 ml of solvent, up to 500 $\mu$moles of sodium chloride (bromide or iodide salts should not be used for obvious reasons) or several milligrams of protein or nuleic acid, is a modification of the system of Patterson and Greene (18). It is prepared from 8 g PPO, 0.2 g POPOP, one liter of Triton X-100 and two liters of toluene.

b. The vials used should be of good optical quality, clean, and of low $^{40}K$ content. Polyethylene vials are often used instead of glass. Quartz and Vycor are generally too expensive.

c. One can sometimes avoid or minimize chemical quenching by keeping the concentration of quenchers down. Often this is not possible, however. For chemical quenching or color quenching when the color is dissolved in the solvent, external standards can be used.

The investigator should estimate the extremes in quenching in his solutions from the quantity of quenching agent—either the material to be counted or a known quencher such as carbon tetrachloride or acetone. Add equal known quantities of an internal standard, such as $^{14}C$, to a set of volumetric flasks containing some scintillation solvent; then add enough quenching agent to cover the range of quencher concentration in the samples. The flasks are brought to the mark with scintillator solvent.

The lower level of the discriminator setting (lower "window edge") of the external standard is put above the point where $\gamma$ rays cause glass or plastic to fluoresce; the upper level is set above the standard spectrum.

Count the samples for the calibration curve; then the unknown samples. Counting efficiencies are read from the curve, and the observed cpm are converted to dpm by simple division.

For color quenching when the color is dissolved in the solvent, balanced counting can help. This is also known as balance-point operation. When the spectrum is shifted by quenching, the maximum counting rate is no longer centered on the "window" (see Figure 8.24).

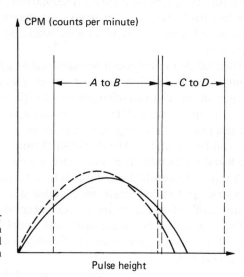

**Figure 8.24**
Use of discriminator settings for "balance-point" counting. (From a brochure prepared by the Packard Instrument Co., and reprinted with their permission.)

If the counter is set to give the maximum counting rate with the unquenched sample, a shift in the pulse height to lower energies doesn't affect the rate very much, since the loss of counts on the high side of the spectrum is partly compensated for by the gain on the low side.

The character of the isotope is important also. $^3$H, $^{14}$C, and $^{32}$P emit $\beta$ particles of different energies. Accordingly, each isotope has its own set of tap, window, and gain settings for maximum counting efficiency. Photomultipliers in any given scintillation counter will "drift," so that the settings for maximum counting efficiency of a given isotope will change. Accordingly, new maximum efficiency settings for an instrument should be determined every six months.

Another procedure for overcoming quenching when a spectral shift occurs is the channels ratio method (19, 20). This is especially helpful when quenching is from yellow color on a solid support.

In this method, the counting efficiency is correlated with the spectral shift.

A single isotope sample is counted using two channels (A + B). These may have the same lower limits or upper limits or may have different upper and lower discriminator limits (20). With an unquenched sample, the count ratio of Channel B to Channel A will be different from that observed for a quenched sample. By counting a series of samples, each prepared with the same known amount of activity but differing quantities of a quenching agent (such as carbon tetrachloride, acetone, or pyridine), a quench calibration curve such as that shown in Figure 8.23b can be constructed. With such a curve, the counting efficiency may be determined by examining the B/A ratio, provided the samples are counted with the same instrument settings used to construct the calibration curve. Alternatively, one of the discriminator limits can be varied until the efficiency versus channels ratio graph is linear (unlike the one in 8.23b); this involves more work, but makes correction for quenching easier.

Dual-labeled samples can be examined on a three-channel (A, B, C) instrument by operating first on the more energetic isotope—with discriminator settings of one channel (channel B) set high enough to exclude the less energetic isotope. A second, narrower monitor channel (C) is also set to exclude the less (number 1) energetic isotope so that quenching of the more energetic one can be calculated. The third (A) channel is set so it doesn't overlap with the B and C channels. It gives a combined count rate of both isotopes.

A series of quenched standards is then prepared for each isotope. With the two series and with the instrument adjusted so that the area of overlap is eliminated, calibration curves are then constructed.

Once these curves are available, the samples are counted, and after using the C/B ratio to find the required efficiency figures the following computations are performed:

$$dpm_2 = cpm_C/Eff._{2C} \qquad (8\text{-}26)$$

$$dpm_1 = [cpm_A - (dpm_2 \times Eff._{2A})]/Eff._{1A} \qquad (8\text{-}27)$$

Generally, if all samples of an experiment have been treated similarly, extensive quench analysis is not necessary. If different assay methods are to be compared or if exact data are required, quench analysis with a set of quench standards should be employed.

d. If your samples for counting are in strong acid or base, it is best to neutralize them before putting them in the scintillation fluid. The neutralized samples are also less likely to quench the fluor. To check for chemiluminescence, treat some of a nonradioactive control in the same way and count it.

### 8.4e  *Setting the Scintillation Counter for Dual Label Counting*

A setting is chosen for optimum performance of the photomultiplier tube. Amplification is set full, and the gain is controlled by attenuation of gain.

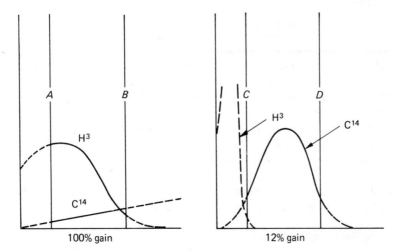

**Figure 8.25**   Showing the use of the gain controls for counting of two isotopes in the same sample. (From a brochure prepared by the Packard Instrument Co., and reprinted with their permission.)

The gain control operates on individual channels. Thus one can set one channel to count optimally one isotope and the other to later count a second (during the night for example) without resetting the instrument.

This use of the gain controls is illustrated in Figure 8.25. However, the ratio of energies of the two isotopes should be at least four for effective dual counting.

### 8.4f   Calculation of Relative Counting Efficiency

The overall efficiency of a counter may be determined by measurement of a standard sample.

$$\frac{cps}{\mu c} = \frac{observed\ cps}{known\ \mu c} \tag{8-28}$$

or

$$\%\ efficiency = \frac{observed\ cps}{\mu c}(3.7 \times 10^4) \tag{8-28a}$$

This is for comparing the results from one counter to those of another.

## 8.5   PRACTICAL ASPECTS

### 8.5a   Units of Radiation

Radioactivity is measured in terms of curies. One curie is that amount of material which gives $3.7 \times 10^{10}$ decompositions per second. Because different substances have different half-lives, this does not represent the same weight.

**TABLE 8.4**

|                | Activity<br>milligram-atom |            | Activity<br>milligram |            |
| -------------- | -------------------------- | ---------- | --------------------- | ---------- |
| $^3$H          | 29                         | curies     | 9.6                   | curies     |
| $^{14}$C       | 62                         | millicuries| 4.4                   | millicuries|
| $^{32}$P       | 9120                       | curies     | 285                   | curies     |
| $^{35}$S       | 1488                       | curies     | 42.8                  | curies     |
| $^{36}$Cl      | 1.1                        | millicuries| 0.03                  | millicuries|
| $^{131}$I      | 16,000                     | curies     | 123                   | curies     |

At 100% enrichment the values shown in Table 8.4 hold. Most isotope is sold by activity in millicuries or microcuries.

### 8.5b  Radiation Safety (5)

Cleanliness in tracer work is important for two reasons: the background count will go up in a poorly managed lab, rendering precise experimentation more difficult; and, although "maximum permissible doses" have been defined, one should never forget that all radiation is harmful and should be minimized. You owe it to your fellow workers—as well as to yourself—to be neat and orderly.

The set of rules presented for the handling of the $\beta$ emitters $^{32}$P, $^{14}$C, and $^3$H in micro-curie amounts is derived in part from "General Procedures for Protection from Radiation Hazards," University of Illinois, Feb. 1959, in part from conversations with the University of Illinois Health Physicist, and in part from past experience in isotopes handling.

In general, low activity beta emitters provide no significant health hazard unless ingested. What is of primary importance is that workers and working areas not become contaminated with long-lived emitters because of careless handling. Follow these guidelines to the letter. Check with an authority about any particular problems that may arise.

#### (i) Transferring and manipulation

1. Pipette only with the aid of some autopipetting device—NEVER by mouth.

2. Work over a tray lined with an absorbent blotter paper. Change this blotter *every day*.

3. Perform all work in a fume hood if volatile radioactive compounds are being used.

4. Wear disposable plastic gloves when handling gross (1 mC) amounts of radioactive material.

5. *Always* wear a lab coat.

6. *Always* wash your hands thoroughly before leaving the laboratory.

### (ii) Treatment of glassware and equipment

1. Label with the commercially available yellow tape all glassware, pipettes, etc., that come in contact with isotopes.

2. When used, place pipettes in a graduate partially filled with water to soak. Place glassware in a plastic bucket. Keep these "hot" dirty dishes separate from "cold" ones.

3. Wash the contaminated glassware with warm soapy water in the "hot" sink designated and *nowhere* else. Wash dishes every day. Do not accumulate "hot" glassware.

4. Rinse off broken glassware and place in solid waste disposal containers.

### (iii) Waste disposal

1. Pour all liquid waste into the central collecting vessel designated, *not* down any drain.

2. Place all solid waste (filter paper, tissues, etc.) into the designated central waste container, *not* in the regular waste cans.

3. If any large volumes of liquid waste will be generated in an experiment, make special arrangements for its disposition ahead of time.

### (iv) Isotope storage

1. Keep all samples, stocks, etc., in sealed containers which are clearly marked as to the type of isotope and carrier compound, amount of radioactivity, specific activity, date assayed and the name of the experimenter.

2. Store $^{32}P$ in foil-wrapped heavy glass vials, *not* in plastic containers.

### (v) Spills and emergencies

1. For minor spills on hands, wash thoroughly with soap and water in the "hot" sink. If $^{32}PO_4$ is the contaminant, soak your hands briefly in a "cold" 0.1 $M$ phosphate solution first.

2. For minor spills on your lab coat, turn in the coat to the health officer or instructor.

3. For spills on benches and floors, sponge (rinsing sponges in the "hot" sink) and rinse areas thoroughly with warm water.

4. Accidental ingestions and gross spills should be reported immediately to the instructor or health officer along with information as to amount and type of radioisotope involved.

### 8.5c  Specific Hazards from Radiation

The hazards from radiation can be divided into those arising from internal and external exposure (Appendix 17). An amount of radioactive material posing no special problem externally may become very dangerous internally. Isotopes with long half-lives which favor specific tissues and are excreted slowly present special hazards. This will depend not only on the isotope but particularly on the chemical form of the isotope: $^3H$ in $^3H_2O$, being volatile, is more of a hazard than $^3H$ in $^3H_3PO_4$, for example. But $^{32}PO_4$ is probably more hazardous internally than $^3H_2O$.

Damage to living organisms by radiation comes from the ionization it produces. Officially, 1 roentgen (R) is the exposure dose of X-or $\gamma$-rays such that the associated emission per 0.001293 of air (1 ml) produces ions carrying 1 esu of electricity of either sign. In terms of ionization energy this is:

$$1R = \frac{1.602 \times 10^{-12} \times 32}{4.803 \times 10^{-10} \times 1.293 \times 10^{-3}} = 82.6 \frac{\text{ergs}}{\text{g air}} \qquad (8\text{-}29)$$

This is with: electron charge $= 4.803 \cdot 10^{-10}$ esu, 32 ev per ion pair, and 1 electron volt $= 1.602 \cdot 10^{-12}$ erg. Although one speaks of external exposure, it is only the radiation absorbed which is damaging. The dose rate, exposure time, and the exposed area of the body are all important.

So one roentgen produces 82.6 ergs per gram of air. The Rep (roentgen equivalent physical) is defined in terms of the dissipation of 93 ergs of *ionizing* radiation per gram of biological tissue (93 ergs is the value associated with one roentgen in water). A Rad is 100 ergs of ionizing radiation when dissipated in one gram of radiated material.

Since different kinds of radiation can have different degrees of effect for the same amount of exposure, the relative biological effectiveness (RBE) for $\gamma$-rays, X-rays and $\beta$ particles is defined as 1; for $\alpha$ particles and protons, the RBE is 10. For heavy ions, the RBE is 20. With these definitions, we can state permissible doses of ionizing radiation. We use the roentgen equivalent man (REM), which is:

$$\text{REM} = \text{Rad} \times \text{RBE} \qquad (8\text{-}30)$$

The maximum permissible doses of radiation by current U. S. Government regulations are 1.25 REM in any 13 consecutive weeks for the whole body, head, trunk, blood forming organs, eye lenses, or gonads. For hands, forearms, feet or ankles, the maximum does is 18.75 REM in any 13 consecutive weeks. The limit is 7.5 REM in any 13 consecutive weeks for the skin of the whole body.

The following empirical equation can be used to determine the exposure from an external $\gamma$-emitting source. The exposure rate, in mR/hr, at 1 foot, is

$$R = 6 \times C \times E \qquad (8\text{-}31)$$

where C is millicuries of activity and $E$ is the $\gamma$-irradiation energy per disintegration. This relation holds quite well for $\gamma$ energies of 0.3 to 3.0 Mev.

Assume a point source of 1 mC disintegrating with an energy of 1 Mev per disintegration. In one hour there will be delivered to each square centimeter of surface of a sphere of 1 ft radius surrounding this point source:

$$\frac{3.7 \times 10^{10} \times 60^2}{10^3 \times (12 \times 2.45)^2 \times 4\pi} = 1.12 \times 10^7 \text{ photons (each with an energy of 1 Mev)}$$

Since $\gamma$ rays lose 3.5 kev/cm/gm of air and there are approximately $10^{-3}$ gm of air per cm of path, 1 Mev will penetrate:[3]

$$\frac{10^{-3} \times 10^6}{3.5 \times 10^3} = \frac{1}{3.5} \text{ cm}$$

For $^{131}$I, $\gamma$ radiation of several energies is released:

| % | Mev | | |
|---|---|---|---|
| 0.028 | 0.724 | = | 0.020272 |
| .093 | .638 | = | .059334 |
| .874 | .365 | = | .318280 |
| .007 | .164 | = | .001148 |
| | | | 0.399034 Mev/disintegration $= E$ |

$$6E = R/C = 0.399034 \times 6 = 2.394 \text{ mR/hr/mC}$$

A dose of 1.25 Rem over 13 weeks (2184 hours) is 1.25/2184, or 0.57 mR/hr. So $C = R/6E = 0.57/2.394 = 0.24$ mC at 1 ft continuously. 0.24 mC is equivalent to:

$$\frac{0.24}{10} \times 1.5 \times 10^5 = 3.6 \times 10^3 \text{ mg or 3.6 gm of } \gamma \text{ globulin}$$

$$(M = 1.5 \times 10^5) \text{ labelled at } 10 \frac{\text{mC}}{\text{mg}}$$

## 8.6 TREATMENT OF DATA: COUNTING STATISTICS (12, 21)

This section is included since counting statistics involve counts in a given time, that is reasonably large, discrete numbers. They are Poisson statistics and consequently are different from the statistics covered in Chapter 2.

The value of the arithmetic mean $m$ of a set of $n$ measurements is obtained from the usual equation:

$$m = \frac{\sum\limits_{n}^{n} x_i}{n} \tag{8-32}$$

Any measurement deviating from the mean is given by $x - m$.

---

[3] Or less—this calculation errs on the safe side.

### 8.6a  Reliability of Results

We are interested first in a way of determining the reliability of a given result, whether a single counting of a sample or several countings of a particular sample. Several measures of precision or "errors" are used to indicate the reliability of a result. The standard deviation ($\sigma$) is defined as:

$$\left(\frac{\sum_{}^{n}(x-m)^2}{n-1}\right)^{1/2} \tag{8-33}$$

For a single observation $\sigma$ may be estimated from:

$$\sigma = (N)^{1/2} \tag{8-34}$$

$N$ is the count of a sample. For a number of observations a more exact value is: $\sigma = (\bar{N})^{1/2}$, and the standard deviation of the mean is

$$\sigma_{\bar{N}} = \left(\frac{\bar{N}}{n}\right)^{1/2} \tag{8-35}$$

where $n$ is the number of observations.

It is customary to use $\sigma$ to express error. We define $k$ as a proportionality constant between "error," $E_N$, and the standard deviation. Thus the error of an activity is

$$E_N = k(N)^{1/2} = k\sigma \tag{8-36}$$

The value of $k$ is selected from Table 8.5. $P(x)$ is the probability that $x$ will exceed $k\sigma$; that is, that the error $E_N$ will be greater than $k\sigma$.

**TABLE 8.5**

| $k = x/\sigma$ | $P(x)$ |
|---|---|
| 0.6745 | 0.500 ("probable error") |
| 1.0000 | 0.317 ("standard error") |
| 1.6449 | 0.100 ("reliable error") |
| 1.4500 | 0.050 ("95/100" error) |
| 2.5758 | 0.010 ("99/100" error) |

It is frequently desired to express the error as a percentage. The error for a single observation is

$$E_N\% = \frac{100k(N)^{1/2}}{N} = \frac{100k}{(N)^{1/2}} \tag{8-37}$$

where $k$ may be selected to give the desired probability.

Since any counting rate is

$$R = \frac{N}{t} \tag{8-38}$$

Then the error in the rate is

$$E_R = \frac{E_N}{t} = \frac{k(N)^{1/2}}{t} = \frac{k(Rt)^{1/2}}{t} = k\left(\frac{R}{t}\right)^{-1/2} \tag{8-39}$$

The percent error of a rate is

$$E_R\% = \frac{100k}{(Rt)^{1/2}} \tag{8-40}$$

And from this we may find the time required for a desired percent error:

$$t = \frac{10^4 k^2}{R(E\%)^2} \tag{8-41}$$

Occasionally, in a series of replicate samples, one of the values may vary so widely from the others that it disproportionately affects the mean. Such a value may be rejected on the basis of Chauvenet's criterion provided the set of observations is distributed normally (Chapter 2). If that is the case, such a value may be rejected when the magnitude of its deviation from the mean of all measurements is such that the probability of occurrence is less than $1/2n$. To use the criterion, we calculate the standard deviation of the mean as:

$$\sigma_N = \left(\frac{\bar{N}}{n} - 1\right)^{1/2} \tag{8-42}$$

And the deviation in question as

$$X = (\bar{N} - N) \tag{8-43}$$

The ratio of these quantities

$$P = \frac{X}{\sigma_N} \tag{8-44}$$

is referred to a table of critical ratios (see Figure 8.26), and the suspected observation is rejected if the calculated ratio exceeds that shown for the $n$ observations used to calculate the mean.

The total counting rate is the sum of the background counting rate and the sample counting rate, so the errors in the two are propagated:

$$e_t = \sqrt{e_1^2 + e_2^2 + \ldots + e_n^2 +} \tag{8-45}$$

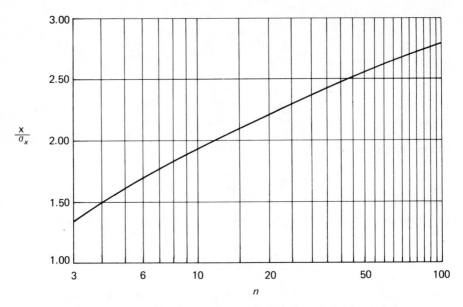

**Figure 8.26**   Chauvenet's criterion for rejection of a value with a deviation $x$. [The values are taken from R. T. Overman and H. M. Clark, *Radioisotope Techniques*, McGraw-Hill, New York (1960).]

The total error, $e_t$, is the square root of the sum of the squares of the individual errors.

The net activity of a sample (that is, minus "background" counts) is the desired quantity in most instances.

$$N_s = N_c - N_b, \quad \text{or} \quad R_s = R_c - R_b \qquad (8\text{-}46,\ 8\text{-}46a)$$

Since $\sigma_s = (\sigma_c^2 + \sigma_b^2)^{1/2}$, and $\sigma = (N)^{1/2}$, the error of a net activity may then be written:

$$E_{N_s} = (E_{N_c}^2 + E_{N_b}^2)^{1/2} = k(N_c + N_b)^{1/2} \qquad (8\text{-}47)$$

And the per cent error of a net activity is

$$E_{N_s}\% = \frac{100k(N_c + N_b)^{1/2}}{N_c - N_b} \qquad (8\text{-}48)$$

These formulas apply only when the sample and background are counted for equal times, i.e., $t_c = t_b$.

When the sample and background are counted for different times:

$$E_{R_s} = (E_{R_c}^2 + E_{R_b}^2)^{1/2} = k\left(\frac{R_c}{t_c} + \frac{R_b}{t_b}\right)^{1/2} \qquad (8\text{-}49)$$

And the percent error of the net rate is

$$E_{R_s}\% = \frac{200k\left(\dfrac{R_c}{t_c} + \dfrac{R_b}{t_b}\right)^{1/2}}{R_c - R_b} \tag{8-50}$$

The most efficient distribution of counting times between sample and background is

$$\frac{t_c}{t_b} = \left(\frac{R_c}{R_b}\right)^{1/2} \tag{8-51}$$

And

$$t_b = \frac{10^4\left[\left(\dfrac{R_c}{R_b}\right)^{1/2} + 1\right]k^2}{\left[\left(\dfrac{R_c}{R_b}\right)^{1/2} - 1\right]^2(E_{R_s}\%)^2 R_b} \tag{8-52}$$

And

$$t_c = t_b\left(\frac{R_c}{R_b}\right)^{1/2} \tag{8-53}$$

If the values of $R_c$ and $R_b$ are known approximately from a preliminary determination, then the required times for the desired error can be calculated using these formulas.

There are two statistical problems which may arise from the counters: the instrument is not counting randomly, or there is a drift in the instrument so that there is a trend of measurements towards higher or lower values. Both of these possibilities may be checked by statistical methods (20, 21). The procedure involves comparison of the expected $\sigma$ and actual $\sigma$ of a set of counts from a single sample. It is a powerful and easy way to detect trouble with the counter.

### 8.6b Limits of Detection of Radioactivity

We are also interested in how little radioactivity can be reliably detected, since some samples may have low activities.

The precision of a counting setup (i.e., least detectable amount of radioactivity) may be calculated if we define precision as that count corresponding to the standard deviation. It may be expressed in units of radioactivity if the calibration factor (c. f.) or overall efficiency of the counting system is known:

$$\text{Precision} = \frac{\sigma_s}{\text{c.f.}} = \frac{\left(\dfrac{R_c}{t_c} + \dfrac{R_b}{t_b}\right)^{1/2}}{\text{c.f.}} \tag{8-54}$$

Since, however, precision depends upon time this latter factor must be arbitrarily assigned. With limiting samples $R_c = R_b$, therefore:

$$\text{Precision} = \frac{\left(\dfrac{2R_b}{t_b}\right)^{1/2}}{\text{c.f.}} \tag{8-55}$$

which is the activity detectable on the average two out of three times with a given counting time, $t_b$. If we desire to increase the reliability of detection we may introduce an appropriate value of $k$:

$$\text{Precision} = \frac{k\left(\dfrac{2R_b}{t_b}\right)^{1/2}}{\text{c.f.}} \tag{8-56}$$

The precision of the counting setup is helpful in determining the order of magnitude of the smallest sample that can be detected.

## 8.7 SUMMARY

The advantages of the use of isotopes are well known. Among these advantages is great sensitivity—quantities of material down to $10^{-12}$ moles can be measured. Isotopes are almost indispensable agents in modern biochemical investigations. Their disadvantages are primarily questions of safety, and this is strongly influenced by the isotope and its chemical form. A sometimes major disadvantage is the existence of "isotope" effects; these are usually important only with $^3H$ and $^{14}C$ compounds.

As far as the advantages of proportional versus scintillation counting, these depend somewhat on the isotope to be counted—$^{32}P$ can be counted easily with a proportional counter but not $^{14}C$. Generally, scintillation counting is preferred because it can be used to count almost any isotope with reasonable efficiency and fewer problems are found with self-absorption, etc. This seems to outweigh the generally greater time that must be spent in sample preparation, though some substances may be troublesome enough in this respect to warrant the use of a proportional counter.

## REFERENCES

1. Carswell, D. J., *Introduction to Nuclear Chemistry*, Elsevier, New York (1967).
2. Friedlander, G., J. W. Kennedy and J. M. Miller, *Nuclear and Radiochemistry*, 2nd ed., Wiley, New York (1964).
3. Overman, R. T., *Basic Concepts of Nuclear Chemistry*, Rheinhold, New York (1963).
4. Weidner, R. T. and R. L. Sells, *Elementary Modern Physics*, Allyn and Bacon, Boston (1960).

5. Wang, C. H. and D. L. Willis, *Radiotracer Methodology in Biological Science*, Prentice-Hall, New Jersey (1965).
6. Chase, G. D. and J. L. Rabinowitz, *Principles of Radioisotope Methodology*, Burgess, Minneapolis (1962).
7. Korsunsky, M., *The Atomic Nucleus* (translated by G. Yankovsky), Gordon, New York (1964).
8. Johnson, W. H., K. S. Quisenberry and A. O. Nier in *Handbook of Physics*, Part 9, E. U. Condon and H. Odishaw, Eds., McGraw-Hill, New York (1958), Ch. 2, p. 55.
9. Landau, B. R., G. E. Bartsch, J. Katz and H. G. Wood, *J. Biol. Chem.*, **239**, 686 (1964).
10. Weinberg, R. A., U. Loening, M. Willems and S. Penman, *Proc. Nat. Acad. Sci. U. S.*, **58**, 1088 (1967).
11. Dice, J. F. and R. T. Schimke, *J. Biol. Chem.*, **247**, 98 (1972).
12. Overman, R. T. and H. M. Clark, *Radioisotope Techniques*, McGraw-Hill, New York (1960).
13. Hanson, D. L. and E. T. Bush, *Anal. Biochem.*, **18**, 320 (1967).
14. Bush, E. T., *Anal. Chem.*, **36**, 1082 (1964).
15. Birks, J. B., *The Theory and Practice of Scintillation Counting*, Pergamon Press, New York (1964).
16. Ziegler, C. A., D. J. Chleck and J. Brinkerhoff, *Anal. Chem.*, **29**, 1774 (1957).
17. Drysdale, G. N., *J. Biol. Chem.*, **234**, 2399 (1959).
18. Patterson, M. D. and R. C. Greene, *Anal. Chem.*, **37**, 854 (1965).
19. Hendler, R. W., *Anal. Biochem.*, **7**, 110 (1964).
20. Herberg, R. J., "Channels Ratio Method of Quench Correction in Liquid Scintillation Counting," Packard Technical Bulletin No. 15, Packard Instrument Co., Downers Grove, Illinois, December 1965.
21. Young, H. D., *Statistical Treatment of Experimental Data*, McGraw-Hill, New York (1962).
22. Bennet, C. A. and N. L. Franklin, *Statistical Analysis in Chemistry and Chemical Industry*, Wiley, New York (1954).

**APPENDICES**

# 1. PROPERTIES OF PLASTICS

(a) *Resistance of Plastics to Chemicals*

| Chemical | Cellulose Nitrate | Conventional Polyethylene | Linear Polyethylene | Polyallomer | Polypropylene |
|---|---|---|---|---|---|
| Acids, inorganic | E* | E | E | E | E |
| Acids, organic | E* | E | E | E | E |
| Alcohols | N | E | E | E | E |
| Aldehydes | E | G (N) | G (F) | G (N) | G (N) |
| Amines | — | G | G | G | G |
| Bases | E* | E | E | E | E |
| Dimethyl Sulfoxide | — | E | E | E | E |
| Esters | N | E | G | G | G |
| Ethers | N | G | E | E | E |
| Foods | E | E | E | E | E |
| Glycols | E | E | E | E | E |
| Hydrocarbons, aliphatic | N | G (N) | G | G | G |
| Hydrocarbons, aromatic | N | G (N) | G | G | G |
| Hydrocarbons, halogenated | N | G | G | G | G |
| Ketones | N | G | G | G | G |
| Mineral Oil | E | E | E | E | E |
| Oils, essential | E | G | G | G | G |
| Oils, lubricating | E | G | E | E | E |
| Oils, vegetable | E | E | E | E | E |
| Proteins, unhydrolyzed | G | E | E | E | E |
| Salts | E | E | E | E | E |
| Silicones | — | G | E | E | E |
| Water | E | E | E | E | E |

*N for strong, concentrated acids or bases; E for 10% solutions.

| Chemical | Cellulose Nitrate | Conventional Polyethylene | Linear Polyethylene | Polyallomer | Polypropylene |
|---|---|---|---|---|---|
| *Physical Properties of Plastics* | | | | | |
| Temperature limit, °C | — | 80 | 120 | 130 | 135 |
| Specific gravity | — | 0.92 | 0.95 | 0.90 | 0.90 |
| Tensile strength, psi | — | 2000 | 4000 | 2900 | 5000 |
| Brittleness temperature | — | −100 | −100 | −40 | 0 |
| Water absorption, % | — | <0.01 | <0.01 | <0.02 | <0.02 |
| Flexibility | Good | Excellent | Rigid | Slight | Rigid |
| Transparency | Clear | Translucent | Translucent | Translucent | Translucent |
| Autoclavable | — | No | With caution | Yes | Yes |

*Symbols:*

E: Excellent. Exposure for at least 30 days at room temperature has no effect.

G: Good. Short exposures (less than 24 hours) at room temperature cause no damage.

F: Fair. Short exposures at room temperature cause little or no damage under unstressed conditions.

N: Not recommended. Short exposures may cause permanent damage.

The second letter where given indicates results of exposure at 52°; the first letter refers to results at room temperature.

| Chemical | TPX | Teflon ETFE, FEP, TFE | Polycarbonate | General purpose polystyrene | Styrene-acrylonitrile | Polyvinyl-chloride† |
|---|---|---|---|---|---|---|
| Acids, inorganic | E | E | E | N | E | G |
| Acids, organic | E | E | G | G | E | G |
| Alcohols | E | E | G | G | G (N) | G |
| Aldehydes | G (N) | E | F (N) | N | F (N) | F (N) |
| Amines | G | E | N | G | G | N |
| Bases | E | E | N | G | E | E |
| Dimethyl Sulfoxide | E | E | N | N | N | N |
| Esters | E | E | F | N | N | F |
| Ethers | G | E | F | F | N | F |
| Foods | E | E | E | E | G | G |
| Glycols | E | E | G (N) | G | G | F (N) |
| Hydrocarbons, aliphatic | G | E | F (N) | N | E | F |
| Hydrocarbons, aromatic | F | E | N | N | N | N |
| Hydrocarbons, halogenated | F | E | N | N | N | N |
| Ketones | G | E | N | N | N | N |
| Mineral Oil | E | E | E | G | G | E |
| Oils, essential | G | E | G | N | F | N |
| Oils, lubricating | E | E | G | G | G | E |
| Oils, vegetable | E | E | E | G | E | E |
| Proteins, unhydrolyzed | E | E | E | G | E | G |
| Salts | E | E | E | E | E | E |
| Silicones | E | E | E | G | G | G |
| Water | E | E | E | E | E | E |

## Physical Properties of Plastics

| Chemical | TPX* | Teflon* ETFE, FEP, TFE | Polycarbonate | General purpose polystyrene | Styrene-acrylonitrile | Polyvinyl-chloride† |
|---|---|---|---|---|---|---|
| Temperature limit, °C | 175 | 180 | 135 | 70 | 95 | 70 |
| Specific gravity | 0.83 | 1.70 | 1.20 | 1.07 | 1.07 | 1.34 |
| Tensile strength | 4000 | 6500 | 8000 | 6000 | 11000 | 6500 |
| Brittleness temperature | — | -100 | -135 | ‡ | -25 | -30 |
| Water absorption, % | <0.01 | 0.1 | 0.35 | 0.05 | 0.23 | 0.06 |
| Flexibility | Rigid | Moderate | Rigid | Rigid | Rigid | Rigid |
| Transparency | Clear | Translucent | Clear | Clear | Clear | Clear |
| Autoclavable | Yes | Yes | Yes | No | No | No |

*Symbols continued:*
*TPX is the trademark of Imperial Chemical Industries Limited for its brand of methylpentane polymer. Teflon is a fluorinated hydrocarbon made by DuPont.
†"Tygon" is a fluorinated PVC.
‡Normally somewhat brittle at room temperature.

Factors affecting chemical resistance include temperature, pressure and other stresses, length of exposure and concentration of the chemical. As the maximum useful temperature of the plastic is approached, its resistance to attack decreases. Combinations of chemicals can be more harmful than the individual compounds. If in doubt about the resistance of a plastic item to some particular situation, make a test before beginning an experiment.

(*b*)   *Physical, Chemical, and Biological Properties of Plastics*

The polyolefins are a group of resins which includes conventional and linear polyethylene, polyallomer, and TPX. All are unbreakable and are the only plastics lighter than water. Strong oxidizing agents eventually cause embrittlement. All polyolefins can be damaged by long exposure to ultraviolet light.

Conventional polyethylene is translucent and flexible and is used for squeeze-type wash bottles. Linear polyethylene is translucent, more rigid, and somewhat harder than conventional polyethylene. Polypropylene is translucent and autoclavable. Because its coefficient of thermal expansion approximates that of water near room temperature, polypropylene is used for calibrated ware. TPX is the newest of polyolefins. It is transparent, rigid, and resistant to impact and high temperatures, so it is used for graduated cylinders and beakers. It can be autoclaved even at 150°C, and sterilized by dry heat at 160°C. Polyallomer is translucent and rigid. It is more abrasion-resistant than polypropylene, and has a higher impact strength.

Teflon is a fluorinated hydrocarbon (Figure A.1). Both common forms, TFE and FEP, have excellent adhesion resistance and are autoclavable and unbreakable. Teflon TFE is opaque, white, and has the lowest coefficient of friction of any solid. It is used for stopcock and separatory funnel plugs because of its low friction and tight seal. Teflon FEP is flexible and translucent. It has a slight bluish cast and a heavy feel because of its higher density. FEP withstands temperatures from $-270°$ to $+205°C$, and may be sterilized repeatedly by all known chemical or thermal methods.

Both high-impact polystyrene, and the copolymer, styrene-acrylonitrile, are rigid and have excellent dimensional stability.

Polycarbonate is window clear, very strong and rigid. It is autoclavable, nontoxic, unbreakable, and the toughest of all thermoplastics. Its strength and dimensional stability allow its use for high-speed centrifuge ware.

Polyvinyl chloride is transparent and has a slight bluish cast. The narrow mouth bottles are relatively thin-walled and can be flexed slightly. PVC bottles are used for shipping oil or water samples, being strong, somewhat flexible and relatively impermeable to most gases.

In general, most of these plastics are biologically inert.

This material was adopted largely from a brochure copyrighted by Nalge (Sybron Corp.), Rochester, N.Y., and is reproduced in part through the courtesy of that firm.

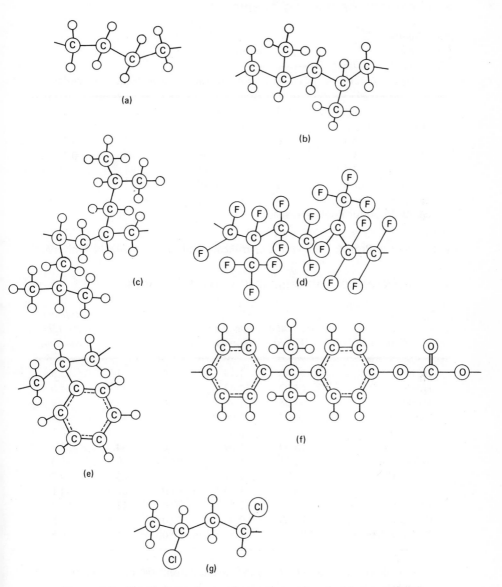

**Figure A.1** Chemical structures of repeating units of: (a) conventional polyethylene; (b) polypropylene; (c) TPX; (d) Teflon FEP; (e) styrene and copolymer; (f) polycarbonate; (g) polyvinylchloride.

## 2. CAPACITIES OF DIALYSIS TUBING (1)

| Code Number (Diameter in 1/32's of an inch) | Width, cm | ml/cm Calculated | ml/cm Tested | cm/100 ml | cm/ml |
|---|---|---|---|---|---|
| 8 | 1 | 0.31 | 0.31 | 322.58 | 3.2 |
| 18 | | | | | |
| 20 | 2.4 | 1.83 | 1.83 | 54.64 | 0.55 |
| 27 | 3.3 | 3.46 | 3.46 | 28.90 | 0.3 |
| 36 | 4.4 | 6.17 | 6.31 | 15.84 | |
| $1\frac{7}{8}$ (60) | 7.6 | 18.40 | 18.79 | 5.32 | |

Add 10–25 cm for knots and space.

Reference:
1. Courtesy of Prof. I. Fridovich, Department of Biochemistry, Duke University.

## 3. CONCENTRATION OF COMMON ACIDS AND BASES

| Compound | Molecular Weight | Specific Gravity | Percent | N | ml/Liter for 1N* Solution |
|---|---|---|---|---|---|
| HCl | 36.46 | 1.19 | 36.0 | 11.7 | 85.5 |
| $HNO_3$ | 63.02 | 1.42 | 69.5 | 15.6 | 64.0 |
| $H_2SO_4$ | 98.08 | 1.84 | 96.0 | 35.9 | 28.4 |
| $CH_3COOH$ | 60.03 | 1.06 | 99.5 | 17.6 | 56.9 |
| $NH_4OH$ | 35.04 | 0.90 | 58.6 | 15.1 | 66.5 |
| $H_3PO_4$ | 98.00 | 1.69 | 85.0 | 44.1 | 22.7 |
| Thioglycolic Acid | 92.12 | 1.26 | 80.0 | 10.9 | 91.3 |
| HCOOH | 46.03 | 1.21 | 97.0 | 25.5 | 39.2 |
| | 46.03 | 1.19 | 88.0 | 22.7 | 44.1 |
| $HClO_4$ | 100.5 | 1.67 | 70.0 | 11.65 | 85.7 |
| Pyridine | 79.10 | 0.98 | 100.0 | 12.4 | 80.6 |
| 2-Mercaptoethanol | 78.13 | 1.14 | 100.0 | 14.6 | 68.5 |

To calculate concentration ($c$) from the weight percent ($w$) of a compound, use the formula:

$$\frac{10\ ws}{M} = c$$

$M$ is the molecular weight, and $s$ the specific gravity.

*Remember, the normality ($N$) is not the same as the molarity ($M$) for sulfuric and phosphoric acid.

## 4. p*K* VALUES OF SOME IMPORTANT COMPOUNDS IN BIOCHEMISTRY (1-5)

| Substance (*Molecular Weight*) | p$K_1$ | p$K_2$ | p$K_3$ | p$K_4$ |
|---|---|---|---|---|
| Acetaldehyde (44.05) | 1.10 | — | — | — |
| Acetic acid (60.05) | 4.76 | — | — | — |
| β-Alanine (189.09) | 3.60 | 10.19 | — | — |
| 2-Amino-2-methyl-1-propanol (89.14) | 9.9 | — | — | — |
| Ammonium hydroxide (35.05) | 9.26 | — | — | — |
| Ascorbic acid (176.12) | 4.17 | 11.57 | — | — |
| Arsenic acid (1/2 $H_2O$) (150.93) | 2.3 | 4.4 | 9.2 | — |
| Aspartic acid (133.1) | 2.10 | 3.86 | 9.82 | — |
| Adenosine 5′ triphosphate ("ATP") ($Na_2H_2$-$4H_2O$) (623) | 4.1 (amino) | 6.1 | 6.3 | 6.5 |
| Barbituric acid (128.09) | 4.00 | 12.5 | | |
| Benzoic acid (122.12) | 5.20 | — | — | — |
| N, N bis (2-hydroxyethyl) glycine ("bicine") (163.2) | 8.35 | — | — | — |
| Boric acid (61.84) | 9.23 | — | — | — |
| Carbonic acid (63.0) | 6.1 | 9.8 | — | — |
| Citric acid ($H_2O$) (210.14) | 3.08 | 4.75 | 5.40 | — |
| Cysteine (HCl, hydrate) (175.6) | 1.71 | 8.33 | 10.78 | — |
| Diphosphopyridine nucleotide ("NAD") ("DPN") ($4H_2O$) (735) | 3.7 | — | — | — |
| Ethylenediaminetetracetic acid ("EDTA") ($Na_4$-$2H_2O$) (416.2) | 0.26 | 0.96 | 2.67 | 2.70 |
| | p$K_5(N)$, 6.16 | | p$K_6(N)$, 10.26 | |
| Formic acid (46.03) | 3.77 | — | — | — |
| Fumaric acid (116.07) | 3.00 | 4.52 | — | — |
| Glutamic acid (HCl) (183.6) | 2.10 | 4.07 | 9.47 | — |
| Glutaric acid (132.11) | 4.3 | 5.54 | — | — |
| Glycine (75.07) | 2.35 | 9.78 | — | — |
| Glycylglycine (132.12) | 3.14 | 8.07 | — | — |
| N-2-hydroxyethyl-piperazine-N′-2-ethanesulfonic acid ("HEPES") (238.3) | 7.55 | — | — | — |
| Histidine (HCl-$H_2O$) (209.65) | 1.77 | 6.10 | 9.18 | — |
| Hydrazine (32.05) | 8.07 | — | — | — |
| Hydrochloric acid (36.46) | −0.47 | — | — | — |
| Hydroxylamine (HCl) (69.50) | 5.97 | — | — | — |
| Imidazole (68.08) | 7.07 | — | — | — |
| Lactic acid (90.08) | 3.86 | — | — | — |
| Lysine (HCl) (182.66) | 2.18 | 8.95 | 10.53 | — |
| Maleic acid (116.07) | 1.83 | 6.58 | — | — |
| Malic acid | 3.40 | 5.02 | — | — |
| Malonic acid (104.06) | 2.80 | 5.68 | — | — |
| Mercaptoacetic acid (92.1) | 3.67 | 10.31 | — | — |
| Mercaptoethanol (78.1) | 9.5 | — | — | — |
| 2-(N-morpholino) ethanesulfonic acid ("MES") (195) | 6.15 | — | — | — |

| Substances (Molecular Weight) | $pK_1$ | $pK_2$ | $pK_3$ | $pK_4$ |
|---|---|---|---|---|
| Nitric acid (63.02) | −1.3 | — | — | — |
| Oxalic acid ($2H_2O$) (126.07) | 1.19 | 4.21 | — | — |
| Phenol (94.11) | 10.0 | — | — | — |
| Phosphoenolpyruvate (cyclohexylammonium$_3$) (465) | — | 3.5 | 6.4 | — |
| 2-Phosphoglycerate (cyclohexylammonium$_3$) (483) | 1.42 | 3.6 | 7.1 | — |
| Phosphoric acid (98.04) | 1.96 | 7.12 | 12.32 | — |
| Phthalic acid (166.13) | 2.90 | 5.51 | — | — |
| Piperazine-N, N′ bis (2-ethanesulfonic acid) ("PIPES") (151) | 6.8 | — | — | — |
| Propionic acid (74.08) | 4.8 | — | — | — |
| Pyridine (79.1) | 5.14 | — | — | — |
| Pyruvic acid (Na) (110.1) | 0.59 | — | — | — |
| Riboflavin (376.4) | 9.93 | — | — | — |
| Salicylic acid (138.12) | 3.0 | — | — | — |
| Sulfuric acid (98.08) | −3.0 | 1.9 | — | — |
| Tartaric acid (150.09) | 2.95 | 4.16 | — | — |
| N-tris (hydroxymethyl) methyl-2-aminoethane-sulfonic acid ("TES") (229.2) | 7.5 | — | — | — |
| Triphosphopyridine nucleotide ("NADP") ("TPN") ($NaH_2$-$4H_2O$) (837) | 3.7 | — | — | — |
| N-tris (hydroxymethyl) methylglycine ("Tricine") (179) | 8.15 | — | — | — |
| Tris (hydroxymethyl) aminomethane (121.14) | 8.14 | — | — | — |
| Urea (60.06) | 0.18 | — | — | — |
| Veronal (Diethylbarbituric acid) (184.19) | 7.43 | — | — | — |

When making up buffers or other solutions containing these substances, remember that apparent $pK$'s are functions of temperature and ionic strength and composition of the solutions. Therefore, all pH's should be adjusted using a pH meter, and not simply by calculations based on the above values. In any case, the pH of any buffer should not be more than one pH unit from the $pK$.

On dilution, the pH of a concentrated buffer tends to shift towards its $pK$.

*References:*
1. Some data supplied courtesy of Jencks, W. P. "$pK$'s of Acids and Bases" (unpublished).
2. *Handbook of Biochemistry—Selected Data for Molecular Biology*, H. A. Sober, Editor, Chemical Rubber Co., Cleveland, Ohio (1968).
3. Reference 1, Appendix 3.
4. Good, N. E., G. D. Winget, W. Winter, T. N. Connolly, S. Izawa, and R. M. M. Singh, *Biochem.*, **5**, 467 (1966).
5. Chaykin, S., *Biochemistry Laboratory Techniques*, John Wiley and Sons, New York, N. Y. (1966).

## 5. TABLES FOR ADJUSTMENT OF CONCENTRATION OF AMMONIUM SULFATE SOLUTIONS

An equation for estimating how much solid ammonium sulfate or solution should be added to a given volume to obtain a given degree of saturation is given in reference 1; data for these estimations at temperatures from 0° to 30° is found in reference 2. At 0° the grams of solid ammonium sulfate to be added to a volume V (in milliliters) at a percent saturation of ammonium sulfate $S_1$ to obtain a final percent saturation $S_2$ is: grams = Volume/1000 ×

$$\frac{\text{concentration of saturated solution at } 0°, \text{g/ml}}{\left(1\text{-concentration of saturated solutions} \times \text{partial specific volume of salt at } 0° \times \frac{S_2}{100}\right)}$$

$$\times \frac{S_2 - S_1}{100} = \frac{0.0515(S_2 - S_1)}{100 - 0.27\, S_2} \times V$$

For the sake of convenience, tables for these adjustments for 0° (3) and 25° (4) are provided.

*(a)* *Values for 0°*

The values are in grams of ammonium sulfate to be added per liter of solution. The figures down the left side of the table are initial percent ($S_1$) and the figures across the top are final percent saturations.

Final Percent Saturation ($S_2$)

| $S_1$ | 10 | 15 | 20 | 25 | 30 | 33 | 35 | 40 | 45 | 50 | 55 | 60 | 65 | 67 | 70 | 75 | 80 | 85 | 90 | 95 | 100 |
|---|---|---|---|---|---|---|---|---|---|---|---|---|---|---|---|---|---|---|---|---|---|
| 0 | 53 | 80 | 106 | 134 | 164 | 187 | 194 | 226 | 258 | 291 | 326 | 361 | 398 | 421 | 436 | 476 | 516 | 559 | 603 | 650 | 697 |
| 5 | 27 | 56 | 79 | 108 | 137 | 162 | 166 | 197 | 229 | 262 | 296 | 331 | 368 | 390 | 405 | 444 | 484 | 526 | 570 | 615 | 662 |
| 10 | | 28 | 53 | 81 | 109 | 133 | 139 | 169 | 200 | 233 | 266 | 301 | 337 | 358 | 374 | 412 | 452 | 493 | 536 | 581 | 627 |
| 15 | | | 26 | 54 | 82 | 87 | 111 | 141 | 172 | 204 | 237 | 271 | 316 | 327 | 343 | 381 | 420 | 460 | 503 | 547 | 592 |
| 20 | | | | 27 | 55 | 75 | 83 | 113 | 143 | 175 | 207 | 241 | 276 | 296 | 312 | 349 | 387 | 427 | 469 | 512 | 557 |
| 25 | | | | | 27 | 46 | 56 | 84 | 115 | 146 | 179 | 211 | 245 | 264 | 280 | 317 | 355 | 395 | 436 | 488 | 522 |
| 30 | | | | | | | 28 | 56 | 86 | 117 | 148 | 181 | 214 | 233 | 249 | 285 | 323 | 362 | 402 | 445 | 488 |
| 33 | | | | | | | | 40 | 70 | 101 | 133 | 166 | 200 | 214 | 235 | 271 | 309 | 347 | 387 | 429 | 472 |
| 35 | | | | | | | | 28 | 57 | 87 | 118 | 151 | 184 | 201 | 218 | 254 | 291 | 329 | 369 | 410 | 453 |
| 40 | | | | | | | | | 29 | 58 | 89 | 120 | 153 | 170 | 182 | 212 | 258 | 296 | 335 | 376 | 418 |
| 45 | | | | | | | | | | 29 | 59 | 90 | 123 | 138 | 156 | 190 | 226 | 263 | 302 | 342 | 383 |
| 50 | | | | | | | | | | | 30 | 60 | 92 | 107 | 125 | 159 | 194 | 230 | 268 | 308 | 348 |
| 55 | | | | | | | | | | | | 30 | 61 | 76 | 93 | 127 | 161 | 197 | 235 | 273 | 313 |
| 60 | | | | | | | | | | | | | 31 | 44 | 62 | 95 | 129 | 164 | 201 | 239 | 279 |
| 65 | | | | | | | | | | | | | | | 31 | 63 | 97 | 132 | 168 | 205 | 244 |
| 67 | | | | | | | | | | | | | | | | 52 | 85 | 120 | 156 | 194 | 233 |
| 70 | | | | | | | | | | | | | | | | 32 | 65 | 99 | 134 | 171 | 209 |
| 75 | | | | | | | | | | | | | | | | | 32 | 66 | 101 | 137 | 174 |
| 80 | | | | | | | | | | | | | | | | | | 33 | 67 | 103 | 139 |
| 85 | | | | | | | | | | | | | | | | | | | 34 | 68 | 105 |
| 90 | | | | | | | | | | | | | | | | | | | | 34 | 70 |
| 95 | | | | | | | | | | | | | | | | | | | | | 35 |

Initial Percent Saturation ($S_1$)

(b) *Values for 25° (4)*

The table is otherwise identical to the previous one.

*Final Percent Saturation ($S_2$)*

| $S_1$ | 10 | 15 | 20 | 25 | 30 | 33 | 35 | 40 | 45 | 50 | 55 | 60 | 65 | 67 | 70 | 75 | 80 | 85 | 90 | 95 | 100 |
|---|---|---|---|---|---|---|---|---|---|---|---|---|---|---|---|---|---|---|---|---|---|
| 0 | 56 | 84 | 114 | 144 | 176 | 196 | 209 | 243 | 277 | 313 | 351 | 390 | 430 | 446 | 472 | 516 | 561 | 608 | 662 | 709 | 767 |
| 10 | | | 57 | 86 | 118 | 137 | 150 | 183 | 216 | 251 | 288 | 326 | 365 | 380 | 406 | 449 | 494 | 537 | 592 | 634 | 694 |
| 15 | | | 28 | 58 | 88 | 107 | 119 | 151 | 185 | 219 | 255 | 292 | 331 | 347 | 371 | 412 | 456 | 501 | 547 | 596 | 647 |
| 20 | | | | 29 | 59 | 78 | 91 | 123 | 155 | 189 | 225 | 262 | 300 | 313 | 340 | 382 | 424 | 466 | 520 | 559 | 619 |
| 25 | | | | | 30 | 49 | 61 | 93 | 125 | 158 | 193 | 230 | 267 | 280 | 307 | 348 | 390 | 429 | 485 | 522 | 583 |
| 30 | | | | | | 19 | 30 | 62 | 94 | 127 | 162 | 198 | 235 | 247 | 273 | 314 | 356 | 394 | 449 | 485 | 546 |
| 33 | | | | | | | 12 | 43 | 74 | 107 | 142 | 177 | 214 | 227 | 252 | 292 | 333 | 372 | 426 | 462 | 522 |
| 35 | | | | | | | | 31 | 63 | 94 | 129 | 164 | 200 | 213 | 238 | 278 | 319 | 358 | 411 | 447 | 506 |
| 40 | | | | | | | | | 31 | 63 | 97 | 132 | 168 | 180 | 205 | 245 | 285 | 322 | 375 | 410 | 469 |
| 45 | | | | | | | | | | 32 | 65 | 99 | 134 | 147 | 171 | 210 | 250 | 287 | 339 | 373 | 431 |
| 50 | | | | | | | | | | | 33 | 66 | 101 | 113 | 137 | 176 | 214 | 250 | 302 | 336 | 392 |
| 55 | | | | | | | | | | | | 33 | 67 | 80 | 103 | 141 | 179 | 215 | 264 | 298 | 353 |
| 60 | | | | | | | | | | | | | 34 | 47 | 69 | 105 | 143 | 179 | 227 | 261 | 314 |
| 65 | | | | | | | | | | | | | | | 34 | 70 | 107 | 143 | 190 | 224 | 275 |
| 67 | | | | | | | | | | | | | | | | 55 | 91 | 129 | 177 | 209 | 251 |
| 70 | | | | | | | | | | | | | | | | 35 | 72 | 107 | 153 | 186 | 237 |
| 75 | | | | | | | | | | | | | | | | | 36 | 72 | 115 | 149 | 198 |
| 80 | | | | | | | | | | | | | | | | | | 36 | 77 | 112 | 157 |
| 85 | | | | | | | | | | | | | | | | | | | 36 | 75 | 114 |
| 90 | | | | | | | | | | | | | | | | | | | | | 79 |
| 95 | | | | | | | | | | | | | | | | | | | | | 38 |

*Initial Percent Saturation ($S_1$)*

*References*
1. Kunitz, M., *J. Gen Physiol.*, **35**, 423 (1952).
2. Taylor, J. F., *The Proteins*, H. Neurath and K. Bailey, Editors, Vol I, Part A, Academic Press, New York (1953), p. 55.
3. Modified from Dawson, R. M. C., D. C. Elliott, W. H. Elliott, and K. M. Jones, *Data for Biochemical Research*, 2nd Edition, Oxford University Press, New York (1969), p. 616.
4. Modified from Green, A. A. and W. L. Hughes, *Methods in Enzymology*, S. P. Colowick and N. O. Kaplan, Editors, Vol I, Academic Press, New York (1955), p. 67.

## 6. METHODS OF DETERMINATION OF PROTEIN CONCENTRATION

I. $OD_{215}$–$OD_{225}$. This is essentially a measure of peptide bond concentration (1, 2). Sensitivity: 10–100 $\mu$g/ml.
1. Measure the absorbence of a protein solution at 215 and 225 nm.
2. Subtract $OD_{225}$ from $OD_{215}$, obtaining $\Delta OD$.
3. Determine $F$ so that $F \times \Delta OD = \mu$g protein/ml. For bovine serum albumin, $F$ is 145. This method is best used with solutions containing only protein and water since many substances strongly absorb in this region.

II. $OD_{280}/OD_{260}$. This is a measure of tyrosine and tryptophan concentration. Sensitivity: from 0.01 OD.
1. Measure the absorbence of a protein solution versus a buffer blank at both 260 and 280 nm.
2. Calculate protein content per ml by the formula of Kalckar (3):
$1.45OD_{280} - 0.74OD_{260} = $ mg protein/ml
or from the table of Warburg and Christian (4, 5). Again, scattering from turbidity interferes with this method as does the presence of extraneous material which absorbs in the ultraviolet. Note also that different proteins have different tyrosine and tryptophan contents, so this method will give different values for equal concentrations of different proteins.

III. Microbiuret Method of Goa. This method measures peptide bond concentrations chemically (6).

*Reagents:*
A. Biuret (modified Benedict's). Dissolve 173 g sodium citrate and 100 g sodium carbonate in about 500 ml of water with gentle warming if necessary. Dissolve 17.3 g copper sulfate pentahydrate in 100 ml of water and add to the first solution. Make up to 1000 ml with water. This reagent has a long shelf life (over 1 year) but should be discarded if a reddish precipitate forms.
B. Sodium hydroxide—6% w/v.
*Sample:* 0.1—4.0 mg protein in 4 ml 3% NaOH.

*Test:*
1. To a sample containing 0.1—4 mg protein in up to 2 ml volume add 2 ml 6% NaOH.
2. Add 0.2 ml reagent, mix and allow to stand 15 minutes at room temperature.
3. Read at 330 or 540 nm. The color is stable for several hours.

*Standard:* Crystalline bovine serum albumin is widely used, but dialyzed, lyophilized crude serum is probably better since the variety of proteins present in serum should give a better average value.

IV. Biuret-Phenol Method (5). This also measures peptide bonds, together with tyrosine residues. Sensitivity: 10—300 $\mu$g protein.

*Reagents:*
1. 4% $Na_2CO_3$ w/v
2. 2% $CuSO_4 \cdot 5H_2O$ w/v
3. 4% sodium tartrate w/v
4. Folin phenol reagent (Eimer and Amend) (this measures tyrosines).
5. Protein reagent: to 100 ml 4% $Na_2CO_3$ add 1.0 ml 2% $CuSO_4 \cdot 5H_2O$ and 1.0 ml sodium tartrate solution. Prepare this solution fresh daily.
6. Dilute protein reagent: 10 ml $Na_2CO_3$ + 0.5 ml copper solution + 0.5 ml sodium tartrate.

*Test:*
1. To 0.5 ml sample containing 10–300 $\mu$g protein add 5.0 ml protein reagent.
2. Let stand 45 minutes at room temperature or at 45°C for 10 minutes.
3. Add 0.5 ml dilute Folin reagent (1 volume commercial reagent + 2 volumes $H_2O$).
4. Mix, and let stand for 15 minutes at room temperature.
5. Read at 700 nm in a spectrophotometer or in a Klett with filter #66 (red). The color is stable for several hours.

*Note:* The dilute protein reagent is used for protein estimation in the 1—10 $\mu$g range.

*Standard:* See above.

Any substance that can reduce the copper interferes with this reagent: ammonium salts (such as Tris), organic amines and sulfhydryl reagents are especially troublesome.

*References*
1. Murphy, J. B. and M. W. Kies, *Biochim. Biophys. Acta*, **45**, 382 (1960).
2. Waddel, W. J., *J. Lab. Clin. Med.*, **48**, 311 (1956).
3. Kalckar, H., *J. Biol. Chem.*, **167**, 461 (1947).
4. Warburg, O. and W. Christian, *Biochem. J.*, **310**, 384 (1942).
5. Layne, E. in *Methods in Enzymology*, S. P. Colowich and N. O. Kaplan, Eds., Vol. III, Academic Press, New York (1957), p. 447.
6. Goa, J., *Scand. J. Clin. Lab. Invest.*, **5**, 218 (1953).

## 7. PROCEDURE FOR PHOSPHATE DETERMINATION (1, 2)

All glassware should first be cleaned with dichromate cleaning solution to remove traces of phosphate adsorbed to the glass.

*A. Total Phosphate Determinations* (1). The sample is placed in a 13 by 100 mm Pyrex test tube. Add 0.05 ml of 10% Mg $(NO_3)_2 \cdot 6H_2O$ in ethanol and evaporate the solution to dryness over a Bunsen burner. Keep shaking the tube rapidly until the sample is dry and no more brown fumes come off. The sample is now "ashed."

Let the tube cool. Then add 0.3 ml of $1M$ HCl. Put a clean marble over the mouth of the test tube and heat the tube for 15 minutes in a boiling water bath. This hydrolyzes any prophosphates present. The sample can now be assayed for phosphate (see below).

*B. Acid-labile Phosphate Determination.* Add HCl to $1M$ to the sample in a test tube, put a clean marble onto the test tube, and heat the contents 10 minutes in a boiling water bath. Assay the sample for phosphate.

*C. Inorganic Phosphate Determination* (2). The sample is transferred to a 15 ml volumetric or a 18 by 150 mm Pyrex test tube. Be sure to rinse all the sample into the new container. Water is added to bring the sample volume to about 10 ml. 1 ml of 72% perchloric acid (or 1.2 ml of 60%) is added, then 1 ml of 5% ammonium molybdate, and then 0.5 ml of the reducing agent. This is prepared from 0.5 g of 1, 2, 4-aminonaptholsulfonic acid, 30 g of sodium bisulfite and 6 g of sodium sulfite, diluted to 250 ml with water. This solution should be filtered to clarify it; if the first filtration doesn't work, let the solution stand a day and then filter it again. This solution is stable for two weeks at room temperature.

Dilute the sample to 15 ml with water. Wait 5 minutes and read the absorbance of the solution at 660 nm. The sensitivity range is 0.05 to 1 $\mu$mole of phosphate.

*References*
1. Ames, B. N. and D. T. Dubin, *J. Biol. Chem.*, **235**, 769 (1960).
2. King, E. J., *Biochem. J.*, **26**, 292 (1932).

# 8. STATISTICAL TABLES

*(a)  Arguments of n*

| $n$ | $\sqrt{n}$ | $\sqrt{\dfrac{1}{n}}$ | $\sqrt{\dfrac{1}{n(n-1)}}$ | $n!$ |
|---|---|---|---|---|
| 1 | 1.00000 | 1.00000 | 1.00000 | 1.00000 |
| 2 | 1.41421 | .70711 | .70711 | 2.00000 |
| 3 | 1.73205 | .57735 | .40825 | 6.00000 |
| 4 | 2.00000 | .50000 | .28867 | $2.40000 \times 10$ |
| 5 | 2.23607 | .44721 | .22361 | $1.20000 \times 10^2$ |
| 6 | 2.44949 | .40825 | .18257 | $7.20000 \times 10^2$ |
| 7 | 2.64575 | .37796 | .15430 | $5.04000 \times 10^3$ |
| 8 | 2.82843 | .35355 | .13363 | $4.0320 \times 10^4$ |
| 9 | 3.00000 | .33333 | .11785 | $3.6288 \times 10^5$ |
| 10 | 3.16228 | .31623 | .10541 | $3.6288 \times 10^6$ |
| 11 | 3.31662 | .30151 | .09535 | $3.9917 \times 10^7$ |
| 12 | 3.46410 | .28867 | .08704 | $4.7900 \times 10^8$ |
| 13 | 3.60555 | .27735 | .08006 | $6.2270 \times 10^9$ |
| 14 | 3.74166 | .26726 | .07412 | $8.7178 \times 10^{10}$ |
| 15 | 3.87298 | .25820 | .06901 | $1.3077 \times 10^{12}$ |
| 16 | 4.00000 | .25000 | .06455 | $2.0923 \times 10^{13}$ |
| 17 | 4.12311 | .24254 | .06063 | $3.5569 \times 10^{14}$ |
| 18 | 4.24264 | .23570 | .05717 | $6.4024 \times 10^{15}$ |
| 19 | 4.35890 | .22942 | .05407 | $1.2165 \times 10^{17}$ |
| 20 | 4.47214 | .22361 | .05130 | $2.4329 \times 10^{18}$ |
| 21 | 4.58258 | .21822 | .04879 | $5.1091 \times 10^{19}$ |
| 22 | 4.69042 | .21320 | .04652 | $1.1240 \times 10^{21}$ |
| 23 | 4.79583 | .20851 | .04445 | $2.5852 \times 10^{22}$ |
| 24 | 4.89898 | .20412 | .04256 | $6.2045 \times 10^{23}$ |
| 25 | 5.00000 | .20000 | .04082 | $1.5511 \times 10^{25}$ |

| d.f. | Probability of Larger Value of t, Sign Ignored | | |
|---|---|---|---|
| | 0.05 | 0.02 | 0.01 |
| 1 | 12.706 | 31.821 | 63.657 |
| 2 | 4.303 | 6.0965 | 9.925 |
| 3 | 3.182 | 4.541 | 5.841 |
| 4 | 2.776 | 3.747 | 4.604 |
| 5 | 2.571 | 3.365 | 4.032 |
| 6 | 2.447 | 3.143 | 3.707 |
| 7 | 2.365 | 2.998 | 3.499 |
| 8 | 2.306 | 2.896 | 3.355 |
| 9 | 2.262 | 2.821 | 3.250 |
| 10 | 2.228 | 2.764 | 3.169 |
| 11 | 2.201 | 2.718 | 3.106 |
| 12 | 2.179 | 2.681 | 3.055 |
| 13 | 2.160 | 2.650 | 3.012 |
| 14 | 2.145 | 2.624 | 2.977 |
| 15 | 2.131 | 2.602 | 2.947 |
| 16 | 2.120 | 2.583 | 2.921 |
| 17 | 2.110 | 2.567 | 2.898 |
| 18 | 2.101 | 2.552 | 2.878 |
| 19 | 2.093 | 2.539 | 2.861 |
| 20 | 2.086 | 2.528 | 2.845 |
| 21 | 2.080 | 2.518 | 2.831 |
| 22 | 2.074 | 2.508 | 2.819 |
| 23 | 2.069 | 2.500 | 2.807 |
| 24 | 2.064 | 2.492 | 2.797 |
| 25 | 2.060 | 2.485 | 2.787 |

(c)  *Error Functions*

| t | erf *t* | t | erf *t* | t | erf *t* |
|---|---|---|---|---|---|
| .00 | .00000 | .45 | .47548 | .75 | .71116 |
| .01 | .01128 | .46 | .48466 | .80 | .74210 |
| .02 | .02256 | .47 | .49375 | .85 | .77067 |
| .03 | .03384 | .48 | .50275 | .90 | .79691 |
| .04 | .04511 | .49 | .51167 | .95 | .82089 |
| .05 | .05637 | .50 | .52050 | 1.00 | .84270 |
| .06 | .06762 | .51 | .52924 | 1.05 | .86244 |
| .07 | .07886 | .52 | .53790 | 1.10 | .88021 |
| .08 | .09008 | .53 | .54646 | 1.15 | .89612 |
| .09 | .10128 | .54 | .55494 | 1.20 | .91031 |
| .10 | .11246 | .55 | .56332 | 1.25 | .92290 |
| .11 | .12362 | .56 | .57162 | 1.30 | .93401 |
| .12 | .13476 | .57 | .57982 | 1.35 | .94376 |
| .13 | .14587 | .58 | .58792 | 1.40 | .95229 |
| .14 | .15695 | .59 | .59594 | 1.45 | .95970 |
| .15 | .16800 | .60 | .60386 | 1.50 | .96611 |
| .16 | .17901 | .61 | .61168 | 1.55 | .97162 |
| .17 | .18999 | .62 | .61951 | 1.60 | .97635 |
| .18 | .20094 | .63 | .62705 | 1.65 | .98038 |
| .19 | .21184 | .64 | .63459 | 1.70 | .98379 |
| .20 | .22270 | .65 | .64203 | 1.75 | .98667 |
| .21 | .23352 | .66 | .64938 | 1.80 | .98909 |
| .22 | .24430 | .67 | .65663 | 1.85 | .99111 |
| .23 | .25502 | .68 | .66378 | 1.90 | .99279 |
| .24 | .26570 | .69 | .67084 | 1.95 | .99418 |
| .25 | .27633 | .70 | .67780 | 2.00 | .99532 |
| .26 | .28690 | .71 | .68467 | | |
| .27 | .29742 | .72 | .69143 | | |
| .28 | .30788 | .73 | .69810 | | |
| .29 | .31828 | .74 | .70468 | | |
| .30 | .32863 | | | | |
| .31 | .33891 | | | | |
| .32 | .34913 | | | | |
| .33 | .35928 | | | | |
| .34 | .36936 | | | | |
| .35 | .37938 | | | | |
| .36 | .38933 | | | | |
| .37 | .39921 | | | | |
| .38 | .40901 | | | | |
| .39 | .41874 | | | | |
| .40 | .42839 | | | | |
| .41 | .43797 | | | | |
| .42 | .44747 | | | | |
| .43 | .45689 | | | | |
| .44 | .46623 | | | | |

# 9. PAPER CHROMATOGRAPHY OF AMINO ACIDS, NUCLEOTIDES, CARBOHYDRATES AND LIPIDS

(a)  $R_f$  Values in Various Solvents

1. Amino Acids (1–3)

|  | Solvent Systems | | | | | |
|---|---|---|---|---|---|---|
| Amino Acid | 1 | 2 | 3 | 4 | 5 | 6* |
| Alanine | 0.58 | 0.30 | 0.11 | 0.09 | 0.45 | 0.22 |
| Arginine HCl | 0.56 | 0.138 | 0.03 | 0.05 | 0.22 | 0.07 |
| Asparagine | 0.44 | 0.172 | 0.02 | — | — | — |
| Aspartic Acid | 0.18 | 0.106 | 0.03 | 0.10 | 0.27 | 0.09 |
| Cystine | — | — | — | 0.01 | 0.04 | 0.11 |
| Glutamic Acid | 0.31 | 0.12 | 0.05 | 0.01 | 0.37 | 0.11 |
| Glutamine | 0.60 | 0.21 | 0.04 | — | — | — |
| Glycine | 0.39 | 0.22 | 0.06 | 0.05 | 0.32 | 0.19 |
| Histidine | 0.64 | 0.25 | 0.03 | 0.09 | 0.22 | 0.22 |
| Isoleucine | 0.84 | 0.55 | 0.36 | 0.40 | 0.62 | 0.47 |
| Leucine | 0.84 | 0.55 | 0.34 | 0.46 | 0.64 | 0.52 |
| Lysine HCl | 0.48 | 0.08 | 0.02 | 0.02 | 0.11 | 0.16 |
| Methionine | — | — | — | 0.05 | 0.48 | 0.40 |
| Phenylalaine | 0.84 | 0.58 | 0.29 | 0.38 | 0.55 | 0.58 |
| Proline | 0.86 | 0.31 | 0.14 | — | 0.47 | 0.29 |
| Serine | 0.36 | 0.23 | 0.05 | 0.02 | 0.32 | 0.29 |
| Threonine | 0.49 | 0.30 | 0.09 | 0.04 | 0.45 | 0.51 |
| Tryptophan | 0.80 | 0.66 | 0.22 | 0.30 | 0.35 | 0.41 |
| Tyrosine | 0.67 | 0.64 | 0.16 | 0.19 | 0.41 | 0.34 |
| Valine | 0.78 | 0.44 | 0.24 | 0.15 | 0.63 | 0.37 |

*Solvent System:*
1. Phenol saturated with water (pH 5.0—5.5) (1).
2. Collidine-lutidine (1 : 3) saturated with water (1).
3. 1-butanol-acetic acid (9 : 1) saturated with water (1).
4. 1-butanol saturated with 2 $N$ NH$_4$OH (2).
5. Methanol-water-pyridine (80 : 20 : 4) (3).
6. t-butanol-methylethyl ketone-water-diethylamine (40 : 40 : 20 : 4) (3).

*Solvent 6 will give a strong ninhydrin background, unless the diethylamine is removed. This can be accomplished by spraying the chromatogram with 0.1 $M$ borate buffer, pH 9.3 and drying over a hot plate or in an oven at 100°.

## 2. Nucleotides and Related Compounds (4–8)

| Compound | Solvent Systems | | | | | | | |
|---|---|---|---|---|---|---|---|---|
| | 1 | 2 | 3 | 4 | 5* | 6* | 7 | 8 |
| Adenine | — | — | — | — | — | 1.06* | 0.48 | 0.33 |
| Cytosine | — | — | — | — | — | — | 0.45 | 0.45 |
| Guanine | — | — | — | — | — | 0.0 | 0.23 | 0.22 |
| Thymine | — | — | — | — | — | — | 0.62 | 0.80 |
| Uracil | — | — | — | — | — | 1.13 | 0.50 | 0.69 |
| Orthophosphate | 0.61 | — | — | — | 1.00* | 1.00* | — | — |
| Adenosine-2'-phosphate | — | 0.35 | 0.74 | — | 0.62 | — | — | — |
| Adenosine-3'-phosphate | — | 0.35 | 0.67 | — | 0.62 | 0.29 | — | — |
| Adenosine-5'-phosphate | 0.37 | 0.28 | 0.69 | 1.00 | 0.57 | — | 0.08 | 0.42 |
| Adenosine-2,3'-phosphate | — | — | 0.57 | — | — | — | — | — |
| Cytidine-5'-phosphate | 0.34 | 0.26 | 0.85 | — | 0.61 | 0.34 | 0.07 | 0.50 |
| Deoxycytidine-5'-phosphate | — | — | — | — | — | — | 0.15 | 0.64 |
| Deoxyuridine-5'-phosphate | — | — | — | — | — | — | 0.21 | 0.83 |
| Guanosine-2'-phosphate | 0.20 | 0.67 | 0.79 | — | — | — | — | — |
| Guanosine-3'-phosphate | 0.20 | 0.67 | 0.79 | — | — | — | — | — |
| Guanosine-5'-phosphate | — | — | 0.74 | — | 0.47 | — | 0.05 | 0.39 |
| Riboflavin-5'-phosphate | — | — | — | — | 0.45–0.72 | — | — | — |
| Thymidine-5'-phosphate | — | — | — | — | — | — | 0.24 | 0.88 |
| Uridine-5'-phosphate | 0.51 | 0.43 | 0.85 | — | 1.10 | — | 0.12 | 0.70 |
| Uridine diphosphoglucose | — | — | — | — | 0.26 | — | 0.00 | 0.69 |
| Adenosine diphosphate | 0.10 | 0.07 | 0.77 | 0.80 | 0.27 | — | 0.02 | 0.40 |
| DPN | — | — | — | — | 0.06 | 0.21 | — | — |
| Adenosine triphosphate | 0.04 | 0.08 | 0.83 | 0.66 | 0.10 | 0.07 | 0.00 | 0.39 |
| TPN | — | — | — | — | 0.03 | — | — | — |

*Relative to orthophosphate.

*Solvent Systems and Conditions*
1. Acetone-25% trichloracetic acid (3:1). Acid washed Whatman #1 paper. 4°C Ascending (4).
2. *Iso*amyl alcohol-tetrahydrofurfuryl alcohol-0.08 $M$ potassium citrate buffer, pH 3 (1:1:1). Descending (4).
3. 5% aqueous $Na_2HPO_4$ saturated with *iso*amyl alcohol (both phases added to the trough) (5).
4. Isobutyric acid—1 $N$ $NH_4OH$—0.1 $M$ EDTA (100:60:1.6) (5).
5. Methyl Cellosolve-pyridine-glacial acetic acid-water (8:4:1:1) (6).
6. Ethanol-1 $M$ ammonium acetate, pH 7.5 (5:2) (7).
7. t-butanol-methylethyl ketone-formic acid-water (8:6:3:3) (8). Ascending.
8. Isopropanol-water-HCl (65:18.4:16.6) (8). Ascending.

## 3. Carbohydrates (8–11)

| Carbohydrate | Solvent Systems | | | | |
|---|---|---|---|---|---|
| | 1* | 2 | 3 | 4 | 5 |
| N-acetylglucoseamine | 0.81 | — | — | — | — |
| Arabinose | 0.70 | 0.22 | 0.22 | 0.28 | 0.70 |
| 2-deoxyribose | — | — | 0.46 | 0.53 | 0.85 |
| Dihydroxyacetone | — | — | 0.48 | 0.58 | 0.83 |
| Erythrose | — | — | 0.44 | — | — |
| Fructose | 0.69 | — | 0.21 | 0.25 | 0.68 |
| Fucose | 0.82 | — | 0.30 | 0.35 | 0.75 |
| Galactose | 0.53 | 0.14 | 0.14 | 0.17 | 0.64 |
| Galacturonic acid | — | 0.13 | — | 0.18 | 0.58 |
| Glyceraldehyde | — | — | 0.52 | 0.26 | 0.64 |
| Glucose | 0.59 | 0.17 | 0.15 | 0.18 | 0.66 |
| Glucoseamine-HCl | 0.43 | — | — | 0.16 | 0.35 |
| Glucuronic acid | 0.18 | — | — | — | — |
| Lactose | 0.28 | — | 0.02 | 0.05 | 0.46 |
| Maltose | — | — | 0.06 | 0.08 | 0.50 |
| Mannose | 0.68 | 0.195 | 0.21 | 0.22 | 0.70 |
| Rhamnose | 1.00 | 0.34 | 0.36 | 0.41 | 0.75 |
| Ribose | — | — | 0.29 | 0.34 | 0.74 |
| Ribulose | — | — | 0.33 | 0.28 | 0.65 |
| Sedoheptulose | — | — | 0.20 | 0.26 | 0.67 |
| Sorbose | — | — | 0.18 | — | — |
| Sucrose | — | — | 0.09 | 0.14 | 0.65 |
| Xylose | 0.81 | 0.265 | 0.25 | 0.30 | 0.73 |

*Solvent Systems:*
1. 1-butanol-pyridine-water (5:3:2) (9).
2. Ethyl acetate-acetic acid-water (3:1:3) (10).
3. 1-butanol-ethanol-water (52:32:16) (11).
4. t-butanol-methylethyl ketone-formic acid-water (8:6:3:3) (8).
5. Isopropanol-water-HCl (68:18.4:16.6) (8).

*Rhamnose is used as the reference compound.

## 4. Sugar Phosphates (6, 8, 12, 13)

| Sugar Phosphate | Solvent Systems | | | | |
|---|---|---|---|---|---|
| | 1 | 2* | 3 | 4 | 5 |
| Dihydroxyacetone phosphate | — | 0.31 | — | — | — |
| 2, 3 diphosphoglycerate | — | — | — | — | 0.07 |
| Fructose 1, 6-diphosphate | 0.03 | 0.28 | 0.24 | 0.82 | 0.08 |
| Fructose 6-phosphate | 0.08 | 0.76 | 0.44 | 0.80 | 0.36 |
| Glucose 1-phosphate | 0.11 | 0.58 | 0.60 | 0.68 | 0.36 |
| Glucose 6-phosphate | 0.06 | 0.63 | 0.48 | 0.74 | 0.29 |
| Glycerol-1-phosphate | — | — | — | — | 0.39 |
| Mannose 6-phosphate | — | 0.67 | — | — | — |
| Phosphate | — | 1.00 | 0.28 | — | 0.21 |
| Phosphocreatine | — | 0.35 | — | — | — |
| Phosphoenolpyruvate | — | 0.98 | 0.46 | — | — |
| 6-Phosphogluconic acid | 0.06 | — | — | 0.81 | — |
| 3-Phosphoglyceraldehyde | — | 0.30 | — | — | 0.19 |
| 3-Phosphoglycerate | — | 0.62 | 0.35 | — | 0.22 |
| 2-Phosphoglycerate | — | — | 0.18 | — | 0.41 |
| Pyrophosphate | — | 0.32 | — | — | — |
| Ribose 5-phosphate | — | 0.79 | — | — | — |
| Ribose 1-phosphate | 0.11 | — | — | 0.82 | 0.40 |
| Ribulose 5-phosphate | — | 0.94 | — | — | — |
| Sedoheptulose 7-phosphate | — | 0.72 | — | — | — |
| Xylulose 5-phosphate | — | 0.90 | — | — | — |

*Solvent Systems:*
1. t-butanol-methylethyl ketone-formic acid-water (8: 6: 3: 3) (8).
2. Methyl Cellosolve-pyridine-glacial acetic acid-water (8: 4: 1: 1) (6).
3. Methanol-ammonia (0.9 specific gravity)-water (6: 1: 3) (12).
4. Isopropanol-water-HCl (65: 18.4: 16.6) (8).
5. Methyl Cellosolve-methylethyl ketone-3 $M$ ammonia (7: 2: 3) (13).

*Phosphate is used as the reference compound.

## 5. Lipids (14)

| Compound | $R_f$ | Color of Fluorescent Spots (See "Spray Reagents" Below) |
|---|---|---|
| Lysolecithin | 0.42 | Yellow |
| Sulfoquinovosyl diglyceride | 0.32 | Blue |
| Phosphatidyl inositol | 0.25–0.45 | Blue |
| Phosphatidyl glycerol aminoacyl ester | 0.40 | Yellow |
| Sphingomyelin | 0.40–0.48 | Yellow |
| Digalactosyl diglyceride | 0.41 | Yellow |
| Lecithin | 0.44–0.55 | Yellow |
| Phosphatidyl glycerol | 0.51 | Blue |
| Phosphatidyl serine | 0.53–0.59 | Blue |
| Phosphatidyl ethanolamine | 0.59–0.63 | Yellow |
| Diphosphatidyl glycerol (cardiolipin) | 0.69 | Blue |
| Monogalactosyl diglyceride | 0.75 | Yellow |
| Phosphatidic acid | 0.81–0.96 | Blue |
| Neutral lipids | 0.90–1.00 | Yellow |

*Mobile Phase:*
Diisobutyl ketone-acetic acid-water (40: 25: 5)
*Stationary Phase:* Silicic acid impregnated paper

The paper is dipped in a solution made by adding 310 g of silicic acid slowly with stirring to 1 liter of 7.2 $M$ NaOH, allowing the solution to cool to room temperature, and adding water to 1500 ml. Then the paper is blotted and soaked 30 minutes in 6 $N$ HCl with occasional agitation, washed with running water 2–3 hours, and then with several changes of distilled water over 1–2 hours until free of chloride. The papers are allowed to dry in a hood, then washed by chromatography with chloroform-methanol (1: 1) and dried in a hood.

*(b)   Spray Reagents and Color Yield*

*1.  Amino Acids*

Ninhydrin-collidine spray reagent for amino acids (15):
   50 ml 0.1 % ninhydrin in ethanol
    2 ml collidine
   15 ml glacial acetic acid
Spray and heat for 1–3 minutes over a hot plate. The spots fade rapidly.

Color of the Spots of the Various Amino Acids (15)

| Amino Acid | Color | Amino Acid | Color |
|---|---|---|---|
| Alanine | Blue-Purple | Lysine | Blue-Purple |
| Arginine | Blue-Purple | Methionine | Grey |
| Aspartic acid | Blue | Phenylalanine | Turquoise Blue |
| Cystine | Brown | Proline | Yellow |
| Glutamic acid | Blue-Purple | Serine | Grey-Yellow |
| Glycine | Reddish Purple | Threonine | Grey |
| Histidine | Turquoise Blue | Tryptophan | Yellow-Brown |
| Isoleucine | Blue-Purple | Tyrosine | Turquoise Blue |
| Leucine | Blue-Purple | Valine | Blue Purple |

Preservation of amino acid spots (16).
*Reagent:* 1 ml saturated $Cu(NO_3)_2$
         0.2 ml 10 % (v/v) $HNO_3$ in 95 % ethanol
         Make up to 100 ml.
Spray, momentarily expose the paper to $NH_3$, and spray with Krylon.

*2.  Nucleotides*

  (i)  Phosphate Spray (17)
      $HClO_4$ (60 %), 5 ml
      HCl (1 $N$), 10 ml
      Ammonium molybdate (4 %), 28 ml
      Water, 55 ml
Spray the dry chromatogram and heat in oven (80°) for 5–7 minutes. Develop spots in UV light.
Inorganic phosphate gives a yellow spot after heating, turning greyish-blue after UV irradiation. Organic phosphate gives a blue spot after exposure to UV light.

  (ii)  Periodate Spray (18)
      0.5 % $NaIO_4$ in water
      0.5 % benzidine* in ethanol-acetic acid (4: 1)
Spray lightly with periodate, wait 2–3 minutes and spray with benzidine. Any compound having a glycol group (or any other group which is oxidized by periodate) will give a white spot in a background of oxidized benzidine (blue changing to brown).

*Benzidine is a CARCINOGEN; be careful!

### 3. Carbohydrates (7)

The paper is dipped into a solution made by dilution of 0.1 ml of saturated aqueous $AgNO_3$ to 20 ml with acetone (water is added dropwise with stirring until the solution clears). The paper is dried and sprayed with a solution made by diluting saturated aqueous NaOH with ethanol to a final sodium hydroxide concentration of 0.5 $M$. Reducing sugars immediately form black spots; other sugars react more slowly. The paper is washed a few minutes in 6 $N$ $NH_4OH$, then at least an hour in running water and dried in an oven.

### 4. Sugar Phosphates (19)

The chromatograms are first dipped in a solution made from 1.5 g $FeCl_3 \cdot 6H_2O$, 30 ml 0.3 $N$ HCl and 970 ml of acetone. They are dried in a hood, then dipped in a solution of 12.5 g of sulfosalicylic acid in 1 liter of acetone. Sensitivity: 0.05 $\mu$ moles of phosphate.

### 5. Lipids (14)

The dried chromatogram is soaked 1–3 minutes in 0.0012% rhodamine 6G in water, rinsing off the excess in distilled water. The stained chromatogram is viewed while wet with an ultraviolet lamp.

### References

1. Lederer, E. and M. Lederer, *Chromatography*, 2nd edition, Elsevier, New York (1957), p. 312, 328.
2. Block, R. J., E. L. Durrum, and G. Zweig, *A Manual of Paper Chromatography and Paper Electrophoresis*, 2nd edition, Academic Press, New York (1956), p. 108.
3. Redfield, R. R., *Biochim. Biophys. Acta*, **10**, 344 (1953).
4. Reference 1, p. 374.
5. Krebs, H. A. and R. Hems, *Biochim. Biophys. Acta*, **12**, 172, (1953).
6. Runeckles, V. R. and G. Krotkov, *Arch. Biochem. Biophys.*, **70**, 442 (1957).
7. Paladini, A. C. and L. F. Leloir, *Biochem. J.*, **51**, 426 (1952).
8. Fink, K., R. E. Cline, and R. M. Fink, *Anal. Chem.*, **35**, 389 (1963).
9. Masamune, H. and Z. Yosizawa, *Tokoku J. Exptl. Med.*, **59**, 1 (1953).
10. Jermyn, M. A. and F. A. Isherwood, *Biochem. J.*, **44**, 402, (1949).
11. Putnam, E. W. in *Methods in Enzymology*, Vol. III, S. P. Colowick and N. O. Kaplan, Eds., Academic Press, N. Y. (1957), p. 62.
12. Bandurski, R. S. and B. Axelrod, *J. Biol. Chem.*, **193**, 405 (1951).
13. Mortimer, D. C., *Can. J. Chem.*, **30**, 653 (1952).
14. Kates, M., in *Lipid Chromatographic Analysis*, Vol. I, G. V. Marinetti, Ed., Marcel Dekker, N.Y. (1957), p. 1.
15. Cowgill, R. W. and A. B. Pardee, *Experiments in Biochemical Research Techniques*, Wiley, New York (1957).
16. Kewerau, E. and T. Wieland, *Nature*, **168**, 77 (1951).
17. Hanes, C. S. and F. A. Isherwood, *Nature*, **164**, 1107 (1949).
18. Viscontini, M., D. Hock and P. Karrer, *Helv. Chim. Acta*, **38**, 643 (1955).
19. French, D., D. W. Knapp, and J. H. Pazur, *J. Am. Chem. Soc.*, **72**, 5150 (1950).

## 10. $R_f$'s OF DNP- AND DANSYL AMINO ACIDS IN VARIOUS SOLVENTS

(a)  *Paper Chromatography of DNP-Amino Acids (1, 2)*

|  | *Solvent Systems* | | | |
| *Compound* | 1* | 2 | 3 | 4* |
|---|---|---|---|---|
| DNP-alanine | 0.45 | 0.44 | 0.58 | 0.53 |
| α-DNP-arginine | 0.48 | 0.74 | 0.57 | — |
| DNP-asparagine | 0.24 | 0.20 | — | 0.62 |
| DNP-aspartic acid | 0.08 | 0.05 | 0.05 | 0.93 |
| DNP-cysteic acid | 0.02 | — | — | — |
| Di-DNP-cystine | 0.32 | 0.20 | 0.41 | 0.17 |
| DNP-glutamic acid | 0.09 | 0.05 | 0.06 | 0.85 |
| DNP-glycine | 0.35 | 0.33 | 0.45 | 0.42 |
| Di-DNP-histidine | 0.43, 0.70 | 0.32, 0.70 | 0.47, 0.70 | — |
| DNP-isoleucine | 0.64 | 0.68 | 0.80 | 0.51 |
| DNP-leucine | 0.64 | 0.68 | 0.80 | 0.51 |
| Di-DNP-lysine | 0.72 | 0.75 | 0.78 | 0.08 |
| ε-DNP-lysine | 0.40 | 0.65 | 0.40 | — |
| DNP-methionine | 0.59 | 0.57 | 0.71 | 0.48 |
| DNP-phenylalanine | 0.66 | 0.68 | 0.80 | 0.33 |
| DNP-proline | 0.43 | 0.50 | — | 0.68 |
| DNP-serine | 0.30 | 0.23 | 0.45 | 0.66 |
| DNP-threonine | 0.36 | 0.31 | 0.53 | 0.75 |
| DNP-tryptophan | 0.62 | 0.63 | 0.75 | 0.15 |
| Di-DNP-tyrosine | 0.74 | 0.70 | 0.83 | 0.04 |
| DNP-valine | 0.56 | 0.60 | 0.70 | 0.58 |
| 2, 4 dinitrophenol | 0.44 | 0.25 | 0.62 | 0.42 |
| 2, 4 dinitroaniline | 0.98 | 0.98 | 0.98 | 0.08 |

*Solvents:*
1. Toluene-pyridine-glycol monochlorohydrin-0.8 $N$ $NH_4OH$ (5: 1: 3: 3) (1).
2. Phenol-isoamyl alcohol-water (1: 1: 1) (1).
3. Pyridine-isoamyl alcohol-1.6 $N$ $NH_4OH$ (1).
4. 1.5 $M$ phosphate buffer, pH 6 (2).

  *Systems 1 and 4 are well suited for two-dimensional chromatography (2).

## (b) Thin-Layer Chromatography of Dansyl (DNS-) Amino Acids (3, 4)

| Compound | Solvent Systems | | | | | |
|---|---|---|---|---|---|---|
| | 1 | 2 | 3 | 4 | 5 | 6 |
| DNS-isoleucine | 0.92 (0.87) | 1.0 | 1.00 | 0.89 | 0.65 | 0.19 |
| DNS-leucine | 0.89 (0.83) | 1.0 | 0.99 | 0.63 | 0.63 | 0.10 |
| DNS-valine | 0.85 (0.77) | 1.0 | 0.97 | 0.75 | 0.54 | 0.13 |
| DNS-proline | 0.79 (0.70) | 0.93 | 0.96 | 0.50 | 0.24 | 0.07 |
| DNS-phenylalanine | 0.69 (0.59) | 0.94 | 0.99 | 0.42 | 0.51 | 0.07 |
| DNS-methionine | 0.65 (0.52) | 0.92 | 0.99 | 0.33 | 0.43 | — |
| DNS-alanine | 0.61 (0.47) | 0.89 | 0.96 | 0.29 | 0.37 | — |
| Di-DNS-tyrosine | 0.55 (0.46) | 0.82 | 0.92 | 0.23 | 0.48 | — |
| Di-DNS-lysine | 0.48 (0.40) | 0.93 | 0.96 | 0.33 | 0.54 | — |
| DNS-glycine | 0.38 (0.32) | 0.65 | 0.93 | 0.12 | 0.32 | — |
| DNS-tryptophan | 0.34 (0.34) | 0.87 | 0.95 | 0.25 | 0.45 | — |
| DNS-histidine | 0.27 (0.24) | 0.59 | 0.91 | 0.09 | 0.32 | — |
| DNS-glutamic acid | 0.18 (0.16) | 0.42 | 0.85 | 0.08 | 0.05 | — |
| DNS-threonine | 0.18 (0.18) | 0.25 | 0.57 | 0.02 | 0.28 | — |
| DNS-serine | 0.15 (0.15) | 0.20 | 0.52 | 0.02 | 0.25 | — |
| DNS-aspartic acid | 0.06 (0.06) | 0.02 | 0.60 | 0 | 0.04 | — |
| DNS-(CM) cysteine | 0.05 — | 0.02 | 0.70 | 0 | — | — |
| DNS-glutamine | 0.03 (0.03) | 0.07 | 0.14 | 0 | 0.15 | — |
| DNS-asparagine | 0.02 (0.03) | 0.02 | 0.10 | 0.03 | 0.15 | — |
| $\epsilon$-DNS-lysine | 0.02 (0) | 0.08 | 0.05 | 0.04 | 0.16 | — |
| $\alpha$-DNS-lysine | 0.02 — | 0 | 0 | 0.06 | 0.12 | — |
| DNS-arginine | 0 (0) | 0 | 0 | 0 | 0.16 | — |

The $R_f$ values in parenthesis (solvent 1) and those for solvents 5 and 6 are from reference 4; the others are from reference 3. The chromatography is done on Silica Gel G-coated glass plates.

*Solvents:*
1. Benzene-pyridine-acetic acid (80: 20: 2).
2. Chloroform- t-amyl alcohol-acetic acid (70: 30: 3).
3. Chloroform- t-amyl alcohol-formic acid (70: 30: 3).
4. Chloroform- t-amyl alcohol-acetic acid (140: 60: 1).
5. n-butanol saturated with 0.2 N NaOH.
6. n-butanol-chloroform (3: 97, v/v).

*References*
1. Reference 1, Appendix 9, p. 332.
2. Levi, A. L., *Nature*, **174**, 126 (1954).
3. Morse, D. and B. L. Horecker, *Anal. Biochem.*, **14**, 429 (1966).
4. Cole, M., J. C. Fletcher, and A. Robson, *J. Chromatog.*, **20**, 616 (1965).

# 11. CHROMATOGRAPHIC MATERIALS

*(a) Characteristics of Ion Exchange Resins*

*1. Cation Exchange Resins*

| Designation | Polymer Type | Supplier | Functional Group | Temp. Limit | Capacity Meq/g | Particle Size Wet Mesh | Cross-Linkage or Porosity | Notes |
|---|---|---|---|---|---|---|---|---|
| | | | *Strong* | | | | | |
| Dowex 50 (AG) | Styrene | D, B | $SO_3H$ | 150 | 4.9–5.2 | 20–400 | 1–16 | |
| Zerolit 225 | Styrene | Z | $SO_3H$ | — | — | 14–<200 | 1–20 | |
| Rexyn 101 | Styrene | F | $SO_3H$ | — | 4.5–4.6 | 16–400 | Medium | |
| Amberlite IR-120 | Styrene | M | $SO_3H$ | 250 | 5.0 | 20–50 | 8 | |
| Amberlite-CG-120 | Styrene | M | $SO_3H$ | 250 | 4.5 | 100–400 | 8 | |
| Amberlyst-15 | Styrene | M | $SO_3H$ | 302 | 4.9 | 16–50 | — | Non-aqueous |
| Bio Rex 40 | Phenolic | B | $—CH_2SO_3H$ | 40 | 2.9 | 20–325 | Coarse | |
| CGC-240-250 | Styrene | J | $SO_3H$ | — | 4–5 | 16–400 | 8–12.5 | |
| MFC 6, 10 | Styrene | J | $SO_3H$ | — | 4 | <35 micron | 6–10 | Micro-fine |
| C-350 | Styrene | J | $SO_3H$ | — | 4.6 | 16–50 | 20 | Macro-porous |
| | | | *Weak* | | | | | |
| Zerolite 216 | Phenolic | Z | —OH, —COOH | — | — | — | — | |
| Bio Rex 63 | Styrene | B | $\phi—PO_3^=$ | — | 6.6 | 16–325 | 5.5 | |
| Bio Rex 70 | Acrylic | B | $RCOO^-$ | — | 10.2 | 16–325 | 10 | |
| Zerolit 226 | Acrylic | Z | —COOH | — | 10.2 | 14–100 | 2.5–4.5 | |
| Amberlite IRC-50 | Acrylic | M | $RCOO^-$ | 250 | 10.0 | 20–50 | — | |
| Rexyn 102 | Acrylic | F | $RCOO^-$ | — | 7.5–9 | 16–400 | Medium | |
| Rexyn 105 | Coal | F | $RSO_3^-$; $RCOO^-$ | — | 1.5 | 16–50 | — | |
| Rexyn 104 | Alumino-silicate | F | $AlO_2$ | — | 1.4 | 16–50 | — | |
| Amberlite CG-50 | Acrylic | M | $R—COO^-$ | 250 | 10.0 | 100–400 | — | |
| Chelex 100 | Styrene | B | $\phi CH_2N(CH_2COO^-)_2$ | — | 2.9 | 100–400 | Coarse | |
| CGC-270-271 | Acrylic | J | $R—COO^-$ | — | 9.0 | 100–400 | 2.5 | |

343

## 2. Anion Exchange Resins

| Designation | Supplier | Polymer Type | Functional Group | Temp. Limit | Capacity Meq/g | Particle Size Wet Mesh | Cross-Linkage or Porosity | Notes |
|---|---|---|---|---|---|---|---|---|
| | | | *Strong* | | | | | |
| Dowex 1 (AG) | D, B | Styrene | $\phi CH_2 N(CH_3)_3$ | 50(OH); 150(Cl) | 3–3.5 | 50–<400 | 1–10 | |
| Dowex 21K (AG) | D, B | Styrene | $\phi CH_2 N(CH_3)_3$ | 50(OH); 150(Cl) | 4.5 | 16–100 | — | |
| Zerolit FF-1P | Z | Styrene | $\phi CH_2 N(CH_3)_3$ | — | — | 14–<200 | 2–9 | |
| Zerolit KM-P | Z | Styrene | $R_3 N$ | 100(Cl) | — | — | — | |
| A-540, CGA 540, 541 | J | Styrene | $\phi CH_2 N(CH_3)_3$ | 60(OH) | 3.2–4.0 | 16–400 | 7–9 | |
| Dowex 2 (AG) | D, B | Styrene | $\phi CH_2 N(CH_3)_2$ $C_2 H_4 OH$ | | 3.0 | 20–400 | 4–10 | |
| A-2, A-550 | J | Styrene | $\phi CH_2 N(CH_3)_2$ $C_2 H_4 OH$ | 40(OH); 100(Cl) | 3.3 | 16–50 | 4, 9 | |
| Rexyn 201 | F | Styrene | $\phi CH_2 N(CH_3)_3$ | — | 3.3–4.0 | 16–400 | Medium | |
| Rexyn 202 | F | Styrene | $\phi CH_2 N(CH_3)_2$ $C_2 H_4 OH$ | — | 3.3 | 16–100 | Medium | |
| Rexyn 203 | F | Styrene | $RN(C_2 H_2)_2$ | — | 4.0 | 100–400 | Medium | |
| Rexyn 204 | F | Styrene | $RN(C_5 H_6 N)_3$ | — | 3.7 | 16–100 | Medium | |
| Rexyn 205 | F | Styrene | $R_3 N, R_4 N$ | — | 5.5 | 16–50 | Medium | |
| Amberlite IRA 400 | M | Styrene | $-N(CH_3)_3$ | 140 | 3.3 | 20–50 | 8 | |
| Amberlite CG 400 | M | Styrene | $-N(CH_3)_3$ | 140 | 3.3 | 100–400 | 8 | |
| Amberlite IRA 401S | M | Styrene | $-N(CH_3)_3$ | 140 | — | 20–50 | 4 | |
| Amberlite IRA 410 | M | Styrene | $-N(CH_3)_2 C_2 H_4 OH$ | 105 | 3.3 | 16–50 | 8 | |
| Amberlite IRA 938 | M | Styrene | $-N(CH_3)_3$ | 140 | 3.7 | 20–50 | — | |
| Amberlyst A-26 | M | Styrene | $-N(CH_3)_3$ | 140 | 4.2 | 16–50 | — | Non-aqueous |

| Designation | Polymer Type | Supplier | Functional Group | Temp. Limit | Capacity Meq/g | Particle Wet Size Mesh | Cross-Linkage or Porosity | Notes |
|---|---|---|---|---|---|---|---|---|
| A302 | Styrene | J | Aliphatic Polyamine | 40 | 4.2 | 16–50 | — | |
| CGA-300, 301 | Styrene | J | $-N(CH_3)_2$ | 40 | 3.5 | 100–400 | 7 | Macro-porous |
| A-641 | Styrene | J | $-N(CH_3)_3$ | 70 | 3.7 | 16–50 | 16 | Micro-fine |
| MFA-6 | Styrene | J | $-N(CH_3)_3$ | 100(Cl); 60(OH) | 2.2 | <35 micron | 6 | |
| Bio Rex 9 | Styrene | B | $-N(C_5H_6N)_3$ | — | 3.7 | 16–325 | — | |
| Bio Rex 5 | Epoxy-Polyamine | B | $RN(CH_3)_3$, $RN(CH_3)_2$ | — | 8.8 | 20–400 | 4 | |
| Rexyn 207 | Phenolic | F | $R_3N$ | — | 5.5 | 16–50 | — | |
| *Weak* | | | | | | | | |
| Dowex 3 (AG) | Styrene | D, B | $\phi N(CH_3)_2H$ | 65 | 2.8 | 20–400 | 4 | |
| Amberlite IR-4B | Phenolic | M | Polyamine | 105 | 8.0 | — | — | |
| Amberlite CG-4B | Phenolic | M | Polyamine | 105 | 8.0 | 100–400 | — | |
| Amberlite IR-45 | Styrene | M | Polyamine | 205 | 5.0 | 20–50 | — | |
| Amberlite IRA-68 | Polyacrylic | M | $N(R_2)$ | 176 | 5.6 | 16–50 | — | |
| Amberlyst A-21 | Styrene | M | $N(CH_3)_2$ | 212 | 4.5 | 16–50 | — | |
| Rexyn 206 | Aliphatic Polyamine | F | $R_2NH$, $R_3N$ | — | 5.5 | 16–50 | — | |
| Rexyn 208 | Styrene | F | $RN(CH_3)_2$ | — | 4 | 100–400 | — | |
| Zerolit H-1P | Styrene | Z | $N(CH_3)_2$ | — | — | — | — | |
| Zerolit M-1P | Styrene | Z | Polyamine | — | — | — | — | |
| A-260 | Aliphatic Polyamine | M | Polyamine | 40 | 4.0 | 16–50 | — | |
| CGA 315–316 | Styrene | M | Polyamine | 40 | 3.5 | 100–400 | 7 | |
| ANGA 316 | Styrene | M | Polyamine | 40 | 5.0 | 16–50 | 9 | |

(b) Substituted Ion-Exchange Celluloses, Polyacrylamide and Dextran Gels.

| Designation | Chemical Type | Functional Group | Acidity or Basicity | Exchange Capacity Meq/g | Vol. per g | Manufacturer |
|---|---|---|---|---|---|---|
| **1. Cation Exchangers (Cellulose)** | | | | | | |
| CM-Cellulose | Carboxymethyl | $-CH_2COOH$ | Weak | 0.6 | 7.5 | a, b, c, d, f |
| P-Cellulose | Phosphorylated | $-OPO_3H_2$ | Strong | 0.8 | 7.5 | a, b, c, f |
| SE-Cellulose | Sulfoethyl | $-OC_2H_4SO_3Na$ | Strong | 0.2 | 7.5 | a, b |
| Hydroxylapatit-Cellulose | Calcium phosphate | | — | — | 7.0 | a |
| **2. Anion Exchangers (Cellulose)** | | | | | | |
| DEAE-Cellulose | Diethylaminoethyl | $-OC_2H_4NH(C_2H_5)_2Cl$ | Strong | 0.9–1.0 | 6.0–9.0 | a, b, c, d, e, f |
| TEAE-Cellulose | Triethylaminoethyl | $-OC_2H_4N(C_2H_5)_3Cl$ | Strong | 0.5–0.8 | 7.5 | a, b |
| GE-Cellulose | Guanidoethyl | $-OC_2H_4NHC(NH_2){:}NH_2Cl$ | Strong | 0.2–0.3 | 7.5 | a |
| Ecteola-Cellulose | Mixed Amines | Undefined | Intermediate | 0.3 | 7.0 | a, b, d, f |
| PAB-Cellulose | Para-aminobenzoyl | $-OCH_2-C_6H_4-NH_2$ | Weak | 0.2 | 8.0 | a, b |
| AE-Cellulose | Aminoethyl | $-O-C_2H_4NH_2$ | Weak | 0.2–0.3 | 8.0 | a, b |
| BD-Cellulose | Benzoylated-DEAE | | Intermediate | 0.8 | 7.0 | a |
| BND-Cellulose | Benzoylated, Naphthoylated-DEAE | | Intermediate | 0.8 | 7.0 | a |
| PEI-Cellulose | Polyethyleneimine | $NHCH_2CH_2$ | Weak | 0.1 | 7.0 | a |

Although resins are grouped in the tables for part (a) according to cross-linkage and particle size, most manufacturers offer selections of mesh ranges in a desired cross-linkage. The most useful cross-linkage and mesh size of Dowex 2, for example, is 8% in the 100–200 mesh particle size; of Dowex 50, are 2% and 8% in the 100–200 mesh particle size.

*Manufacturers (Part a)*

D: Dow Chemical Co., 2030 Dow Center, Midland, Michigan 48640

B: Bio-Rad Laboratories, 32nd and Griffin Ave., Richmond, California 94804

Z: Zerolit, Ltd., (Gallard-Schlesinger Chemical Co., 584 Mineola Avenue, Carle Place, Long Island New York 11514)

F: Fisher Scientific Co., 711 Forbes Ave., Pittsburgh, Pennsylvania 15219

M: Mallinckrodt Chemical Works, 2nd and Mallinckrodt Sts., St. Louis, Missouri 63160 (Products of Rohm and Haas Co.)

J: J. T. Baker Chemical Co., 222 Red School Lane, Phillipsburg, N.J. 08865

*Manufacturers (Part b)*

a. Serva-Entwicklungs Laboratory, Heidelberg, Germany (Gallard-Schlesinger Chemical Co., Long Island, N.Y.)

b. Bio-Rad Laboratories, Richmond, California

c. Whatman (H. Reeve Angel & Co., Ltd., 9 Bridewell Place, Clifton, N.J. 07014)

d. Brown Co., 550 Main Street, Berlin, New Hampshire

e. Eastman Organic Chemicals, 343 State St., Rochester, New York 14650

f. Mann Research Laboratories, 136 Liberty St., New York, N.Y. 10006

Note: Suppliers a, b, c, and f offer resins in forms suitable for thin-layer applications.

## 3. Sephadex and Bio-Gel Exchangers

| Designation | Functional Group | Particle Size Mesh | Exchange Capacity Meq/g | Bed Volume ml/g |
|---|---|---|---|---|
| *i. Cation Exchangers* | | | | |
| CM-Sephadex C-25 | —O—$CH_2COOH$ | 40–120 | 4.5 | 15–20 |
| CM-Sephadex C-50 | —O—$CH_2COOH$ | 40–120 | — | 15–20 |
| SP-Sephadex C-25 | $C_3H_6SO_3H$ | 40–120 | 2.3 | |
| SP-Sephadex C-50 | $C_3H_6SO_3H$ | 40–120 | 2.3 | |
| Bio-Gel CM-2 | | 100–200 | 6.0 | 5.6 |
| *ii. Anion Exchangers* | | | | |
| DEAE-Sephadex A-25 | —O—$C_2H_4N(C_2H_5)_2Cl$ | 40–120 | 3.5 | — |
| DEAE-Sephadex A-50 | —O—$C_2H_4N(C_2H_5)_2Cl$ | 40–120 | 3.5 | 15–20 |
| QAE-Sephadex A-25 | —O—$C_2H_4N(C_2H_5)_2$ | 40–120 | 3.0 | |
| QAE-Sephadex A-50 | —O—$C_2H_4N(C_2H_5)_2$ | 40–120 | 3.0 | |

Sephadex Ion Exchangers are derivatives of Sephadex G-25 and G-50. Bio-Gel exchangers are derivatives of Bio-Gel P polyacrylamide gels.

## (c) Characteristics of Molecular Sieving Materials

| Type | Exclusion Limit | Water Retain (g/g gel) | Bed Volume per g | Particle Size (microns) |
|---|---|---|---|---|
| *1. Dextran Types* | | | | |
| Sephadex G-10 | 700 | 1.0 | 2–3 | 40–120 |
| Sephadex G-15 | 1500 | 1.5 | 2.5–3.5 | 40–120 |
| Sephadex G-25 Fine | 5000 | 2.5 | 5 | 20–80 |
| Sephadex G-25 Coarse | | | | 100–300 |
| Sephadex G-50 Fine | 10,000 | 5.0 | 10 | 20–80 |
| Sephadex G-50 Coarse | | | | 100–300 |
| Sephadex G-75 | 50,000 | 7.5 | 12–15 | 40–120 |
| Sephadex G-100 | 100,000 | 10.0 | 15–20 | 40–120 |
| Sephadex G-150 | 150,000 | 15.0 | 20–30 | 40–120 |
| Sephadex G-200 | 200,000 | 20.0 | 30–40 | 40–120 |
| Sephadex LH-20* | 100–4000 in EtOH | 2.1 $H_2O$ | 4($H_2O$) | |
| | 100–2000 in $CHCl_3$ | 1.8 $CHCl_3$ | 3–3.5 ($CHCl_3$, EtOH) | 25–100 |
| *2. Polyacrylamide Types* | | | | |
| Bio-Gel P-2 | 1600 | 1.5 | 3.8 | 50–400 |
| Bio-Gel P-4 | 3600 | 2.4 | 5.8 | 50–400 |
| Bio-Gel P-6 | 4600 | 3.7 | 8.8 | 50–400 |
| Bio-Gel P-10 | 10,000 | 4.5 | 12.4 | 50–400 |
| Bio-Gel P-30 | 30,000 | 5.7 | 14.8 | 50–400 |
| Bio-Gel P-60 | 60,000 | 7.2 | 19.0 | 50–400 |
| Bio-Gel P-100 | 100,000 | 7.5 | 19.0 | 50–400 |
| Bio-Gel P-150 | 150,000 | 9.2 | 24.0 | 50–400 |
| Bio-Gel P-200 | 200,000 | 14.7 | 34.0 | 50–400 |
| Bio-Gel P-300 | 300,000 | 18.0 | 40.0 | 50–400 |

*Sephadex LH-20 is a hydroxymethyl derivative of Sephadex G-25 and is used for separation of substances in organic solvents.

## 3. Agarose Types*

| Designation | Particle Size (Mesh) | Operating Range or Exclusion Limit | Approximate % Agarose | |
|---|---|---|---|---|
| Bio-gel A-0.5 M | 50–400 | 500,000 | 10 ⎫ | |
| Bio-gel A-1.5 M | 50–400 | $1.5 \times 10^6$ | 8 ⎪ | Sold in three mesh |
| Bio-gel A-5 M | 50–400 | $5 \times 10^6$ | 6 ⎪ | ranges, 50–100, |
| Bio-gel A-15 M | 50–400 | $15 \times 10^6$ | 4 ⎬ | 100–200, and 200–400 |
| Bio-gel A-50 M | 50–400 | $50 \times 10^6$ | 2 ⎪ | |
| Bio-gel A-150 M | 50–400 | $150 \times 10^6$ | 1 ⎭ | |
| Sepharose 2B | 60–250 (wet) | $10^5 - 20 \times 10^6$ | 2 | |
| Sepharose 4B | 40–190 (wet) | $10^5 - 5 \times 10^6$ | 4 | |
| Sepharose 6B | 40–210 (wet) | $10^5 - 1 \times 10^6$ | 6 | |
| Sagavac 2 | 66–250 (wet) | $181 \times 10^6$ | 2 ⎫ | |
| Sagavac 3 | 66–250 (wet) | $500,000{-}100 \times 10^6$ | 3 ⎪ | |
| Sagavac 4 | 66–250 (wet) | up to $15 \times 10^6$ | 4 ⎪ | Sold in two mesh |
| Sagavac 5 | 66–250 (wet) | $200,000{-}10 \times 10^6$ | 5 ⎪ | ranges (British) particle |
| Sagavac 6 | 66–250 (wet) | $2.6 \times 10^6$ | 6 ⎬ | sizes 66–142 micron, |
| Sagavac 7 | 66–250 (wet) | $50,000{-}1.5 \times 10^6$ | 7 ⎪ | 142–250 micron |
| Sagavac 8 | 66–250 (wet) | to 930,000 | 8 ⎪ | |
| Sagavac 9 | 66–250 (wet) | 25,000–500,000 | 9 ⎪ | |
| Sagavac 10 | 66–250 (wet) | to 400,000 | 10 ⎭ | |

*Agarose is a linear polysaccharide with alternating residues of D-galactose and 3,6 anhydro-L-galactose. All are sold as swollen wet suspensions. Stability, pH 4–9.

## 4. Porous Polystyrene Bead Filtration Media*

| Designation | Particle Size (Mesh) | Exclusion Limit | Bed Vol/g (Benzene) |
|---|---|---|---|
| Bio-Beads S-X1 | 200–400 | 14,000 | 9.8 |
| Bio-Beads S-X2 | 200–400 | 2,700 | 6.2 |
| Bio-Beads S-X3 | 200–400 | 2,000 | 5.1 |
| Bio-Beads S-X4 | 200–400 | 1,400 | 4.2 |
| Bio-Beads S-X8 | 200–400 | 1,000 | 3.9 |
| Bio-Beads S-X12 | 200–400 | 400 | 2.6 |
| Bio-Beads SM-2 | 20–50 | 14,000 | 2.9 |

*Bio-Beads S are cross-linked, nonionic, polystyrene polymers. The S-type functions in organic solvents, while type SM functions in aqueous media.

### 5. Porous Glass Bead Filtration Media*

| Designation | Pore Diameter, A° | Exclusion Limit | Mesh Sizes |
|---|---|---|---|
| Bio-Glas 200 | 200 | 30,000 | All available in |
| Bio-Glas 500 | 500 | 100,000 | 4 mesh ranges, |
| Bio-Glas 1000 | 1000 | 500,000 | 50–100, 100–200, |
| Bio-Glas 1500 | 1500 | 2,000,000 | 200–325, >325 |
| Bio-Glas 2500 | 2500 | >9,000,000 | |
| CPG-10-75 | 75 | 28,000 | |
| CPG-10-125 | 125 | 48,000 | |
| CPG-10-175 | 175 | 68,000 | All available in 3 |
| CPG-10-240 | 240 | 95,000 | mesh ranges, |
| CPG-10-370 | 370 | 150,000 | 80–120, 120–200 |
| CPG-10-700 | 700 | 300,000 | 200–400 |
| CPG-10-1250 | 1250 | 550,000 | |
| CPG-10-2000 | 2000 | 1,200,000 | |

*A series of highly porous glass materials with rigidly controlled pore sizes ranging from 200 to 2500 A°. The matrix is of silicate glass with a network of interconnecting pores. [It is advisable to include salt in eluting buffers to prevent protein and nucleic acid adsorption to the glass.]

### Manufacturers and Suppliers of Materials in Tables 1–5

Ion exchange Sephadexes, Sephadexes and Sepharoses are products of Pharmacia Fine Chemicals, Inc., 800 Centennial Ave., Piscataway, New Jersey 08854.

Ion exchange Bio-Gels Bio-gels, Bio-gel A, Bio-Beads and Bio-Glas, are products of Bio-Rad Laboratories, 32nd and Griffin Ave., Richmond, California 94804.

Sagavac is a product of Seravac Laboratories, Maidenhead, England (Gallard–Schlesinger, Chemical Co., 584 Mineola Avenue, Carle Place, Long Island, New York 11514).

CPG porous glass beads are products of Corning Glass Works, Corning, New York 14830.

## 12. SOLUTIONS FOR ELECTROPHORESIS EXPERIMENTS

(a) *Solutions for the "Standard" Analytical Disc System (1).*

| *Lower Gel* | | *Upper Gel* | |
|---|---|---|---|
| A. 1 $N$ HCl | 24.0 ml | B. 1 $M$ H$_3$PO$_4$ | 25.6 ml |
| Tris | 18.2 g | Tris | 5.7 g |
| TEMED* | 0.23 ml | TEMED* | 0.46 ml |
| Water to 100 ml (pH 8.9) | | Water to 100 ml (pH 6.9) | |
| C. Acrylamide | 30.0 g | D. Acrylamide | 10.0 g |
| Bisacrylamide | 0.8 g | Bisacrylamide | 2.5 g |
| Water to 100 ml | | Water to 100 ml | |
| F. Ammonium persulfate | 0.14 g† | E. Riboflavin | 4.0 mg |
| Water to 100 ml | | Water to 100 ml | |

H. 50X concentrated Upper Buffer:

| Tris | 30.0 g | | |
|---|---|---|---|
| Glycine | 144.0 g | 1 $N$ HCl = 83 ml conc HCl/liter | |
| Water to 1000 ml | | 1 $M$ H$_3$PO$_4$ | |
| Use 20 ml/liter final buffer (pH 8.3) | | = 68.3 ml 85% H$_3$PO$_4$/liter | |

Tracking dye: 0.005% Bromophenol blue in water.
Staining solution: 1% Amido black in 7% acetic acid.

---

*TEMED is N, N, N′, N′ tetramethylethylenediamine. TEMED, acrylamide, methylene bisacrylamide and Tris can be obtained from Eastman Kodak. The other chemicals were purchased from the Fisher Chemical Company.

†Riboflavin and light may be used instead. Use the same concentration of riboflavin as is used in polymerizing upper gels; an excess inhibits polymerization. Note that polymerization is not as complete in riboflavin polymerized gels, and protein $R_f$'s will be greater.

(b) *Some Other Disc Systems (1)*

*1. Running pH 4.3; 7-1/2% gel: (stacks at pH 5.0)*

Stock solutions: as in standard, except for:

| A. 1 $N$ KOH | 48.0 ml | B. | 48.0 ml |
|---|---|---|---|
| Glacial acetic acid | 17.2 ml | | 2.87 ml |
| TEMED | 1.0 ml | | 0.46 ml |
| Water to 100 ml (pH 4.3) | | | Water to 100 ml (pH 6.7) |

H. 10 X concentrated upper buffer

| β-alanine | 31.2 g |
|---|---|
| Glacial acetic acid | 8.0 ml |
| H$_2$O to 1 liter (pH 5.0) | |

*Lower gel:* 1 part A to 2 parts C to 1 part water to 1 part F.
*Upper gel:* 1 part B to 2 parts D to 1 part E to 4 parts water.
*Tracking Dye:* 0.2 ml of 0.02% methyl green.

NOTE: RUN WITH ELECTRODES REVERSED.

2. *Running pH; 2.3; 7-1/2% gel: (stacks at pH 4.0)*

Stock solutions: as in standard, except for:

| | | | |
|---|---|---|---|
| A. 1 $N$ KOH | 48.0 ml | B. | 48.0 ml |
| Glacial acetic acid | 213.0 ml | | 2.95 ml |
| TEMED | 4.6 ml | | 0.47 ml |
| $H_2O$ to 400 ml (pH 2.9) | | | $H_2O$ to 100 ml (pH 5.9) |

G. Ammonium persulfate  1.4 g
   $H_2O$ to 50 ml
H. 10 X concentrated upper buffer
   Glycine          28.1 g
   Glacial acetic acid   3.06 ml
   $H_2O$ to 1 liter (pH 4.0)

*Lower gel:* 2 parts A to 1 part C to 1 part G.
*Upper gel:* as above.
*Tracking dye:* as above.
NOTE: RUN WITH ELECTRODES REVERSED

3. *Running pH; 8.5; 7-1/2% gel: (stacks at pH 7.3) (2)*

Stock solutions: as in standard except for:

A. 1 $N$ HCl          48.0 ml
   TEMED             0.46 ml
   Tris and water to pH 7.9 and 100 ml
B. 1 $N$ HCl          48.0 ml
   TEMED             0.46 ml
   Imidazole and water to pH 5.7 and 100 ml
H. Upper buffer: 0.034M asparagine; brought to pH 7.3 with Tris.

*Lower gel:* 1 part A to 1 part C to 1 part water to 1 part F.
*Upper gel:* 1 part B to 2 parts D to 1 part E to 4 parts water.

Several additional systems with running pH's from 5.2 to 9.0 are described by Richards et al. (3). We have not tried these. Still other systems can be devised if the ion mobilities and pH's are known (4). A table of buffer ion mobilities is given in reference 5, p. 110. Jovin et al. (6) have compiled data on a large number of disc systems.

(c)  *Reagents and Solutions for SDS Acrylamide Gel Electrophoresis (7).*

A. *Gel buffer:*   7.8 g $NaH_2PO_4 \cdot H_2O$
                38.6 g $NaHPO_4 \cdot 7H_2O$
                 2.0 g SDS
             Water to 1 liter

B. *22% acrylamide solution:* 22.2 g acrylamide
                               0.6 g methylenebisacrylamide
                               Water to 100 ml

C. *Ammonium persulfate: 4 mg/ml*

For gel solutions, mix (per gel): 1.25 ml gel buffer
                                     1.13 ml acrylamide solution
                                     0.13 ml ammonium persulfate (4 mg/ml)
                                     $4\lambda$ TEMED

For upper and lower buffer, dilute the gel buffer with an equal volume of water.

Protein sample: $3\lambda$ 0.05% bromphenol blue
                   1 drop glycerol
                   $5\lambda$ mercaptoethanol
                   $50\lambda$ 0.01 $M$ phosphate, pH 7.0 in 0.1% SDS and 1% mercaptoethanol (dialysis buffer)
                   10–50$\lambda$ protein

Staining solution: 0.25% Coomassie Brilliant Blue and 9% acetic acid in methanol.

Stain 2–10 hours.

Destaining solution: 5% methanol and 7% acetic acid in water.

(d)  *Solution for Isoelectric Focusing in an Acrylamide Gel.*

| Reagent | ml/ml gel solution | ml/ml gel solution |
|---|---|---|
| C | 0.25 | 0.17 |
| E | 0.125 | — |
| F | — | 0.25 |
| TEMED | 0.5$\lambda$ | 0.5$\lambda$ |
| 40% ampholyte (any appropriate pH range) | 0.05 ml | 0.05 ml |
| Protein | see Chapter 5 | see Chapter 5 |
| $H_2O$ | (to 1.0 ml) | (to 1.0 ml) |

Ampholyte is sold by LKB, Inc.

(e) *Solutions for 8M Urea Gels*

| Lower Gel | ml/ml of gel solution |
|---|---|
| 2.5 × A | 0.10 |
| 2.5 × C | 0.08 |
| 25.0 × F | 0.02 |
| 10M urea | 0.80 |
| *Upper Gel* | |
| 5 × B | 0.025 |
| 2 × D* | 0.125 |
| 2 × E* | 0.06 |
| 10M urea | 0.80 |

Protein samples should be made dense with urea rather than sucrose.

(f) *Dehydrogenase Stain (8)*

| Reagent | Amount/gel |
|---|---|
| 1M Tris, pH 9.0 | 0.5 ml |
| 10 mM DPN† | 0.1 |
| 10 mg/ml NBT‡ | 0.1 |
| 0.4 µg/ml PMS¶ | 0.1 |
| 10 mM substrate | 0.1 |
| Water | 1.1 |

The reagents should be kept *dark*.

*References*
1. Davis, B. J., *N.Y. Acad. Annals*, **121**, 404 (1964).
2. Hedrick, J. L. and A. J. Smith, *Arch. Biochem. Biophys.*, **126**, 155 (1968).
3. Richards, E. G., J. A. Coll, and M. A. Gratzer, *Anal. Biochem.*, **12**, 452 (1965).
4. Ornstein, L., *N.Y. Acad. Annals*, **121**, 321 (1964).
5. Bier, M., Editor, *Electrophoresis*, Academic Press, N.Y. (1959).
6. Jovin, T. M., M. L. Dante, and A. Chrambach, *Multiphasic Buffer Systems Output*, National Technical Information Service, Springfield, Virginia (1970), P.B. #196085 to 196019 and 203016.
7. Weber, K. and M. Osborn, *J. Biol. Chem.*, **244**, 4406 (1969).
8. Gabriel, O. in *Methods in Enzymology*, W. B. Jakoby, Ed., Vol. XXII, Academic Press, N.Y. (1971), p. 578.

*Solution 2 × E may have to be heated (gently) before everything dissolves. The methylene bisacrylamide concentration of solution 2 × D must be cut to 3% (15% of the acrylamide concentration).
†DPN = diphosphopyridine nucleotide
‡NBT = nitro blue tetrazolium
¶PMS = phenazine methosulfate

# 13. PHYSICAL PROPERTIES OF VARIOUS PROTEINS

These proteins have been sufficiently well characterized so that they can be used as standards in SDS-gel electrophoresis, sucrose density gradient centrifugation, and Sephadex chromatography. All are available commercially. Note that these values will vary somewhat, depending on the source. For the assay conditions of the enzymes, see the appropriate section in *Methods in Enzymology*.

| Protein (Source) | M | $S^{\circ}_{20,w}$ $\cdot 10^{13}$(S) | $\bar{V}$, cc/g | Extinction Coefficient, $\frac{OD_{280}}{mg/ml}$ | Stokes Radius, Å | Number of Subunits |
|---|---|---|---|---|---|---|
| Cytochrome c (bovine heart) | 13,370 | 1.83 | 0.728 | 2.32 (9.65 at 416 nm) | 17.4 | 1 |
| Lysozyme (chicken egg white) | 13,930 | 1.91 | 0.703 | 2.64 | 20.6 | 1 |
| Ribonuclease (bovine pancreas) | 13,700 | 2.00 | 0.707 | 0.73 | 18.0 | 1 |
| Trypsin inhibitor (soybean) | 22,460 | 2.3 | 0.735 | 1.00 | 22.5 | 1 |
| Carbonic anhydrase (bovine B) | 30,000 | 2.85 | 0.735 | 1.90 | 24.3 | 1 |
| Ovalbumin (chicken egg) | 45,000 | 3.55 | 0.746 | 0.736 | 27.6 | 1 |
| Serum Albumin (bovine*) | 67,000 | 4.31 | 0.732 | 0.667 | 37.0 | 1 |
| Enolase (yeast) | 90,000 | 5.90 | 0.742 | 0.895 | 34.1 | 2 |
| Glyceraldehyde 3-phosphate dehydrogenase (rabbit muscle) | 145,000 | 7.60 | 0.737 | 0.815 | 43.0 | 4 |
| Alcohol dehydrogenase (yeast) | 141,000 | 7.61 | 0.740 | 1.26 | 41.7 | 4 |
| Aldolase (rabbit muscle) | 156,000 | 7.35 | 0.742 | 0.938 | 47.4 | 4 |
| Lactic dehydrogenase (bovine heart) | 136,000 | 7.45 | 0.747 | 0.970 | 40.3 | 4 |
| Catalase (bovine liver) | 247,500 | 11.30 | 0.730 | 1.64 (276 nm) | 52.2 | 4 |

*Routinely found containing 5–10% dimer (M = 133,000).

*References*
1. Reference 2, Appendix 4
2. *Worthington Enzymes, Enzyme Reagents, Related Biochemicals*, Catalogue of the Worthington Chemical Co., Freehold, New Jersey (1972).
3. Nelson, C. A., *J. Biol. Chem.*, **246**, 3895 (1971).
4. Tanford, C., "Physical Chemistry of Macromolecules," Wiley, New York (1961).
5. Ackers, G. K., *Biochemistry*, **3**, 723 (1964).

## 14. DENSITIES AND VISCOSITIES OF SOLUTIONS

| Salt | Molarity | $\Delta\rho$ | | | | $\eta/\eta^\circ$ | | | |
|---|---|---|---|---|---|---|---|---|---|
| | | 20° | 25° | 30° | 40° | 20° | 25° | 30° | 40° |
| NaCl | 0.10 | 0.0041 | 0.0041 | 0.0041 | 0.0040 | 1.009 | — | 1.009 | 1.010 |
| | 0.20 | 0.0083 | 0.0082 | 0.0081 | 0.0080 | 1.017 | — | 1.019 | 1.021 |
| | 0.50 | 0.0204 | 0.0202 | 0.0200 | 0.0196 | 1.042 | — | 1.049 | 1.054 |
| | 1.00 | 0.0403 | 0.0399 | 0.0395 | 0.0389 | 1.089 | — | 1.101 | 1.111 |
| | 2.00 | 0.0788 | 0.0780 | 0.0774 | 0.0763 | 1.208 | — | 1.255 | 1.241 |
| | 3.00 | 0.1160 | 0.1149 | 0.1140 | 0.1127 | 1.366 | — | 1.382 | 1.397 |
| KCl | 0.10 | 0.0050 | — | — | — | — | 0.998 | — | — |
| | 0.20 | 0.0090 | — | — | — | — | 0.996 | — | — |
| | 0.50 | 0.0250 | — | — | — | — | 0.994 | — | — |
| | 1.00 | 0.0490 | — | — | — | — | 0.991 | — | — |
| | 2.00 | 0.0990 | — | — | — | — | 1.004 | — | — |
| KH₂PO₄ | 0.01 | — | 0.0010 | — | — | 1.003 | — | — | — |
| | 0.10 | — | 0.0093 | — | — | 1.029 | — | — | — |
| | 0.20 | — | 0.0193 | — | — | 1.058 | — | — | — |
| | 0.50 | — | 0.0469 | — | — | — | — | — | — |
| Na₂HPO₄ | 0.01 | — | — | 0.0014 | — | — | — | 1.007 | — |
| | 0.10 | — | — | 0.0141 | — | — | — | 1.066 | — |
| | 0.20 | — | — | 0.0281 | — | — | — | 1.132 | — |
| NaH₂PO₄ | 0.01 | 0.0008 | — | — | — | 1.003 | — | — | — |
| | 0.10 | 0.0079 | — | — | — | 1.030 | — | — | — |
| | 0.20 | 0.0158 | — | — | — | 1.061 | — | — | — |
| | 0.50 | 0.0404 | — | — | — | 1.182 | — | — | — |
| | 1.00 | 0.0789 | — | — | — | 1.409 | — | — | — |

| Salt | Molarity | $\Delta\rho$ 20° | 25° | 30° | 40° | $\eta/\eta°$ 20° | 25° | 30° | 40° |
|---|---|---|---|---|---|---|---|---|---|
| HOAc | 0.01 | 0.0000 | — | — | — | 1.001 | — | — | — |
|  | 0.10 | 0.0008 | — | 0.0008 | 0.0005 | 1.012 | — | — | — |
|  | 0.20 | 0.0018 | — | 0.0017 | 0.0015 | 1.023 | — | — | — |
|  | 0.50 | 0.0043 | — | 0.0040 | 0.0036 | 1.057 | — | — | — |
|  | 1.00 | 0.0086 | — | 0.0079 | 0.0071 | 1.119 | — | — | — |
|  | 2.00 | 0.0168 | — | 0.0154 | 0.0140 | 1.257 | — | — | — |
|  | 3.00 | — | — | — | — | 1.418 | — | — | — |
| NaOAc | 0.01 | 0.0004 | — | — | — | 1.004 | — | — | — |
|  | 0.10 | 0.0043 | — | — | — | 1.036 | — | — | — |
|  | 0.20 | 0.0086 | — | — | — | 1.073 | — | — | — |
| KOAc | 0.10 | 0.0055 | — | — | — | — | 1.025 | — | — |
|  | 0.20 | 0.0095 | — | — | — | — | 1.055 | — | — |
|  | 0.50 | 0.0255 | — | — | — | — | 1.135 | — | — |
|  | 1.00 | 0.0495 | — | — | — | — | 1.255 | — | — |
|  | 2.00 | 0.103 | — | — | — | — | 1.54 | — | — |
| Na$_2$CO$_3$ | 0.01 | 0.0011 | — | — | — | 1.005 | — | — | — |
|  | 0.10 | 0.0109 | — | 0.0106 | 0.0106 | 1.049 | — | — | — |
| Na$_2$B$_4$O$_7$ | 0.01 | — | — | 0.0019 | — | — | — | 1.009 | — |
|  | 0.10 | — | — | 0.0187 | — | — | — | 1.086 | — |
| HCl | 0.01 | 0.0002 | — | — | — | 1.001 | — | — | — |
|  | 0.10 | 0.0018 | — | — | — | 1.007 | — | — | — |
|  | 1.00 | 0.0179 | — | — | — | 1.061 | — | — | — |

| Salt | Molarity | $\Delta\rho$ | | | | $\eta/\eta°$ | | | |
|---|---|---|---|---|---|---|---|---|---|
| | | 20° | 25° | 30° | 40° | 20° | 25° | 30° | 40° |
| NaOH | 0.10 | 0.0046 | — | — | — | 1.023 | — | — | — |
| | 0.20 | 0.0090 | — | — | — | 1.045 | — | — | — |
| | 0.25 | — | — | — | — | 1.056 | — | — | — |
| Na-citrate | 0.01 | 0.0020 | — | — | — | 1.010 | — | — | — |
| | 0.10 | 0.0195 | — | — | — | 1.101 | — | — | — |
| | 0.20 | 0.0382 | — | — | — | 1.209 | — | — | — |
| Guanidine —HCl | 2.0 | — | 0.0499 | — | — | 1.090 | — | — | — |
| | 4.0 | — | 0.0990 | — | — | 1.273 | — | — | — |
| | 6.0 | — | 0.1448 | — | — | 1.619 | — | — | — |
| | 8.0 | — | 0.1892 | — | — | — | — | — | — |
| Urea | 2.0 | — | 0.0320 | — | — | 1.090 | — | — | — |
| | 4.0 | — | 0.0636 | — | — | 1.220 | — | — | — |
| | 6.0 | — | 0.0926 | — | — | 1.403 | — | — | — |
| | 8.0 | — | 0.1221 | — | — | 1.660 | — | — | — |
| Tris-HCl, pH 8.0 $\frac{\Gamma}{2} =$ | 0.1 | — | 0.0075 | — | — | 1.030 | — | — | — |
| | 0.2 | — | 0.0146 | — | — | 1.077 | — | — | — |
| | 0.5 | — | 0.0390 | — | — | 1.231 | — | — | — |
| $D_2O$, volume percent | 30 | — | 0.0322 | — | — | — | 1.068 | — | — |
| | 50 | — | 0.0537 | — | — | — | 1.118 | — | — |
| | 60 | — | 0.0645 | — | — | — | 1.137 | — | — |
| | 67 | — | 0.0716 | — | — | — | 1.153 | — | — |
| | 75 | — | 0.0806 | — | — | — | 1.172 | — | — |
| | 80 | — | 0.0860 | — | — | — | 1.184 | — | — |
| | 90 | — | 0.0967 | — | — | — | 1.207 | — | — |

Values of density and viscosity of water at various temperatures can be found in references 5 and 6. Note that the effect of high concentrations of two or more components on the viscosity of the solvent is not additive (2).

References

1. Modified from Svedberg, T. and K. O. Pederson, *The Ultracentrifuge*, Oxford University Press, N.Y., (1940).
2. Kawahara, K. and C. Tanford, *J. Biol. Chem.*, **241**, 3228 (1966).
3. International Critical Tables, E. W. Washburn, Editor, Vol. III and V, McGraw-Hill, N.Y (1928).
4. Kirschenbaum, I., *National Nuclear Energy Series*, V46, McGraw-Hill, N.Y. (1949).
5. Reference 2, Appendix 4.
6. Reference 1, Appendix 3.

## 15. COMPUTER PROGRAMS FOR CALCULATION OF SCHLIEREN AREAS AND FOR FITTING LINES BY THE LEAST SQUARES METHOD*

The following two programs represent three of the most used numerical methods of data reduction. They are written in BASIC language as it applies to the Data General Nova general purpose computer, and can be readily transformed into a compatible format for a wide range of general purpose computers which have either BASIC or similar high-level language capability. The algorithms as explained by the program comments (REM) can also be easily adapted to use with manual computation or with programable calculator machines. Detailed explanation and derivation of these numerical methods can be found in most elementary statistics and mathematical handbooks, for instance Korn, G. A. and T. M. Karn, *Mathematical Handbook for Scientists and Engineers*, McGraw-Hill, N.Y. (1968).

*Prepared by Dr. J. E. Wampler of the Department of Biochemistry, University of Georgia.

```
1   REM
2   REM
3   REM        THE MEAN OF INPUT DATA AND ITS STANDARD DEVIATION ARE
4   REM    CALCULATED.
5   REM
1Ø  DIM X[1ØØ]
15  PRINT
2Ø  PRINT "NUMBER OF VALUES ";
25  INPUT N
3Ø  PRINT
32  PRINT
35  PRINT "LIST THE DATA"
36  PRINT "NO. ","VALUE"
4Ø  FOR I=1 TO N
45    PRINT
48    PRINT I,
5Ø    INPUT X[I]
55  NEXT I
57  LET S1= Ø
58  LET S2= Ø
6Ø  FOR I=1 TO N
65    LET S1=S1+X[I]
66    LET S2=S2+X[I]*X[I]
7Ø  NEXT I
75  LET S1=S1/N
8Ø  LET S2=S2/N-S1*S1
85  LET S2= SQR ( ABS (S2))
9Ø  PRINT
95  PRINT "THE MEAN =";S1;" WITH STD. DEV. OF ";S2
1ØØ   PRINT
11Ø   PRINT
12Ø   END
```

```
1    REM                  NUMERICAL INTEGRATION
2    REM
3    REM         QUADRATURE USING SIMPSON'S RULE, IE. TAKING THREE POINT
4    REM     GROUPS AND SUMMING INTEGRALS HAVING VALUES CALCULATED
5    REM     BY THE FORMULA    DELTA X/3 TIMES (Y(1)+ 4 Y(2) + Y(3))
6    REM
1Ø   DIM Y[1ØØ]
15   PRINT
2Ø   PRINT "NUMBER OF VALUES, MUST BE ODD ";
25   INPUT N
26   PRINT
3Ø   PRINT "X INCREMENT";
35   INPUT X
4Ø   PRINT
42   PRINT
45   PRINT "LIST THE DATA"
46   PRINT "NO.","Y VALUE"
5Ø   PRINT
55   FOR I=1 TO N
6Ø      PRINT
65      PRINT I,
7Ø      INPUT Y[I]
75   NEXT I
8Ø   LET S= Ø
85   FOR I=2 TO N STEP 2
9Ø      LET J=I-1
95      LET K=I+1
98      LET S=S+Y[J]+(4*Y[I])+Y[K]
1ØØ  NEXT I
1Ø5  LET S=X/3*S
11Ø  PRINT
12Ø  PRINT "THE INTEGRAL VALUE IS ";S
13Ø  PRINT
135  PRINT
14Ø  END
```

```
1   REM                        LINEAR LEAST SQUARES FIT
2   REM
3   REM           THE INPUT DATA EITHER AS X,Y PAIRS OR AS Y ENTRIES WITH
4   REM      CONSTANT X INCREMENT ARE FIT TO A STRAIGHT LINE FOLLOWING THE
5   REM      LEAST SQUARES CRITERION, IE.  THE BEST FIT  IS DEFINED
6   REM      AS THAT WHICH MINIMIZES THE SUM OF SQUARES OF DEVIATIONS
7   REM      OF THE Y VALUES FROM THE LINE.
8   REM
1Ø  DIM X[25],Y[25],S[25]
15  PRINT
2Ø  PRINT
25  PRINT " HOW MANY SETS, X AND Y";
3Ø  INPUT N
31  PRINT
32  PRINT "DATA PAIRS (1) OR Y WITH CONSTANT X INCR. (2)";
33  INPUT Q2
34  IF Q2=1 GOTO  39
35  IF Q2<>2 GOTO   31
36  PRINT
37  PRINT "LIST STARTING X VALUE AND INCREMENT";
38  INPUT X,D
39  PRINT
4Ø  PRINT " LIST THE DATA"
5Ø  PRINT
6Ø  FOR I=1 TO N
63     PRINT
64     IF Q2=1 GOTO  8Ø
65     LET X[I]=X+((I-1)*D)
66     PRINT X[I];
67     INPUT Y[I]
68     GOTO  85
8Ø     INPUT X[I],Y[I]
85  NEXT I
9Ø  LET M1=1
1ØØ LET M2=N
1Ø5 LET M3=N-1
11Ø GOSUB  1ØØØ
12Ø PRINT
125 PRINT
13Ø PRINT "I","X(I)","Y(I)","DEV. FROM FIT"
135 PRINT
137 LET N=M4
14Ø FOR I=1 TO N
145    PRINT I,X[I],Y[I],S[I]
15Ø NEXT I
16Ø PRINT
165 PRINT
168 PRINT "STD. DEV. OF PTS. FROM FIT IS ";E1
17Ø PRINT
175 PRINT
18Ø PRINT "SLOPE = "; S7;" +- ";E3
19Ø PRINT
195 PRINT "INTERCEPT = ";S8;" +- ";E2
```

```
200    PRINT
205    PRINT "CORL. COEF. FOR LINEAR FIT IS ";S9
210    PRINT
220    PRINT
230    PRINT "DO YOU WISH TO DO ANOTHER ANALYSIS, YES=1, NO=2";
240    INPUT Z
250    IF Z=1 GOTO  15
260    STOP
1000   REM       LEAST SQ. SUBROUTINE REQUIRES X, Y, M1, AND M2.
1010   REM       IT FITS A STRAIGHT LINE TO X(I),Y(I) FOR I =M1 TO M2.
1020   REM       M3=M2-M1. DIM S(I) IN MAIN PROGRAM.
1025   LET M3=M2-M1
1030   LET S1= 0
1040   LET S2= 0
1050   LET S3= 0
1060   LET S4= 0
1070   LET S5= 0
1080   FOR I=M1 TO M2
1090     LET S1=S1+X[I]
1100     LET S2=S2+X[I]*X[I]
1110     LET S3=S3+X[I]*Y[I]
1120     LET S4=S4+Y[I]
1130     LET S5=S5+Y[I]*Y[I]
1132   REM       S1 IS SUM OF X VALUES, S2 SUM OF X SQUARED
1134   REM       S3 IS THE SUM OF X TIMES Y, S4 IS THE SUM OF Y VALUES
1136   REM       S5 IS SUM OF Y SQUARED
1140   NEXT I
1150   LET M4=M3+1
1160   LET S6=M4*S2-S1*S1
1170   LET S7=(M4*S3-S1*S4)/S6
1180   LET S8=(S2*S4-S1*S3)/S6
1190   LET S9=M4*S3-S1*S4
1200   LET S9=S9/( SQR (S6)* SQR (M4*S5-S4*S4))
1202   LET E1=(1/(M4*M4*M4))*(S2*S4*S4-2*S1*S3*S4+S1*S1*S5)
1204   LET E2=(S2/M4)-((S1+S1)/(M4*M4))
1206   LET E1=(1/(M4-2))*((E1/E2)-(S8*S8))
1208   LET E3=(S5/M4)-((S4*S4)/(M4*M4))
1210   LET E3=(1/(M4-2))*((E3/E2)-(S7*S7))
1212   LET E2= SQR ( ABS (E1))
1214   LET E3= SQR ( ABS (E3))
1216   LET S2= 0
1220   LET S1= 0
1230   FOR I=M1 TO M2
1240     LET S[I]=Y[I]-S7*X[I]-S8
1250     LET S2=S2+S[I]*S[I]
1260   NEXT I
1270   LET E1= SQR (S2/M3)
1274   REM   THE OUTPUT IS S7= SLOPE, S8= INTERCEPT, S9= CORL. COEF.
1275   REM                 E1= STD.DEV OF PTS., E2= S. D. OF INTER.,
1276   REM                 E3= S.D. OF SLOPE, S(I) IS EACH PT. DEV.
1280   RETURN
4000   END
*
```

363

## 16. EXTINCTION COEFFICIENTS OF SOME BIOCHEMICALLY IMPORTANT COMPOUNDS

The solvent is water at about pH 7, unless otherwise stated.

(a)  *Coenzymes and Cofactors*

| Compound (Salt) (Molecular Weight) | $\lambda_{max}$, nm | $\epsilon$, $Ml^{-1}cm^{-1}$ |
|---|---|---|
| FMN (Na·2H$_2$O) (514.4) | 445 | 13,000 |
| Reduced FMN (Na·2H$_2$O) (516.4) | featureless above 260 | ca. 1,300 |
| FAD (Na) (807.6) | 445 | 13,000 |
| AMP (2H$_2$O) (383) | 259 | 15,400 |
| ADP (NaH$_2$·2H$_2$O) (485) | 259 | 15,400 |
| ATP (NaH$_2$·4H$_2$O) (623) | 259 | 15,400 |
| Adenosine (H$_2$O) (283) | 259 | 15,400 |
| Adenine (135.13) | 261 | 13,400 |
| CMP (Na$_2$·2H$_2$O) (403) | 271 | 9,000 |
| UMP (Na$_2$·2H$_2$O) (404) | 261 | 7,800 |
| GMP (Na$_2$·H$_2$O) (425) | 252 | 13,700 |
| TMP (322.21) | 267 | 9,500 |
| DPN (2H$_2$O) (717) | 259 | 17,800 |
| DPNH (2H$_2$O) (719) | 338 | 6,220 |
| TPN (NaH$_3$·4H$_2$O) (837) | 259 | 18,000 |
| TPNH (NaH$_3$·4H$_2$O) (839) | 339 | 6,200 |
| Ascorbic acid (176.12) | 265 | 7,000 |
| Pyridoxal (HCl) (203.6) | 318; 390 | 8,200; 200 |
| Pyridoxal 5-phosphate (247.15) | 388 | 4,900 |
| Pyridoxamine (2HCl) (241.12) | 325 | 7,700 |
| Coenzyme B$_{12}$ (1579.57) | 525 | 8,570 |
| Dicyano-B$_{12}$ (1383.4) | 578 (pH 10) | 10,000 |
| Folic acid (441.40) | 346 | 7,200 |
| Tetrahydrofolic acid (445.4) | 297 | 29,100 |
| 5, 10-methenyltetrahydrofolate (456.4) | 350 | 24,900 |
| Thiamine (HCl) (337.28) | 259 | 9,970 |
| Coenzyme Q (849.3) | 275 (95% ethanol) | 15,000 |
| Vitamin A (retinol) (286.44) | 325 | 52,400 |
| Coenzyme A (767.55) | 260 | 16,800 |

| Compound | $\lambda_{max}$ | $\epsilon$ |
|---|---|---|
| Heme (616.48) | 572 | 5,500 |
| Chlorophyll a (893.48) | 663 (acetone) | 84,000 |
| Chlorophyll b (907.46) | 645 (acetone) | 51,800 |
| Tyrosine (181.2) | 275 (water); | 1,420 (water); |
| | 278 (ethanol) | 1,790 (ethanol) |
| Tryptophan (204.2) | 280 (water); | 5,600 (water); |
| | 282 (ethanol) | 6,170 (ethanol) |
| Histidine (155.16) | 213 | 6,000 |
| Nonheme iron (per mole) | ca. 390–420 | ca. 5,000 |
| Peptide Bonds | 190 | 6,380 |

(c)  *Substrates and Protein Reagents*

| | | |
|---|---|---|
| Benzoyl-L-arginine ethyl ester (BAEE) (HCl) (342.8) | 253 | 1,150 |
| Tosyl-L-arginine methyl ester (TAME) (HCl) (378.9) | 247 | 540 |
| N-acetyl-L-tyrosine ethyl ester (ATEE) ($H_2O$) (269.3) | 237 | 230 |
| N-benzoyl-L-tyrosine ethyl ester (BTEE) (HCl) (312.7) | 256 | 964 |
| Benzoyl-L-arginine amide (HCl) ($H_2O$) (331.8) | 253 | 1,150 |
| Benzoyl-L-arginine nitroanilide (HCl) (BAPA) (436.8) | 405 | 9,960 |
| CBZ-alanine-paranitrophenyl ester (307.8) | 348 | 5,500 |
| Paranitrophenylacetate (181.1) | use 348 | 5,000 ($\Delta\epsilon$) |
| 5, 5′-dithiobis-(2-nitrobenzoic acid) (DTNB) (Na) (379.17) | 412 | 13,200 |
| Parachloromercuribenzoate (PCMB) (Na) (379.17) | 232 | 16,900 |
| N-ethylmaleimide (125.12) | 300 | 620 |

(d)  *Redox Dyes*

| | | |
|---|---|---|
| Ferricyanide ($K_3$) (329.25) | 420 | 1,000 |
| 2, 6-Dichlorophenolindophenol (DCPIP) (290.1) | 600 | 20,600 |
| 2, 6, 2′-Trichlorophenol indophenol (TCPIP) (324.5) | 660 | 23,000 |
| Methylviologen (257.2) | 600 | 11,300 |
| Benzylviologen (409.4) | 555 | 12,000 |
| Methylene blue (373.9) | 668 | 28,000 |
| Cytochrome c (12,800) | 529 | 14,000 |
| Reduced cytochrome c (12,800) | 550 | 28,000 (19,700 $\Delta\epsilon$ on Reduction) |

(e) *Dyes and Fluorescence Standards*

| | | |
|---|---|---|
| Quinine Sulfate ($H_2SO_4$) ($2H_2O$)<br>(782.97) | 348 ($1N\ H_2SO_4$) | 10,400 |
| Fluorescein (332.30) | 494 (pH 10) | 75,500 |
| Rhodamine B (479.00) | 552 (ethanol) | 71,700 |
| Rhodamine 6G (479.02) | 530 (ethanol) | ca. 71,000 |
| 1, 8 Anilinonapthalene sulfonic acid<br>(ANS) (1/2 Mg) (310) | 350 | 4,300 |
| Dimethylaminonapthalene sulfonic acid<br>("dansyl") (251.30) | 338 | 3,300 |
| 2, 4 dinitrophenol (184.11) | 360 | 18,000 |

*References*
1. Dawson, R. M. C., D. Elliott, W. H. Elliott, and K. M. Jones, *Data for Biochemical Research*, 2nd Edition, Oxford Univ. Press, N.Y. (1969).
2. Damm, H. C., P. K. Besch, and A. J. Goldwyn, *The Handbook of Biochemistry and Biophysics*, World Publishing Co., Cleveland (1966).
3. Circular OR-10 of Pabst Laboratories (1037 W. McKinley Ave., Milwaukee, Wis.), "Ultraviolet Absorption Spectra of 5′ Ribonucleotides."
4. Circular OR-18 of Pabst Laboratories, "Ultraviolet Absorption Spectra of Pyridine Nucleotides."
5. Wetlaufer, D. B., *Adv. Prot. Chem.*, **17**, 296 (1962).
6. Mitchell, W. M. and W. F. Harrington, *J. Biol. Chem.*, **243**, 4683 (1968).

## 17. CONCENTRATION AND BIOLOGICAL HALF-LIFE OF INGESTED RADIOISOTOPES

| Element | Effective Biological Half-Life | Method of Intake | Principal Organ Affected | Fraction of Amount Present in Body in Organ | Fraction of Amount Taken into Body that Reaches the Organ |
|---|---|---|---|---|---|
| $^{239}$Pu (insol.) | 1 yr. | Breathing | Lungs | 1.0 | 0.12 |
| (sol.) | 120 yrs. | Ingestion | Bone | 0.75 | 0.0001 |
| $^{226}$Ra | 45 yrs. | Ingestion | Bone | 0.99 | 0.015 |
| $^{131}$I | 7.7 days | Ingestion | Thyroid | 0.2 | 0.2 |
| $^{90}$Sr | 8 yrs. | Ingestion | Bone | 0.7 | 0.25 |
| $^{45}$Ca | 151 days | Ingestion | Bone | 0.99 | 0.25 |
| $^{35}$S | 18 days | Ingestion | Skin | 0.19 | 0.08 |
| $^{32}$P | 14 days | Ingestion | Bone | 0.92 | 0.2 |
| $^{24}$Na | 15 hrs. | Ingestion | Total body | 1.0 | 0.95 |
| $^{14}$C ($CO_2$) | 35 days | Breathing | Fat | 0.6 | 0.36 |
| $^{14}$C ($CO_2$) | 180 days | Breathing | Bone | 0.07 | 0.036 |

*Reference*
1. U.S. National Bureau of Standards Handbook 52 (1952).

# INDEX

**373**

Stokes radius (contd.):
    relation to effective volume of macromolecule, 46, 158, 168
Stokes shift, 234
Stopper, insertion of tubing, 85
Storage, isotope, 303
Stray light:
    effect on absorption measurements, 239
    prism vs. grating, 239
Streaming potential, 138
Stream splitter, 62
Styrene-acrylonitrile, 318-321
Substituted celluloses, 79-81
    (see also Cellulose ion exchangers)
    (table), 346
    washing, 4
Sucrose density gradients:
    centrifugation, 164, 185, 197-199
    fractionation, 199
    gradient maker, 198
    handling, 199
    uses:
        isoelectric focusing, 153
        sedimentation velocity, 164, 197-199
Sucrose gradients (see Sucrose density gradients)
Sugar phosphate $R_f$ values, 337
Svedberg ($S$), 197
Swinging bucket rotor, 186, 197
Symmetry:
    center, 219
    effects on extinction coefficient, 223
    electronic, 223
    molecular, 166, 169
Synthetic boundary cell, 186, 203, 205

$t$, values, 332
$t$-test, 16-19
Tailing:
    adsorption chromatography, 70
    electrophoresis, 137, 158
    partition chromatography, 39, 93
    sedimentation velocity, 211
TD/TC containers, 7
Teflon, 318-321
Temperature calibration, rotor control, 191
Temperature effects:
    GLC, 58, 64
    Oudin diffusion, 115
    sedimentation velocity, 208, 212
Temperature maintenance, ultracentrifuge, 175, 208, 212
Theoretical efficiency, GLC, definition, 57
Theoretical frictional coefficient, 166
Theoretical plate theory, 53

Thermal conductivity detector, 60
Thermal motion, 162 (see also Diffusion)
Titer, antibody, 98, 110
Total and specific activity, isotope, 291
TPX, 318-321
Transfer, energy (see Energy transfer)
Transfer units ("plates"), significance of number, 39
Transition probability (see Extinction coefficient)
Transition speed (see Speed of transition)
Transitions:
    forbidden, 223, 270
    radiationless, 227
Transmittance, definition, 243
Tritium ($^3$H) samples, scintillation fluid, 298
Trivial reabsorption, 234
Tubing, dialysis (see Dialysis tubing)
Tubing, inserting stopper, 85
Tungsten lamp, 238
Turbulent diffusion (see also Diffusion), 57
Tygon, 319

Ultracentrifugation:
    cell alignment, 189
    cell assembly, 188
    cleanup procedure, 194
    conversion from schlieren to Rayleigh optics, 201
    force needed, 162
    installation of rotor, 189
    operating controls, 190
    photographic developing procedure, 194
    photography, 192
    preparation, 186-189
    shut down, 194
    theory, 161
    thermal motion, 162
Ultracentrifuge:
    camera, 183
    cells, 184
    components, 175-186
    diffusion pump, 175
    drive unit, 176
    light source, 176
    optical systems, 177-183
    preparative, 186, 197
    refrigeration system, 175
    speed control unit, 176
    temperature maintenance (RTIC), 175
    vacuum system, 175
Uncertainty principle, 217
Universal red shift, 225
Urea, use in acrylamide gel electrophoresis, 154, 354

Vacuum system, ultracentrifuge, 175
Van Deemter equation, 58
Variance, 27

Variances, pooled, 26
Velocity:
    alpha particle, 274
    angular, 163
    beta particle, 273, 274
    gamma radiation, 270, 273
Vibrational states, 220, 224, 225, 227, 230, 234
Viscosity:
    gas, effect on GLC, 58
    solution, effect on sedimentation, 197, 208
    solution (table), 356-359
    solvent, determination, 209
Void volume ($V_o$), 46
    percentage of column volume, 48
Voltage gradient, 131
Volume:
    corrected retention, 55
    effective, of macromolecule, 46 (see also Stokes radius)
    effective column, 44
    elution, 44
    fraction, in column chromatography, 92
Volumetric devices, column chromatography, 92
Volumetric pipettes, use, 8

Waste, isotope, 303
Water:
    acid permanganate method for preparation, 6
    deionized or distilled, 6
    hydration, protein, 167
Wavelength shift, absorption maximum, 222
Weight average molecular weight, 172
Weight, molecular, 45, 119, 139, 158, 163-172, 199-203, 212
Windows, cell (see Cell windows)
Work function, 286

Xenon arc lamp, 176, 238, 245
X-rays, preparation, 276

Yphantis method, 172, 199-203

Zone electrophoresis, 137
Zone of equivalence, 109
Zone method, 32
    chromatography, 34, 41, 43, 52, 84
    electrophoresis, 137
    sedimentation velocity, 164
Zone precipitation, 50-51
Zone sedimentation method:
    calculation of distance traveled by unknown, 165, 199
    use of markers, 165, 199